# Residential Energy Auditing and Improvement

# Residential Energy Auditing and Improvement

By
Stan Harbuck
Donna Harbuck

LONDON AND NEW YORK

**Published 2020 by River Publishers**
River Publishers
Alsbjergvej 10, 9260 Gistrup, Denmark
www.riverpublishers.com

**Distributed exclusively by Routledge**
4 Park Square, Milton Park, Abingdon, Oxon OX14 4RN
605 Third Avenue, New York, NY 10017, USA

First issued in paperback 2023

**Library of Congress Cataloging-in-Publication Data**

Harbuck, Stan
    Residential energy auditing and improvement / by Stan Harbuck,
Donna Harbuck.
    pages cm.
    Includes bibliographical references and index.
    ISBN 0-88173-726-7 (alk. paper) -- ISBN 978-1-4822-4462-5 (Taylor &
Francis : alk. paper) -- ISBN 978-8-7702-2316-4 (electronic) 1. Dwellings-
-Energy conservation. 2. Energy auditing. I. Harbuck, D. D. II. Title.
    TJ163.5.D86H368 2014
    696--dc23                                                    2013049045

*Residential Energy Auditing and Improvement / by Stan Harbuck*
First published by Fairmont Press in 2014.

*Routledge is an imprint of the Taylor & Francis Group, an informa business*

Publisher's Note
The publisher has gone to great lengths to ensure the quality of this reprint but
points out that some imperfections in the original copies may be apparent.

ISBN-978-87-7022-925-8 (pbk)
ISBN-978-1-4822-4462-5 (hbk)
ISBN-978-8-7702-2316-4 (online)
ISBN-978-1-0031-5185-2 (ebook master)

While every effort is made to provide dependable information, the publisher, authors, and
editors cannot be held responsible for any errors or omissions.

The views expressed herein do not necessarily reflect those of the publisher.

# Table of Contents

## Chapter 5

### WeatherizationRequirements and Similarities in the Private Arena . . . . . . . . . . . . . . . . . . . . . . . . . . . . . .187

## Chapter 6

### Sealants, Insulation and Barriers and How to Install Them . .197

## Chapter 7

### Auditing, Planning, and Retrofitting . . . . . . . . . . . . . . .221

## Chapter 8

### Work Order Development by the Auditor . . . . . . . . . . . .345

## Chapter 9

### Heating and Cooling. . . . . . . . . . . . . . . . . . . . . . .353

## Chapter 10

### Baseload and How to Improve It . . . . . . . . . . . . . . . . . . . 469

## Chapter 11

### New Construction Energy Evaluations . . . . . . . . . . . . . . 491

# Preface

This book is for energy auditors or retrofitters whether they work in the weatherization program or in the private arena. This book is intended to help instruct in the following certifications:

- Association of Energy Engineers (AEE). Residential Energy Auditor (REA)

- BPI—Building Analyst (BA), Envelope Professional (EP), and the Residential Building Envelope Air Leakage Control Installer (RBE-ALCI).

- RESNET—HERS Rater, Comprehensive HERS Rater (CHERS), Home Energy Survey Professional (HESP), Diagnostic Home Energy Survey Professional (DHESP), and Building Performance Auditor (BPA).

- DOE/NREL—Energy Auditor, Installer, and Quality Control or Technical Monitor Inspector (with supplemental materials).

If you are in the private arena, you may not have to study Chapter 1 or 5. The book also covers BPI, RESNET, and the Draft DOE/NREL requirements. If you are only interested in BPI, you may not have to study Chapter 11, which covers RESNET. To find out more about certification requirements, see Appendix M on certifications.

Your local or state laws or rules may require that any work done on combustion appliances must be done by a licensed or otherwise qualified technician. Air-conditioning technicians must be certified by the EPA.

*Residential Energy Auditing and Improvement* contains procedures and techniques practiced in industry and the trade. Specific procedures vary with each task and must be performed by a qualified, and in some cases, a certified or licensed person. For optimum safety, always refer to specific manufacturer recommendations, insurance regulations, specific job site and plant procedures, applicable federal, state, and local regulations, including OSHA regulations, and any authority having jurisdiction, such as the municipality that the work is being done in. The material contained in this book is intended to be an educational resource for the user of the book. Standards and Testing Inc., Stan Harbuck, nor Donna Harbuck do not assume and hereby disclaim liability to any party for any claims, losses, damages, or disruption under any legal principle, including property damage or personal injury resulting from errors and omissions, negligence, etc. incurred by reliance on this information. Statements made in this book are the opinions of the author(s).

Remember that anytime we do anything that might "tighten up" a home, we reduce the ventilation in a home, and thereby may put the occupants' health and safety at risk. As a result, we must be sure that the home still has an adequate minimum ventilation rate (MVR or BTL).

Always make sure that CO detector(s) have been installed in the home you are working on before starting your work.

Always "test out" (combustion appliance zone (CAZ) and appliance testing) of a home you have worked on at the end of each day before leaving the site.

If you smell or suspect a gas leak, do not light a match, use any electrical appliances, turn lights on or off, or use the phone. These may produce sparks. Leave the building immediately and tell any others in the house to leave with these same considerations. From a different building, call the gas company or a qualified appliance repairperson or plumber for repairs.

Always follow all applicable laws and regulations, including all equipment manufacturer's instructions (including ladders) and OSHA guidelines. Any work done on refrigerant gas systems must be performed by an OPA approved technician. In addition, some states may require work done on combustion appliances be done by a licensed or otherwise qualified technician.

Before working on roofs, consult a physician or your company's safety contact.

**Caution**: If you are opening an electrical panel, always open and close the panel carefully. Never touch any items in the panel box. Also, be careful in removing the panel cover so that you do not angle it out and touch wires in the process.

This publication is designed to provide accurate information about the subject matter covered. It is sold with the understanding that the author and publisher are not engaged in rendering legal, accounting, or other professional services. If legal advice or other expert assistance is required, the services of a competent professional person should be sought.

The information provided is deemed generally reliable but is not guaranteed as, for example, special conditions may apply as to specific situations or as to particular conditions that exist in a given locality.

Be aware that different state or local municipal building codes may apply to your work and that a permit may be required.

# Acknowledgements

**A Tribute to Weatherizers in the
Low Income Weatherization Program**

This weatherization training series is dedicated to the many weatherization workers, managers, administrators and researchers who exemplified the best in helping the poor and those with fixed incomes to better overcome the challenge of balancing a budget in tough economic times while at the same time helping us as a nation reduce our dependence on fossil fuels. These weatherizers purposefully devoted themselves to helping others despite clearly having an opportunity to work in the private arena for more pay and recognition instead.

It is to these individuals who worked to this end, at a time when it was not popular, that this weatherization training series is dedicated. Thank you for devoting yourselves to a life of helping millions of others over helping yourselves. Our "hats off" to you!

Major credit for the development of our knowledge and understanding of saving energy in buildings and as prominent resources for the information provided must also go to:

— US Department of Energy (DOE)
— Environmental Protection Agency (EPA)
— Lawrence Berkeley National Laboratory (LBL)
— National Renewable Energy Lab (NREL)
— Weatherization Assistance Program Technical Assistance Center (WAPTAC)
— U.S. DOE Energy Efficiency and Renewable Energy (EERE)
— Association of Energy Engineers (AEE)
— National Association for State Community Services Programs (NASCSP)
— Affordable Comfort Inc. (ACI)
— Building Performance Institute (BPI)
— Residential Energy Services Network (RESNET)
— Florida Solar Energy Center

Anthony Cox and John Langford should be given special credit for the careful development of the outline for the curriculum to teach weatherization auditing and retrofitting. Their work over many years represented some of the most significant work and research in the field. At the same time, others have been major ongoing pioneers in the development of energy evaluation and retrofit throughout their lives including Terry Brennan, Anthony Cox, John Langford, John Tooley, Max Sherman, John Snell, Cal Steiner, and Tony Woods. In addition, many, many individuals from various fields over the years have contributed in significant ways to advancing weatherization in both the public and private arenas. While the following list is not intended to be a complete list of all those that have made significant contributions to the field, it is intended to give credit to those who have devoted a significant portion of their lives to advancing energy saving in buildings. Included in this list are the following: Hal Aranson, Steve Baden, Rana Belshe, Michael Blasnik, Terry Brennan, Rees Byars, Tony Colontonio, Stephen Cowell, Anthony Cox, Brett Dillon, Chris Dorsey, Laurel Elam, Philip Fairey, Dr. Wolfgang Feist, Jim Fitzgerald, Don Fugler, Doug Garrett, Colin Genge, Suzanne Harmelink, Hap Haven, Skip Hayden, David Hepinstall, Don Hynek, Rick Karg, David Keefe, Gary Klein, Dick Kornbluth, Jan Kosny, John Krigger, Joe Kuonen, John Langford, Jim LaRue, Jack Laverty, Amory Lovins, Joe Lstiburek, Bruce Manclark, Greg McIntosh, Laura McNaughton, Alan Meier, Neil Moyer, Vikki Murphy, Garty Nelson, Tomas O'Leary, Michael Lubliner, Collin Olson, Helen Perrine, Andrew Persily, Lydia Gill Polley, John Proctor, Lester Shen, John Snell, Robert Sonderegger, Cal Steiner, A. Tamasin Sterner, Nehemiah Stone, Al Thumann, John Tooley, George Tsongas, Michael Vogel, Doug Walter, Linda Wigington, Tony Woods, and Alexander Zhivov.

# Chapter 1

# Introduction

The need for energy savings has never been greater. One area where such savings can come is in the energy auditing and retrofitting of homes and other buildings. Saving money for the homeowner, improving indoor air quality, avoiding moisture problems, promoting building durability, and increasing comfort are some of the basic benefits of weatherization. As for saving energy, almost 40% of energy used in the United States is for heating, cooling, lighting, and otherwise powering up both commercial and residential buildings. About 20% is used for heating and cooling buildings. Approximately half of the expenditure for heating and cooling buildings is wasted due to insufficient insulation, air leaks, and inefficiencies in the heating and cooling equipment. It is estimated that improving the efficiency of cooling and heating systems could lead to a savings of five quads of energy per year in the United States. One quad of energy is the equivalent of one quadrillion Btus or 1015 Btus. To put this in context, the U.S. uses about 98 quads of energy a year.

Appropriately guided weatherization work can help prevent mildew or mold, reduce respiratory ailments, and help eliminate odors, thus improving indoor air quality. It also helps buildings last longer and improves the comfort in homes by creating a more balanced airflow. A number of certification programs exist under the Association of Energy Engineers (AEE), The Building Performance Institute (BPI), the Residential Services Network (RESNET), and the Department of Energy/National Renewable Energy Laboratory (DOE/NREL). See Appendix M for more details.

## HISTORY OF ENERGY AUDITING AND RETROFITTING

Much of the early effort of energy auditing and retrofitting was to help the poor. The Weatherization Assistance Program was created in 1976, after the first oil embargo but before the DOE was formed.

While some local programs with similar goals had been in operation for a while, this was the formal beginning of the program on a national scale. The "Winterization" program in Maine provided a model for the national program. The Community Services Administration originally administered the program at a national level. Later, the Federal Energy Administration, a predecessor to the DOE, took over the management. The program used volunteer labor to staple plastic over windows in homes. Thus, this early program installed only low-cost measures and installed very little insulation. Unfortunately with this initial national program, there was little or no financial or production accountability. In the early 1980s, the program grew. It continued to use volunteer labor under the Comprehensive Employment and Training Act in the Department of Labor. Often only temporary measures were installed and there was little or no diagnostic technology to evaluate homes. Because windows and doors were often viewed as major culprits in energy loss, they were often replaced. Other common practices included adding storm windows, weather stripping and caulking, and adding some attic insulation. Later studies showed that door and window replacement is not nearly as cost-effective as retrofit measures involving air sealing and adding insulation. Project Retrotec used very basic heat transfer calculations to evaluate difference in R-values in the home. At this stage, the program addressed the building envelope and workers blew insulation into attics. Work on houses was completed quickly and with much less improvement than is common today.

In 1991, the program began tracking the cost-effectiveness of the installed retrofit measures. In addition, measures to reduce energy use in the home were expanded to include heating and cooling systems. Furthermore, the program no longer had to rely on volunteer labor alone as it was now allowed to pay for the labor involved in retrofitting homes. During this period, some states began using blower door technology and other diagnostic techniques in very crude ways that have since been refined. Continuous feedback loops and accountability became the standard. When an inspector observed retrofit measures that did not meet the standards, training could be required for the work crew. This training could be at a recognized training facility or it could be on-the-job training. The structured Training and Technical Assistance (TTA) helped address the shortcomings of the program. At this stage, some of the components of the program became computerized. In addition, national and state evaluations were conducted on a regular basis.

From the 1990s to the present day, diagnostics and testing have become more common, better refined, and more effective. Highly trained crews perform the installation procedures. Emphasis is now on cost-effective, permanent weatherization measures and retrofits.

Computerized energy audits can now provide far more cost-effective savings. Now, the whole house is viewed as a system involving mechanicals, the building shell and the baseload. This is far better than the shell measures originally used in the old program. With the use of blower door technology, a home can be better air-sealed. A great deal more attic insulation, including dense pack sidewall insulation, can be added than used to be the case. This not only helps insulate but also air seal the walls. Duct sealing and modifications such as adding returns or reconnecting attic ducts is a regular practice today. Even floor grille replacement is provided. Electric baseload measures are also targeted. These include, but are not limited to, installing compact florescent lamps (CFLs), replacing the refrigerator, and modifying the water heater. Health and safety measures that are energy-related, such as replacing faulty furnaces or relining chimneys, are also provided under the program.

MISSION OF WAP

The mission of the Weatherization Assistance Program (WAP) as established by the U.S. Congress is to help reduce the energy costs of low-income families, especially for the elderly (who occupy 34% of low-income housing), children, and people with disabilities. The law was amended in 1990 to include ensuring health and safety. The US government through the US Department of Energy (DOE) funds the Weatherization Assistance Program. The DOE monies are passed on to "grantees" which include 50 state offices, Native American Tribal Organizations, the District of Columbia, and 5 US territories. Each of these grantees then distributes the funding to more than 900 local weatherization agencies nationwide (also known as the "sub grantees"). The War on Poverty Program created by Lyndon Johnson created an infrastructure of Community Action Programs (CAPs) that cover specific geographic areas within a state. These local CAPs have the right of first refusal to become the local area weatherization agency or subgrantee. Thus, most local weatherization agencies are CAPs, although other nonprofits and local government agencies can serve as the local weatherization agency if the local CAP has turned it down. The money distributed to these local weatherization agencies or sub grantees from the state offices or grantees is used to install cost effective, energy-saving measures in low-income households within the geographical boundaries of the local weatherization agency.

More than 90% of low-income households in the United States have an annual income of less than $15,000. More than 13% of low-income households have an annual income of less than $2000. The average energy expenditure in low-income households is $1871 a year. According to the Energy Information Administration (EIA) within the DOE, low-income households typically spend 17% of their annual income on energy while all other households spend only 4%. Energy burden is the percentage of household income used for energy bills. Therefore, the energy burden on low-income households is easily four times that of other households. These facts show the importance of reducing the energy burden of low-income clients.

Some states have found ways to leverage funds from other federal programs, sometimes even going through utility companies to further expand the reach of the WAP. Furthermore, comprehensive weatherization and rehabilitation is possible through coordination with the US Department of Housing and Urban Development (HUD). Compared to utility-sponsored and local weatherization programs, the DOE program is more effective. Evaluations have found benefits on many levels including the fact that more than 6.4 million homes have been weatherized to date and that there is a favorable benefit to cost ratio of 1.8:1. In addition, an average 32% reduction in energy use for space heating has been achieved. All these benefits are in addition to the roughly 8000 jobs nationwide that WAP supports.

# Chapter 2

# Energy Basics

The purpose of auditing and retrofitting is to save energy. To do so, it is important to understand the concept of energy and what it entails.

## THE CONCEPT OF ENERGY

The following concepts of energy will be discussed in this section: potential and kinetic energy, the first two laws of thermodynamics, the relationship of energy to heat, work and temperature, phase change, and heat transfer.

### Potential Energy

Potential energy is energy that is stored. An example of potential energy is gasoline in a car's gas tank. A full gas tank allows a car to drive for miles while an empty tank has no potential energy. Potential energy is also like propane gas in a home's propane gas storage tank. Energy in the propane heats the home. Potential electrical energy is stored in a battery to be used later for lights, motors, etc. A bobsled sitting at the top of a bobsled track has potential energy. Once it is released or pushed from the top, it begins to convert that potential energy into motion or speed (kinetic energy). Similarly, water stored behind a dam as in Figure 2-1 has potential energy that is converted to kinetic energy as it is released into motion through the pipes and turbines that generate electricity.

### Kinetic Energy

Kinetic energy is energy that is in motion or in transition. Good examples are a bicyclist coasting along on a flat road without moving the pedals, or a car that has run out of gas trying to coast into a gas station on a flat road. Another example may be a very heavy spinning

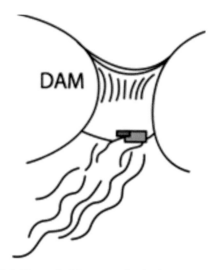

**Figure 2-1. Dams hold a great deal of potential energy.**

metal cylinder that will continue to spin until, over time, friction wears it down to a stop. Electrical energy running along a transmission line has kinetic energy that originated at a dam, coal plant, etc. using the potential energy of the water behind a dam, the fossil fuel energy stored in coal or other sources, to create the electricity.


THE THERMODYNAMIC LAWS

The laws of thermodynamics explain how and why energy is transferred. The First Law of Thermodynamics says that energy can neither be destroyed nor created. In the example of a bicyclist coasting along a flat road without moving the pedals, ultimately the bicyclist slows to a stop because of friction in the bearings of the bicycle wheels. While the bicyclist was only coasting, the kinetic energy the bicyclist had was not actually lost when the bicyclist slowed to a stop. Rather, energy was lost in the form of heat in the wheel friction. Since energy cannot be created, and since there are always losses of energy in any process, including frictional losses to bearings, no one has ever been able to create a perpetual motion machine, Figure 2-2.

In the example of the propane gas tank, the propane is converted to heat by burning it. Even though in reality conversions are not 100%

efficient, the First Law of Thermodynamics says that all the heat in the propane can be accounted for if the byproducts are carefully measured. If you did so, you would find that the energy in the propane has not been destroyed, nor had more energy been produced than was originally in the propane. In other words, no matter where the heat losses are, energy always turns up somewhere, somehow, in one shape, form or another without disappearing and without generating energy from nothing. The Second Law of Thermodynamics as shown in Figure 2-3 says that heat always moves from areas of high temperature to areas of lower temperature. Thus, in a home, heat moves towards the outside of a house in the winter when the home is being heated and towards the inside of the house in the summer when the home is being cooled.

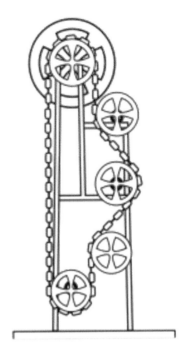

**Figure 2-2. Perpetual motion machines cannot operate perpetually without the input of energy.**

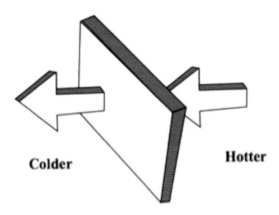

**Figure 2-3. Second Law of Thermodynamics**
Heat always travels from the hotter side to the colder side.

## RELATIONSHIP OF ENERGY TO HEAT, TEMPERATURE, WORK, AND POWER

Energy is a measurable quantity of heat, work, or light. On a microscopic level, molecules in a solid are essentially stationary at lower temperatures. As heat is added, the temperature of the solid increases and the molecules move faster. Temperature is a measure of the amount of heat present. In essence, temperature is measuring the typical motion of molecules or kinetic energy of molecules in a substance. Heat is measured in Btus (British thermal units), calories, joules, watt-hours, barrels of oil, pounds of steam, gallons of propane or therms. A Btu is the amount of energy it takes to raise 1 gallon of water 1°F. Roughly, it is the amount of energy in a kitchen match if it burns completely, Figure 2-4.

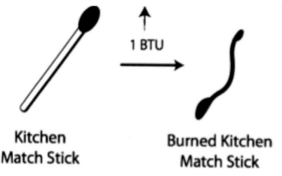

**Kitchen Match Stick**          **Burned Kitchen Match Stick**

**Figure 2-4. A kitchen match has about 1 Btu of energy in it.**

Heat units can be converted from one unit to another using conversion factors. For example, knowing that 1055 Joules equals one Btu, then 2110 Joules represents two Btus. One Btu also represents about 3/10 of a watt-hour or 252 calories. Conversion charts show what number to multiply or divide one unit by to find the equivalent in another unit (see Appendix C).

Temperature is measured in Fahrenheit, Centigrade, or Kelvin. Table 2-1 compares these temperature scales.

Work is energy expended with a result. For instance, raising a one-pound weight one foot off the floor equals one foot-pound of work as shown in Figure 2-5. Work and heat are essentially equivalent. The work expended stirring water in a container actually raises the temperature of the water. In fact, conversion factors show that one Btu is equal to 778

## Table 2-1. Comparison of Temperature Scales

| Water Boils | | 212 | 100 | 100 | 373.15 |
| | 200 | | 90 | | 370 |
| | 190 | | | | 360 |
| | 180 | | 80 | | 350 |
| | 170 | | 70 | | 340 |
| | 160 | | | | |
| | 150 | | 60 | | 330 |
| | 140 | | | | |
| | 130 | | 50 | | 320 |
| | 120 | | | | |
| | 110 | | 40 | | 310 |
| | 100 | | 30 | | 300 |
| | 90 | | | | |
| | 80 | | 20 | | 290 |
| | 70 | | | | |
| | 60 | | 10 | | 280 |
| | 50 | | | | |
| Water Freezes | 40 30 | 32 | 0 | 0 | 273.15 |
| | 20 | | -10 | | 270 |
| | 10 | | | | 260 |
| | 0 | | -20 | | 250 |
| | -10 | | | | |
| Absolute Zero | | -459.67 | -373.15 | | 0 |

Fahrenheit        Celsius        Kelvin

foot-pounds of work. The international energy unit, Joule, is a unit used for both heat and work.

It is good to distinguish energy from power. Power is units of energy used over a specific time frame; power is energy divided by time. Some examples of energy units include Btu, foot-pounds, and joules. Examples of power units include Btu per hour, foot-pounds per minute, joules per second or watts, kilowatts, horsepower, and tons of refrigeration. Some of these units do not show a time unit built because of the way they are defined--watts are defined as joules per second, horsepower is defined as 33,000 foot-pounds per minute, tons of refrigeration is defined as 12,000 Btu per hour. Thus, while it seems counterintuitive, watt-hours (wh), horsepower-minutes (hpmin), and ton-hours (tonhrs) of refrigeration are *energy* units because the time unit comes from the original definition of watts, horsepower, and tons of refrigeration. Table 2-2 compares different unite of energy with units of power.

To calculate how much energy is used, multiply the power by the timeframe over which the power was used. For example, furnaces have

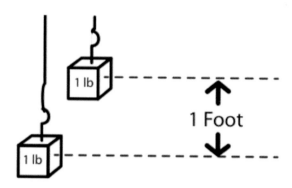

**Figure 2-5. Work: Raising One Pound One Foot =1 Foot-Lb.**

**Table 2-2. Comparison of Energy and Power Units**

| Energy | Power (Energy/time) |
|--------|---------------------|
| Btu    | Btu/hr              |
| Watt   | Watt-hour           |
| Joule  | Joule/sec           |
| Ft lb  | Ft lb/min           |

a Btu rating for power input in Btu/hr. If a furnace has 50,000 Btu/hour of power (a furnace Btu rating of 50,000/hour) and it runs for two hours, it has used 100,000 Btus of energy.

PHASE CHANGE

In dealing with energy and heat, be aware that there are special rules when substances change from one phase or state to another. Changing state simply means that even though it is the same substance (such as water), it is being changed from the solid state (ice) to a liquid state (water), or from a liquid state to a gaseous state (water vapor) or vice versa as in Figure 2-6.

Every material, including water, absorbs heat at a different rate depending upon its state. For instance, the specific heat of solid water (ice) is 0.49 Btu/lb°F. So, if ice is warmed from 0°F to 32°F, 0.49 Btus per pound of ice must be added to raise the temperature each 1 degree.

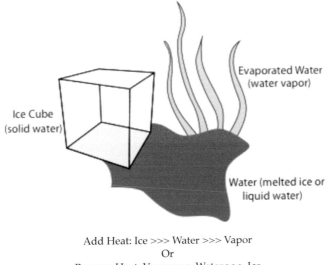

Ice Cube
(solid water)

Evaporated Water
(water vapor)

Water (melted ice or
liquid water)

Add Heat: Ice >>> Water >>> Vapor
Or
Remove Heat: Vapor >>> Water >>> Ice

**Figure 2-6. The Reversible Phases of Water**

Thus, with one pound of ice, approximately 16 Btus must be added to raise the temperature from 0 to 32°F:

$$(32°F\text{-}0°F) \times 0.49 \text{ Btu/lb°F} \times 1 \text{ lb.} = 16 \text{ Btu}$$

Likewise, if the temperature of one pound of water is raised from 32°F to 212°F, 180 Btu must be added to that water. This is true because the definition of 1 Btu is the amount of heat required to raise 1 pound of water 1°F. In other words, the specific heat of liquid water is 1.0 Btu/lb°F. Now, to take this a step further, to raise the temperature of 1 pound of gaseous water, otherwise known as steam, from 212°F to 222°F, multiply these 10°F by 0.47 Btu/lb°F which equals 4.7 Btu of heat. The specific heat of gaseous water (steam) is 0.47 Btu/lb°F. These phase changes are shown in Figure 2-7.

Notice that even water in its various phases or states has a different specific heat for changing temperatures within each phase. Thus, it takes more heat to raise 1 pound of liquid water 1°F than it does to raise 1 pound of either solid (ice) or gaseous (steam) water 1°F. Within a phase or state, adding or removing heat with the subsequent changing temperature is referred to as the *sensible heat*. The specific heat is the

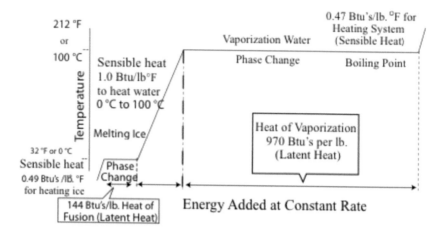

**Figure 2-7. Physical Characteristics of Phase Changes in Water**

number of Btus necessary to raise one pound of the substance by one degree Fahrenheit.

However, in the middle of changing phases, such as going from ice to liquid water or from liquid water to steam, energy or heat has to be added just to get the water to change from one state to another. This is called *latent heat*. Thus, latent heat relates to a change of phase or state while sensible heat only relates to changing the temperature of a substance without changing its phase or state in the process. 144 Btus per pound of water must be expended to convert ice to liquid water. To go from liquid water to steam, 970 Btus per pound of water has to be expended. Different amounts of heat per pound are required to go from one state to another depending on which material it is. These phase change amounts or latent heat amounts are much larger per pound of water than changing the temperature of water between phase changes by even just 1 °F. The heat required per pound of water to go from ice to liquid is called the latent heat of fusion. Likewise, the heat required per pound of water to go from liquid to steam is called the latent heat of evaporation because it is evaporated.

**Phase Change Practical Applications**

So why is learning about specific heat, latent heat of fusion, and latent heat of evaporation so important? This information can be used in real life with homes and buildings. For instance, it is good to know how challenging it is to get liquid water to evaporate out of the house in terms of energy costs. Thus, depending on how much water leaks into a house, much more heat will be required to turn the water to vapor than to raise the temperature of the water or the air. This can factor heavily into the energy usage in a home including, in the reverse process, the additional energy used when an air conditioner has to extract moisture out of high humidity air while cooling the air, as shown in Figure 2-8a.

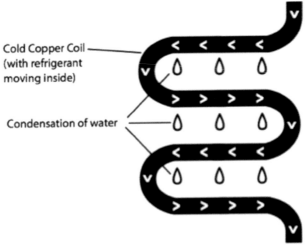

**Figure 2-8a. Condensation Occurring on a Cold Surface**
It takes energy to take moisture out of air.

As another example, in a steam heating system water is evaporated into steam at the boiler. This steam, because of its temperature, rises to all the radiators located above the boiler to heat the home. Thus, steam heating systems require heating the water not just to 212° F, but through the costly latent heat of evaporation of water, to turn it into steam just above 212° F. Using a similar example, an air conditioner uses a special liquid, a refrigerant, to take heat out of the house. At the "evaporator coil," just downstream of the furnace, the refrigerant turns from a liquid into a gas (or evaporates), thus pulling heat out of the air around it. In other words, the refrigerant cools the air around it inside the ductwork. The refrigerant is typically contained in copper coils or pipe in the evaporator. There is a heat exchange from outside the copper pipe to the inside of the copper pipe so refrigerant is not mixed with the air that is breathed in the home. This, now gaseous refrigerant, goes through a pipe to the copper "condenser" coil or pipe outside of the house where the refrigerant passes through the condenser pipe and changes from a vapor to a liquid, see Figure 2-8b. The condenser's copper coils allow the hotter refrigerant vapor inside the pipe to cool down as it passes through its coils, the opposite of what happened in the evaporator coil inside the house.

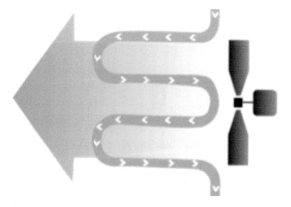

**Figure 2-8b. A Fan Blowing Air Across a Copper Coil**

In some larger buildings, owners are able to take advantage of lower electrical utility fees at night by using an ice or cold-water tank. Water in the ice or cold-water tank is cooled down using electricity at night at the lower electrical rates and then, during the heat of the day when electrical rates are expensive, the "coolness" in the ice or cold-water tank is used to air condition the building, saving on utilities costs.

HEAT TRANSFER

More than 99% of our energy comes from the sun. Once the solar radiation reaches earth, other mechanisms become involved including conduction, convection, radiation, and chemical reactions (plants converting solar energy into chemical energy through photosynthesis). Thus, heat is transferred intentionally or unintentionally through three processes: conduction, convection, and radiation. Heating homes in cold climates and cooling homes in warm climates is sometimes done in a haphazard way. While it is easy to say that the leaks in a home need to be sealed to a cost-effective level, it is a good idea to stop and look at these three very basic methods of heat transfer to better understand how best to build or retrofit a home for cost-effective energy efficiency. A stove burner as shown in Figure 2-9 can be used to illustrate all three transfer processes.

## Heat Flow Types

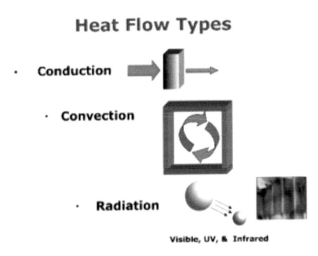

Visible, UV, & Infrared

**Figure 2-9. The various types of heat flow.**

### Conduction

Transfer of heat by conduction involves heat passing through a *solid* or *liquid* object or actual physical contact or touching between two objects: gases are not involved in conduction. When a pot of water is placed on top of an electric stovetop coil, heat is transferred from the burner to the bottom of the pot only where the pot is in direct contact with the electric coil. Conduction also occurs when the stovetop brackets supporting the burner are heated by direct contact with the burner. Both of these examples are demonstrated in Figure 2-10. If you look at a hot piece of iron being worked on by a blacksmith, you will notice that the piece is hotter on the tip than where the blacksmith is holding it. There is heat flowing down the iron from the hotter end towards the cooler end. This heat is being transferred by conduction.

Heat transferred by conduction does not only occur between solids. If a hot beverage is stirred with a cold spoon, the spoon becomes warm from being in direct contact with the hot liquid. When you hold the hot beverage in your hands, heat is transferred to your hands from the cup. In a home, hot or cold are transferred from outside to inside

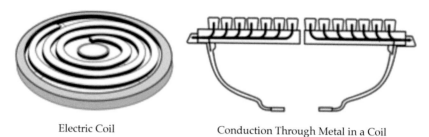

Electric Coil                Conduction Through Metal in a Coil

**Figure 2-10. Examples of Conduction Heating**

through the wall materials by conduction. Heat always flows from the warmer to cooler area, even in conduction. How fast the heat flows is a function of the conduction coefficient (K-value) for the material through which the heat is flowing. More specifically, the K-value is the rate that heat is conducted through a one-inch thick, one square foot slab of the material, Figure 2-11.

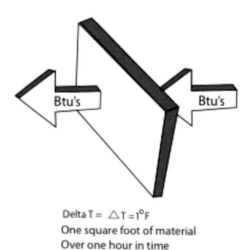

Delta T = $\triangle T = 1°F$
One square foot of material
Over one hour in time

**Figure 2-11. Conductivity Coefficient: Heat Transference Through a Material**

Metals tend to conduct heat very well while insulation, including plastic foam, usually has a very low K-value, or it could not work as insulation. In fact, the R-value of insulation is just a special type of K-value for that insulation. At the same time, as R-value goes up, K-value goes down.

> R ↑ = K ↓
> As R-Value Goes Up, Conductivity Goes Down

## Convection

Heat movement *within liquids* and *gases* is referred to as convection. Convection only applies to liquids and gases because convection requires a medium that moves or can flow. Solids cannot flow.

Referring to the stovetop burner example in Figure 2-12, the hot bottom of the pan warms water in the pan. Once the water becomes warm, it rises to the top of the water in the pan because warm liquids and gases rise. The movement of the warm water to the top and the movement of cold water to the bottom represent convection. In fact, the combination of warm air rising and cold air dropping in a given space is often referred to as a "convective loop." Hot water and air rise because they are less dense (or more buoyant) than cold water or air.

Another common example of convection is blowing across the top of a hot beverage to cool it. Forced convection in this case is being used much like the fan on a furnace blows air past the hot metal heat exchangers to heat the air before it is distributed throughout the house. Another simple example of convection with the stovetop coil is the coil warming the air above it when there is nothing sitting on top of the coil as in Figure 2-13. Once warmed, that air next to the coil rises and can heat the room by this convective process. Cooler air is drawn from below the coil and is heated as it passes by the coil.

As another example, hot water from a water heater is always taken from the top of the tank because that is where the hot water rises by convection inside the tank. The burners at the bottom of the tank heat the water at the bottom of the tank and it rises to the top of the tank.

Another example of convec-

**Figure 2-12. Convection of Water in a Hot Pot**—Warm goes up. After it cools, it goes down.

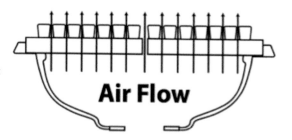

**Figure 2-13. Coil elements heat the air next to them, driving the warmer air up and drawing the cooler air from below.**

tion occurs on a hot summer day looking down a hot asphalt road. Often you can almost see the air warmed by the hot black asphalt rising off the road. Often this hot air is rising so quickly and intensely that it distorts what is seen farther down the road so much that this is often referred to as a "mirage." An example of convection transfer in a heated home is a single pane window in winter, as shown in Figure 2-14. The window cools the air just inside the house and makes the air denser, and therefore heavier. This cold air then flows down towards the floor.

On a cold winter day, the outside wall covering acts much like the single pane window. Thus, because the outside wall covering is cold, the air gap inside the wall is cooled, causing the air to cool and become denser. This colder, heavier air flows towards the bottom of the wall inside the wall cavity, as shown in Figure 2-15.

Because the floor/wall junction or corner and the wall/ceiling junction or corner physically restricts the air next to it from being disturbed, the temperature of the middle section of the wall is closer to the air temperature inside the house because it is more exposed to the air

**Figure 2-14. An example of convection transfer in a heated home is a single pane window in winter. The window cools the air just inside the house and makes the air denser, and therefore heavier. This cold air then flows down towards the floor.**

Net radiation between warm skin and the cold window surface chills the body.

Air cooled by the window drops towards the floor

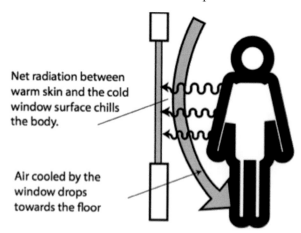

circulation within the house. This difference can be seen in Figure 2-16. Thus, in the winter, the lower junction of the wall with the floor and the upper junction of the wall with the ceiling will appear colder because they have poor mixing of air from the inside of the house. If the house is

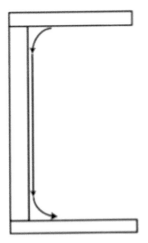

air conditioned in the summer, these upper and lower corners of the walls appear warmer in the image.

Remember heat transferring through a wall is referred to as conduction. However, where there are gaps in the wall, for instance, where there is no

**Figure 2-15. The airflow pattern on a wall in winter shows that little heat exchange occurs at the upper and lower junctions of the wall with the ceiling and the floor where air has a more difficult time affecting the surface temperatures.**

**Figure 2-16. Looking at a typical wall with an infrared camera, you can see convection at work with the most significant temperature differences at wall/ceiling, wall/floor, and wall/wall junctures. A window (dark rectangle) sits in the middle of this short wall. Also a table on the left prevents you from seeing all the floor/wall junctures.**

insulation between the inside and outside coverings of the wall, convection can occur.

At the same time, the inside wall covering such as the drywall in a heated home in winter will be warmer than the air gap or cavity inside the wall. Thus, just inside the wall next to the inside wall covering, the air is being heated, thus expanding it, making it lighter. As a result, in this area, the air flows towards the top of the wall inside the wall cavity. Therefore, in walls without insulation in the wall cavity, a "convective loop" begins to form: the warmer air near the top of the wall cavity begins to flow down along the outside wall as it cools from being exposed to the cold outside wall. The cooler air inside near the bottom of the wall cavity begins to flow up along the inside wall as it warms (and, in the process becomes less dense and thereby more buoyant) from being exposed to the warm inside wall. Thus, it is common to have convective loops in cavities of walls or other spaces where there is a temperature difference between inside and outside, as shown in Figure 2-17. The convective process can be accelerated by forcing air across a surface as happens when air moves past a furnace heat exchanger. This forced airflow increases heat transfer from the heat exchanger so the house heats faster and more efficiently. (see Appendix G on wall R-values to see a similar effect of air films.)

### Radiation

The electric stovetop coil also provides an example of radiation as shown in Figure 2-18. The hot electrical element on a stovetop coil radiates heat. The closer you put your hand to it without touching it, the warmer it feels because of the radia-

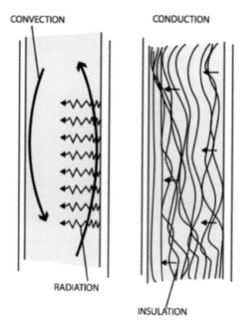

**Figure 2-17. No Insulation In Wall (left), With Insulation in Wall (right) (Radiation applies to both even though it is not shown on the right). Adequate insulation eliminates the convective loop.**

CONVECTION          CONDUCTION

RADIATION

INSULATION

tion from the electrical element. Heat can radiate from the burner all the way to the ceiling, albeit the amount of radiation received at the ceiling will be far less than your hand received closer to the element. Thus, radiation does not require any medium of solids, fluids, or gases for the transfer of heat or any direct physical contact.

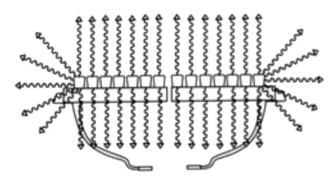

**Figure 2-18. Radiant Heat Extending in All Directions Out From the Coils Regardless of Medium (this happens even in a vacuum)**

A good example of radiation from the weather arena is overnight temperatures that do not drop significantly when there is cloud cover at night. Land and objects that have been warmed by the sun during the day continue to emit radiation into the sky at night. This radiation hits the clouds and is reflected back to the earth as shown in Figure 2-19.

Other examples of radiant heat in homes include warmth from fireplaces, warmth from radiators that are part of hot water or steam boilers, and even warmth from incandescent light bulbs. Everything radiates or absorbs energy, and typically, most solids are both absorbing and radiating energy at the same time. The only question is how much of each. This radiation occurs in addition to any convective or conductive heat transfer going on along with the radiation.

In the home setting, radiant heat is everywhere. For instance, glass windows allow radiation to be transmitted through them, while reflecting some radiation and absorbing some as well, see Figure 2-20. The radiation that is transmitted through glass ends up hitting furniture, carpet, ceilings, etc.

Once that radiation hits these different materials inside the home, the process begins again with some radiation being transmitted through the material, some radiation being reflected, and some being absorbed

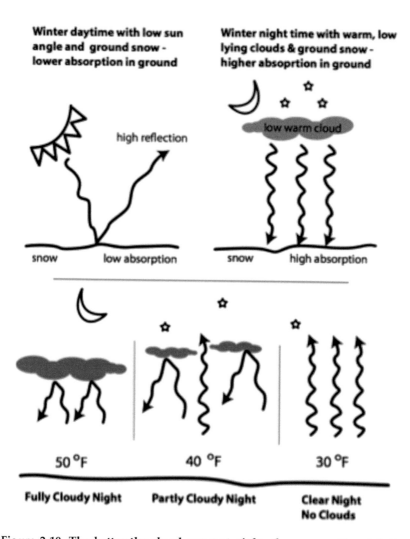

**Figure 2-19. The better the cloud cover at night, the warmer the nighttime temperatures. Nighttime warmth often carries over to allow the next daytime temperatures to be warmer too.**

as heat. This heat energy is typically re-radiated as infrared radiation and the cycle begins again. For instance, some visible light will be reflected off a white carpet and onto the ceiling. The ceiling then reflects some visible light to the wall, etc.

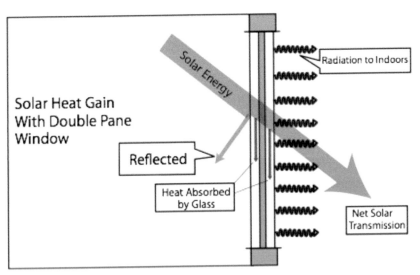

**Figure 2-20. Solar Energy Passing Through a Window. Some energy is reflected; some is absorbed, while some passes through the window (frame missing)**

Outside of the solar range, in the "far infrared" or very long wavelength infrared range, the emissivity can be equal to the amount absorbed, as when a white "cool" roof emits all the far infrared it absorbs, (unlike an unpainted metal roof). Mass walls deliver their heat to the house at night in the far infrared spectrum as well. Thus, for the white roof and mass walls A=E, where A = absorption and E =
. When an object is exposed to solar radiation (for example, by solar exposure), and the total radiation the object is exposed to is assumed to be 100% or 1.0, then, the total of all the radiation transmitted through, absorbed, and reflected from the object will be 1.0 or 100%. Thus R + A + T = 1.0 where R= reflected radiation, A = absorbed radiation, absorption, absorbance, or absorptance, and T = transmitted radiation or transmittance (passing through the object). The corresponding characteristics of the materials involve reflectance, absorbtivity, and transmissivity, respectively as shown in Figure 2-21. All materials have an emittance coefficient. This is also known as a material's emissivity or emissivity coefficient. The closer the coefficient is to 1 or 100%, the easier it is for the material to emit radiation when it is hot. The lower the coefficient, or the closer it is to 0%, the more difficult it is for the material to emit radiation when it is hot.

# Radiation - Visible, UV or Infrared
## Radiation <u>coming in</u> is either:

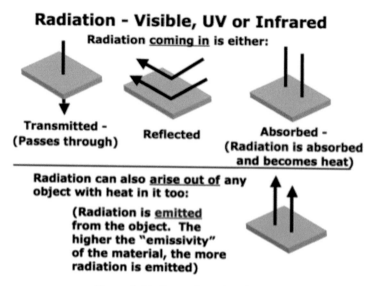

**Transmitted –**
**(Passes through)**          **Reflected**          **Absorbed –**
**(Radiation is absorbed**
**and becomes heat)**

**Radiation can also <u>arise out of</u> any**
**object with heat in it too:**

**(Radiation is <u>emitted</u>**
**from the object. The**
**higher the "emissivity"**
**of the material, the more**
**radiation is emitted)**

**Figure 2-21. Examples of Radiation**

Remember that metals are good conductors of heat. They have a high K-value and a low R-value, and cannot be used as insulators. For instance, if you grab a hot piece of metal, the heat from the metal easily transfers to your hand. It feels hot, and you pull your hand away to avoid being burned. In the case of radiation, metals, especially shiny ones, tend to be poor emitters of radiation energy. They can also reflect solar rays well and that is why some of the best roof coatings are made of aluminum or other appropriate metals. Metals have a low emissivity coefficient. This is the reason a thin metal layer can be on the underside of roof decking to help keep heat from the shingles from transferring into the attic space as easily, as long as no insulation is touching the barrier. Metals are some of the few common substances that reflect both visible and infrared radiation back away from the building. However, nothing can touch that thin metal layer where it faces the inside, or the benefit of being a poor radiation emitter is lost by allowing its high conductivity to transfer heat to whatever it is touching.

Thus, if radiant barriers are placed in walls, floors, ceilings, or attics they cannot touch anything on the side of the radiant barrier that faces the direction of unwanted heat flow. Thus, in a cold-weather area, you could place the barrier on the back of the drywall to keep the heat in. However, to get the radiant barrier to work properly you will need to furr out the drywall away from the vapor retarder/barrier or insulation behind it so that nothing touches the radiant barrier. Similar requirements would exist for ceilings, attics, and floors. Perhaps one of the most fitting uses for radiant barriers would be on the underside of roof rafters in a pitched roof attic in a cooling climate. Thus, extensive heat flow from the shingle surface down through into the attic could be reduced by placing the barriers on the underside of the roof rafters. In this case, it is unlikely that anything would ever be touching the exposed face of the radiant barrier that faces the attic space, and as a result, some heat from the roof deck would be kept from entering the attic by the radiant barrier. You can see why radiant barriers are seldom installed: difficulty in meeting cumbersome application requirements and the risk that materials could fail or fall into a position of touching the radiant barrier and, thereby eliminating the benefits of the radiant barrier. In the end, many consider the installation of radiant barriers as having a low cost effectiveness.

As another example in the building arena, low E-glass windows help reduce the emission of radiation in the unwanted direction. Figure 2-22 shows an example of window label.

Low-e film is placed on the inside pane of a two-pane window in areas that are primarily heating dominated climates, but on the outside pane in ar-

Figure 2-22. Sample NFRC Label for Standardizing Window Characteristics

eas that are primarily cooling dominated climates, see Figure 2-23.

Most radiation, except for visible light, cannot be seen without the use of special equipment. In the case of the infrared spectrum, using an infrared camera can help detect infrared radiation that would otherwise be unseen with the naked eye. On the other hand, ultraviolet light or UV represents only 3 to 5% of the light spectrum because of filtration by the ozone layer, so UV radiation for the most part is ignored. As always, net radiation radiates from the warmer to the cooler area. Figure 2-24 shows how net radiation is determined.

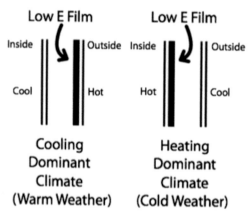

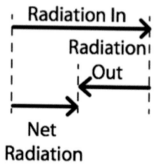

Figure 2-23. Placement of Low-e film in Different Climates

Figure 2-24. Net radiation is the difference between the amount of radiation hitting a surface and the radiation leaving the surface.

### Radiation and Climate Change

Radiation is at the heart of the climate change debate. Pure radiation is how energy reaches the earth from the sun. Fossil fuels such as oil and coal originated from plants grown by the sun, or from dinosaurs that ate the plants grown in the sun. Herein lies the dilemma of climate change: the more we use energy directly from the sun, or so-called renewable energy, the less we need to pollute the earth's atmosphere. The more we use energy that comes indirectly from the sun, as through fossil fuels, the more we use energy at a faster rate than it is being produced by the sun on the planet at the time. Thus, we run the risk that using fuel created in a previous timeframe will overload the atmosphere with pollutants during our time frame. The more we create pollutants that strengthen the "greenhouse effect," the worse the greenhouse effect will become. Figure 2-25 and 2-26 shows how the greenhouse effect occurs. Some also consider nuclear energy to be a renewable energy because it also does not generate climate-changing pollution.

The term "greenhouse effect" is used because of the similarities of this process to the way a greenhouse works. The glass in greenhouses allows almost all the solar radiation to be transmitted into the green house through the glass. That solar radiation then warms objects within the greenhouse as it strikes them. These warm objects then emit infrared radiation. The glass partly absorbs and partly reflects back this radiation into the greenhouse, thus, keeping it from escaping out of the green-house.

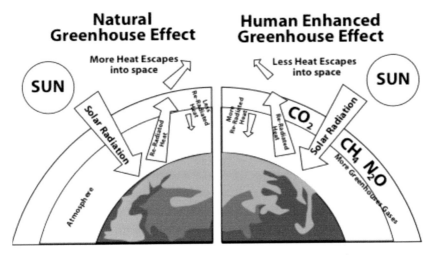

Figure 2-25. The Greenhouse Effect on the Earth

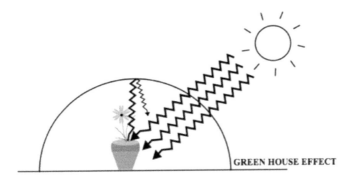

Figure 2-26. The Green House Effect

The closer the sun's rays hit perpendicular to a surface; the more the energy is absorbed by or transmitted through the material. The more the sun's rays hit the material at an angle, the less the material absorbs or transmits energy and the more energy is reflected. This concept is shown in Figure 2-27. For example, a flat roof will be hotter than a sloped roof when the sun is directly overhead because the flat roof sits exactly perpendicular to the sun's rays coming from directly above.

Figure 2-27. Effect of the Sun's Angle on Whether Radiation is Absorbed or Reflected—The closer to parallel the ray is, the more likely it is going to be reflected.

**Applied Heat Transfer**

How do the principles of heat transfer apply to buildings? Heat generated inside a building, say by a furnace, warms the inside surfaces of the building by both radiation and convection. Heat is transferred through the building surfaces themselves mostly by conduction, but also with the help of convection if there are cavities in the wall, and to some degree by radiation as well. On the cold outside of a wall assembly, heat leaves the wall by convection and radiation. An item that is physically touching the building assembly, like a metal stair, will remove heat by conduction. At the same time, when conditioned air leaves the building through a leak, it is replaced by outside air and there is even more significant heat loss. In the end, one of the most significant sources of energy losses in a home is by leakage through "bypasses" in the walls, ceilings, floors, windows, doors, etc. Leakage is considered mass flow or mass transport – we are taking warm air that has mass (albeit a light mass) and it is being transported from, for instance, a warm area inside to a cold area outside. In the process, it cools down, etc. and the heat energy is given up to the outside.

OTHER PERTINENT ENERGY BASICS

This next section discusses other basics in the energy arena, including the combustion process, furnaces, air conditioning and insulation.

**The Combustion Process**

The combustion process is often used to produce heat for homes. Air combines with fuel and a heat source (such as a spark), and oxidation or burning occurs to produce heat. This is often called the combustion triangle—oxygen, heat source (spark), and fuel, Figure 2-28. Combustion products include carbon dioxide, heat, and the often unexpected and overlooked byproduct, water vapor.

$$CH_4 + 2O_2 \rightarrow energy + CO_2 + 2H_2O$$
(Burning of natural gas)

This production of water vapor is why households that use portable, unvented kerosene heaters experience a buildup of moisture, and sometimes may have ice on the inside of windows in the winter in cold weather areas.

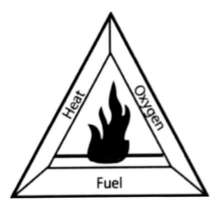

**Figure 2-28. Requirements for Ignition of Burning**

Complete combustion occurs when only carbon dioxide ($CO_2$), water and heat are produced. Complete combustion is very difficult to achieve, although it can come close. Carbon monoxide (CO) can be a byproduct when the ratio of fuel to oxygen is either too high to allow for the complete formation of $CO_2$, or when the temperature of the combustion process is too low to allow for complete burning to occur. If the fuel is not completely burned in the combustion process and it goes out the flue, the "fuel burning efficiency" or "burn efficiency" is not 100%. Even though it is difficult to get all the fuel to burn, only 1% or less of the energy is usually lost to unburned fuel.

**Primary, Secondary, Dilution, and Excess Air**

Air involved in the combustion process can be primary or second-ary air. Primary air is air that is mixed with the fuel before combustion begins. In a furnace, there is an adjustable opening just before the burn-ers to allow primary air to mix with the fuel. Secondary air is additional air that surrounds the flame after the fuel has already been ignited. Thus, the air being drawn in next to the flame is considered secondary air. On a jet burner (Figure 2-29), the air being drawn in through the heat exchanger opening around the jet of blue flame is secondary air.

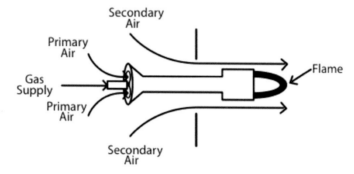

**Figure 2-29. Primary and Secondary Air**

On a gas stovetop burner, the secondary air mixes next to the flame by entering from down below the burner. Part of both the primary and secondary air is used in the combustion process and part passes through as excess air (see "excess air" below). Thus, since flue gases are exhaust-ed to the outside, some energy is wasted by heating gases that are part of the combustion process and the gases that are just "going along for the ride." The losses in energy that occur as a result of the escaped gases that were part of the combustion process are framed in terms of "sensi-ble" and "latent" heat losses. The sensible heat losses due to combustion gases are simply the losses of energy due to heating these gases only to lose them out the flue. The latent heat losses due to combustion gases are the losses of energy due to heating the vapor in the air to the higher temperature only to lose them out the flue. Thus, losses of energy in combustion gases (gases that are actually part of the chemical reaction of combustion) are recognized as either sensible or latent energy losses. Remember, energy is also lost by heating air that is just going along for the ride. Table 2-3 list the various types of heat losses in furnaces.

### Table 2-3. Types of Heat Losses From Furnaces

Sensible

Latent

Excess Air

Dilution

Cycling

Jacket/Cabinet

Duct Leakage/Poor Insulation

Dilution air is additional air from inside the area where the heating unit is located. It is mixed farther downstream of the burners and even past the heat exchangers, where a draft diverter, mixing box, or bell housing exists. Dilution air is drawn into the draft diverter by a negative pressure created by the hot gases being pulled up the flue or chimney by the principle that hot air rises. This is lost energy since heated building air (before it is mixed with the flue gases) is sent out the flue or chimney. Despite attempting to achieve higher efficiencies over time by avoiding the use of dilution air, many new boilers and water heaters still use a draft diverter that must be provided with dilution air. Figures 2-30, 2-31, and 2-32 are diagrams of various draft diverters.

Barometric draft dampers as shown in Figure 2-33 are a special type of mixer that keeps unnecessary dilution air from entering the flue. A special lever on the damper normally keeps the damper closed and only allows it to open when there is negative pressure in the flue. This keeps the damper from being opened when the flue is priming for a while after the heating unit has been started. It does this by only allowing as much dilution air to mix in as is necessary to obtain a good flow of flue gases out the chimney or

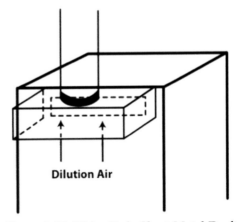

**Dilution Air**

Figure 2-30. Older Style Sheet Metal Draft Diverter on the Outside of a Furnace. These were typically installed on-site by a mechanical contractor.

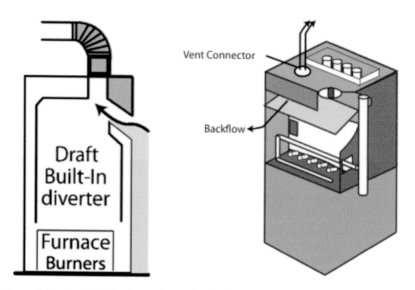

Figure 2-31. (Left) Side view of standard efficiency furnace with built-in draft diverter and proper flow of dilution air mixing into the flue with the combusted gases (Right) Cutaway view of standard (low efficiency) furnace draft diverter with backflow of combusted gases

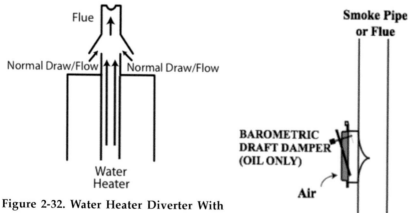

Figure 2-32. Water Heater Diverter With Proper Flow of Dilution Air Mixing Into The Flue

Figure 2-33. A Barometric Draft Damper on a Flue For an Oil-Fired Appliance

flue. Thus, barometric draft dampers are more efficient than the classic "wide open" draft diverters.

Excess air is air that is in excess of what is needed for the combustion process and is released along with the rest of the combustion byproducts above the stovetop burner into open air or out the flue, in the case of a furnace. The excess air is not only both the primary and secondary air that is not used in the combustion process, it also includes the dilution air. Of the primary and secondary air, secondary air represents the larger of the two in terms of excess air because it is not directly mixed with the fuel at ignition, as is primary air. If more than the minimum oxygen required for complete combustion is provided, the unused portion of that oxygen and the rest of the unused air (such as nitrogen) will become an unused portion that exits along with the byproducts of combustion as excess air. Along with it goes the energy it took to heat it. Technically, this unused excess air is only passing through the process unchanged and without participating in the chemical process, and so it really is not a product of combustion. At the same time, if the air is heated in the process and passes out of the flue, it contributes to wasted energy. The higher the percent of excess air in the combustion process, the lower the efficiency of the appliance or furnace. An efficiency diagram is shown below in Figure 2-34.

## HEATING SYSTEM/FURNACE BASICS

The fundamental process of a heating system includes combustion, heat exchange of the produced heat across an exchanger to a separately chambered warm air distributor, and separate ventilation of the byproducts of combustion out the chimney or flue (Figure 2-35).

You will probably come across three categories of heating units: low efficiency, medium efficiency and high efficiency. With each decreasing level of efficiency, the exhaust flue temperatures are higher and more heat is lost when the flue gases are exhausted to the exterior.

### Low Efficiency Furnace

In low or standard efficiency furnaces (Figure 2-36), a high percentage of heat is not extracted from the combustion process. The exhaust air coming from the combustion process is hot enough to safely rise up the chimney or flue on its own. These appliances are typically

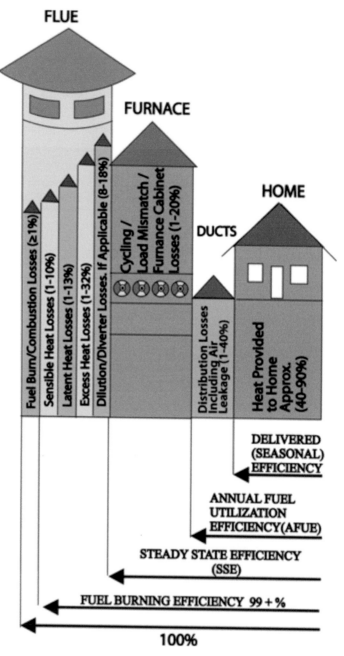

Figure 2-34. Different Energy Losses to Furnace Operation

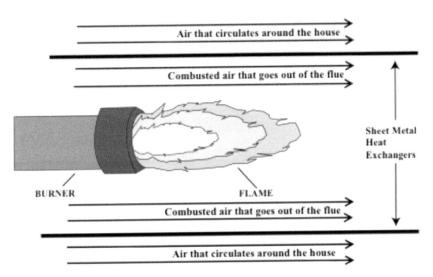

**Figure 2-35. Separation of the Bad Air (Combusted Air) From the Good Air (Circulating Air) by a Heat Exchanger**

called atmospheric or "gravity fed" appliances. Unfortunately, this ability for the hot flue gases to pass out of the house through a flue on their own makes them inefficient. If more heat had been extracted from the combustion process, the exhaust would not be hot enough to be able to rise out of the flue on its own. Another reason these furnaces are so inefficient is that they have a more open clamshell style heat exchanger that extracts less heat than the restricted convoluted heat exchangers found in both medium and high efficiency furnaces.

**Medium Efficiency Furnace**

If a little more heat could be extracted from the combustion process, a fan would have to be used to help exhaust the combustion gases as in a medium efficiency furnace (Figure 2-37). The fan also helps to pull the air through the convoluted heat exchanger. Medium efficiency furnaces have no secondary heat exchanger and are therefore less efficient than high-efficiency furnaces.

**High Efficiency Furnace**

In a high efficiency furnace Figure 2-38), even more heat is extracted from the gases. These furnaces require a flue fan to push the air up the flue because it is no longer hot enough to rise up the flue on its

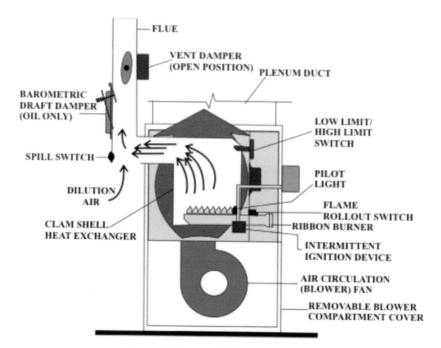

**Figure 2-36. Low Efficiency Furnace**

own. In fact, the temperature is low enough that a plastic PVC sprinkle pipe can serve as the flue for the furnace. So much heat is extracted there is little sensible, latent or excess air heat loss. High-efficiency furnaces are also more efficient because they obtain better heat exchange by using a more convoluted tube or passage way. They are also more efficient because hot gases pass through both a primary and a secondary heat exchanger to extract even more heat. In the secondary heat exchanger, water vapor is extracted from the hot gases in the form of condensate. Medium and low efficiency furnaces do not have secondary heat exchanger that extracts water vapor from the hot gases.

Modern medium and high efficiency furnaces have also become more efficient because they do not use a standing pilot light, and instead use an electronic ignition, such as a "hot plate" or spark ignition system.

**Energy Losses**

All furnaces experience some sensible heat, latent heat and excess air energy losses. There are additional energy losses from the heating equipment cycles. For instance, energy is wasted when the thermostat

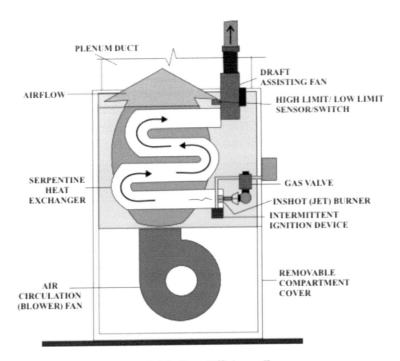

Figure 2-37. Medium Efficiency Furnace

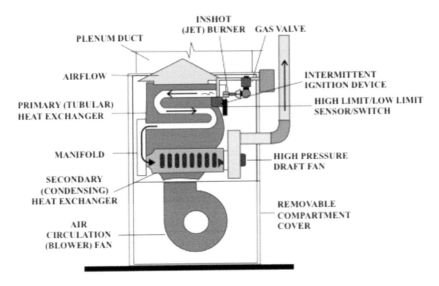

Figure 2-38. High Efficiency Furnace

shuts off a boiler or furnace because there is typically some heated water or air left in the system that will not be circulated. If the furnace or boiler continues to circulate this heated water or air after the burners have been shut off, some, but not all, of that heat can still be utilized to heat the home. Thus, although cycling and jacket losses (see Figure 2-40) can be reduced, they cannot be completely eliminated. Some factors that affect the amount of energy wasted from cycling losses include the heat required to heat the heat exchanger, the number of cycles the furnace will go through in a day, and how efficient the distribution ductwork or piping is. Thus, if there is a thick heat exchanger, as with boilers, then the cycling losses are typically higher than if there is a thin heat exchanger as in furnaces, Figure 2-39.

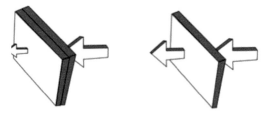

**Figure 2-39. More heat travels through a material over a given timeframe if the material, such as a heat exchanger, is thinner.**

Furthermore, if there is an oversized furnace in a home, the furnace may not even have to heat up to its optimum temperature before it has completely heated the home. In this situation, it will cut its cycle short or "short cycle," resulting in more losses of energy than if an appropriately sized furnace had been placed in the home. Finally, if there is an inefficient ductwork system that allows poor flow in the case of a furnace, then there will be more cycling losses than if the ductwork were more efficient. Note that these are not losses due to poor duct insulation or leaks in the ductwork, only whether they are large enough to accommodate the airflow through them. These considerations assume that the furnace is not located in rooms that are being heated. However, if you are evaluating a room heater, the waste heat resulting from cycling losses will be kept within the room. As a result, room heaters typically have lower "off cycle losses" then do central heating systems.

One final source of energy loss in heating systems comes from "distribution losses." These losses typically are a result of leaks in the pipes or ducts (mass transport) and result from the lack of insula-

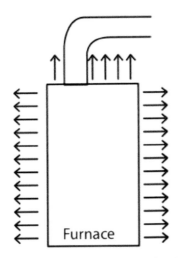

**Figure 2-40. Furnace "Jacket" Energy Losses**

tion around pipes or ducts (radiation, conduction, and convection losses). You can reduce leaks (the most significant source of distribution losses) by sealing them. Ducts and pipes can be insulated and a radiant barrier can be installed to reduce the losses from conduction, convection, and radiation. Another example of distribution losses occurs when the circulation of air or water flows by the heat exchanger too slowly. This kind of a misfit situation, caused by an inadequately sized blower fan or water circulation pump, results in less heat being extracted through the heat exchanger from the burner gases, which, in turn, sends more energy out the flue in the process.

**Efficiency Definitions**

If all the energy losses described above are taken into account, the overall efficiency of the entire heating system can be determined (including the ductwork), which is often termed the "delivered heating efficiency" or "seasonal efficiency." If the distribution losses that occur from the piping or ductwork are eliminated, this is defined as the annual fuel utilization efficiency or AFUE. Thus, while AFUE takes into account unburned fuel, the sensible and latent heat losses of combustion gases, excess air losses of non-combusted gases, off-cycling or jacket losses, and the losses of heat through the furnace wall itself, it does not include losses due to distributing the heat around the house that are caused by leaks or uninsulated pipes or ducts. The AFUE is 90 to 97% for high efficiency furnaces and 80 to 90% for medium efficiency furnaces.

Trim back another aspect of energy losses, (not taking into account cycling losses) to get the steady state efficiency or SSE. It is called steady state because it evaluates efficiency only once the furnace has fully warmed up (on at least 10 minutes) and does not take into account the losses due to cooling down – it does not evaluate transient, but rather only steady state efficiency. This is why you can only check SSE with a combustion analyzer after the furnace is fully heated by taking samples from the flue, as shown in Figure 2-41. Because the SSE is an efficiency test that is done on site with the furnace, rather than in controlled circumstances off-site, it is considered an "in situ" measurement or test. The SSE for high efficiency furnaces is 97 to 98%. The next stage of efficiency does not include the combustion gas losses (sensible and latent losses) or the excess air losses, leaving only the fuel burning efficiency (see Figure 2-34).

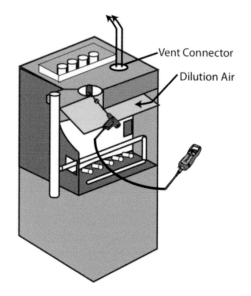

Vent Connector

Dilution Air

**Oil Burners**

High-efficiency oil burners do not condense moisture out of the flue gases with a secondary heat exchanger because the combustion of oil does not involve producing as much moisture as burning natural gas. Instead, they achieve an AFUE of close to 90% using a special flame retention head oil burner (FRHOB) that is capable of almost complete combustion using very little excess air.

**Figure 2-41. Checking Steady State Efficiency on a Low Efficiency Furnace**

In addition, eliminating the use of the barometric draft damper by using a high-pressure burner has also added to higher efficiencies in oil burning boilers and furnaces. Finally, interruptible solid-state igniters and split capacitor motors are now being used to save electrical energy in oil burners.

**Applied Efficiency**

Most homes today are heated with oil, natural gas, or electricity. Because of the inefficiency in combustion furnaces mentioned above, you might ask why not use electric heat since electric heating units are 100% efficient. Electrical energy is much more expensive because all but only a portion of that energy is lost through transmission losses especially if the electricity is used far away from the dam, coal plant, etc. where it is generated, as shown in Figure 2-42.

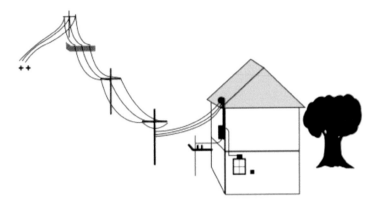

**Figure 2-42. Electric power is lost in the process of transmitting it across power lines.**

One of the benefits of on-site generation of energy such as by solar photovoltaic panels is that there is little transmission loss. Natural gas must also be transported involving expensive pipeline and pumping stations. Fuel oil also can be expensive to transport by truck, etc. Therefore, in addition to combustion efficiencies, all these other factors must be taken into account, Figure 2-43.

**Figure 2-43. Using pipelines and rail to move energy is expensive.**

COOLING UNITS—THE AIR-CONDITIONING PROCESS

There are two basic types of air conditioning units, split and packaged. A split system essentially has a condenser coil outdoors, and an evaporative coil indoors, typically just downstream of the furnace or blower. A packaged system (Figure 2-44) combines the condenser coil and evaporator coil into one unit or box. Single, combined units are typically mounted on the roof or sit on a concrete slab next to the house and have ductwork directly connected to the house.

Figure 2-44. This is an example of a package unit air conditioning system where both the evaporator and condenser are outside. Because both the evaporator and condenser are outside, ductwork must go from the unit to the inside of the house.

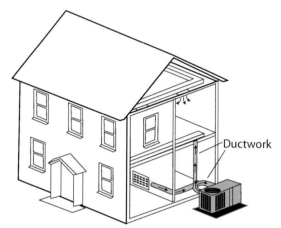

Ductwork

Air conditioners work as follows: Refrigerant gas travels to the compressor as a saturated vapor. At the compressor, it is compressed to a higher pressure and temperature. This hot refrigerant vapor is then routed through the condenser coil or piping, where it is cooled and condenses or changes to a liquid. This process is accomplished by running the hot vapor through the condenser coil with relatively cool air from outside flowing across the condenser coil or pipe, Figure 2-45.

You may have heard an outside condenser unit fan when you were walking around a house. This is the fan that blows relatively cool out-

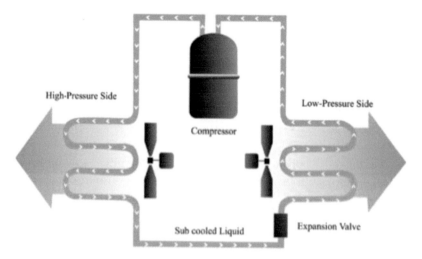

High-Pressure Side

Low-Pressure Side

Compressor

Sub cooled Liquid          Expansion Valve

Figure 2-45. Condenser Coil—left, Evaporator Coil—right

side air across the condenser coil to cool the hot refrigerant vapor and convert it into a liquid.

This condensed liquid refrigerant then passes through an "expansion valve" where it experiences a major reduction in pressure. The reduction in pressure results in evaporation of a portion of the liquid refrigerant. This evaporation cools the refrigerant down to a temperature that is colder than the air temperature inside the home. This cold liquid and vapor refrigerant mixture passes through the evaporator coil or piping located inside the home while a fan blows the warm inside air across the outside surface of the evaporator coil, allowing heat to be transferred from the air to the refrigerant. As the heat is transferred, the remaining liquid part of the cold refrigerant mixture is converted back into vapor as it warms. As the warm, inside circulating air passes through the evaporator coil, it is cooled and the air in the home is conditioned to a lower temperature. The refrigerant vapor exiting the evaporator coil is routed back into the compressor and the refrigeration cycle starts over again.

INSULATION AND R-VALUES

The R-value is a number assigned to a material to quantify its resistance to the transfer of heat. It is typically expressed in terms of resistance per unit of depth. For instance, wood typically has R-1 per inch of thickness. The R-value is determined by guarded hotbox (Figure 2-46) testing. This testing is done by keeping an inner box, the guarded, at a known temperature and placing the material to be tested around it as part of an outer box. The R-value is determined by noting the temperatures on each side of the material being tested.

The U-value is the inverse of the R-value. The U-value is a measure of thermal transmittance, not resistance. It is like the K-value. It is found by measuring how long it takes a known quantity of heat to equalize across a material. Since R-value is the inverse of the U-value, the following equations apply:

$$R = 1/U \text{ and } U = 1/R$$

As an example, a material with a U-value of 0.5 would have an R-value of 2.

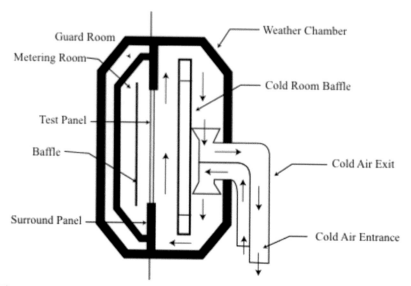

**Figure 2-46. A Hot Box for Testing the R-Value of Materials, Including Windows, Doors, Walls and any Other Materials**

As general guidelines for R-values of specific materials, Table 2-4 gives a list of materials and their generally accepted R-Value.

Thus, of those listed, polyisocyanurate board has the highest R-value, but concrete would have the highest U-value because it does not serve as a very good insulator and has a low R-value.

Determining the actual R-values in a home can be much more complicated. See Appendix G for a discussion of R-values in reality.

**Table 2-4. R-Values of Different Materials**

| Material | R-Value |
|---|---|
| Fiberglass | 3.5 per inch |
| Cellulose | 3.0 per inch |
| Bead Board or expanded polystyrene | 3.6 per inch |
| Styrofoam or extruded polystyrene | 5 per inch |
| Polyisocyanurate Board | 5.6-7.6 per inch |
| Glass | 1 per layer |
| Wood | 1 per inch |
| Concrete | 1 per 8 inches |

# Chapter 3
# House As a System

In the beginning days of weatherization, it was thought that replacing doors and windows would be a major benefit. Over time, it began to be apparent that tightening a home by insulation would be the best improvement. Homes are tighter today in many ways: they have better insulated windows and doors, concrete has replaced stone foundations and floors, drywall and plywood seal better than lath and plaster, and sealing caulks are more durable. Homes of today have more and bigger fans than before: clothes dryers, central vacuums, whole house ventilation fans, air handlers instead of gravity fed circulation systems, stronger kitchen fans, stronger furnace fans, and central air conditioning. Homes of today also have weaker draft appliances: higher efficiency heating appliances require less draft, and sealed combustion units or direct vented appliances bring combustion air directly from outside rather from inside the home. Finally, homes of today have less drying potential: insulation in walls and ceilings slows or stops air movement, single level buildings have a reduced stack effect, etc. People's habits have also changed. Instead of weekly baths, most people now take daily showers. Hot tubs, saunas, indoor pools, and multiple showerheads also add moisture to the interior of the home today.

## WHOLE HOUSE

Unfortunately, as homes have become tighter, safety problems have often developed. Building science experts believe that the key to successful weatherization is to treat the whole house as a system. It is important to understand the interdependency of the components and systems in a home and their combined effect on energy savings and health and safety issues.

Here are some examples of how the components of a home are integrated and how they affect other parts of the home in the process.

1) An uninsulated attic can make the heating and/or cooling system work harder.

2) Leaky canned or recessed light fixtures can increase heat loss and gain, and cause moisture and ice damming problems.

3) Exhaust fans can pump air into crawlspaces or attics. Moisture in this air can condense on the roof deck, floor joists, and other structural features, which in turn, weakens them, as shown in Figure 3-1.

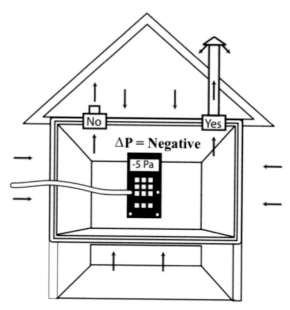

**Figure 3-1. Exhaust air into an attic is not appropriate because it may deposit moisture as condensation. In addition, any exhaust can create a negative pressure in the home with respect to (WRT) outside, which will pull air from the outside into the home.**

4) Attached garages have the potential for carbon monoxide poisoning.

5) Combustion appliances that rely on the gravity feed of the hot air for the exhaust of the combusted gasses out of the flue can potentially cause safety concerns.

6) Moisture from showering, faucet aerators, cooking, and moisture that gets into the house through concrete, soil in a crawlspace, landscaping sprinklers, poor grading, inadequate gutters, and down spouts can be a problem.

7) Air infiltration at windows, doors, and bypasses and air exfiltration at cathedral ceilings and recessed light fixtures or canned lights can cause significant problems.

These and other potential problems must be considered when weatherizing a home. Although the goal of weatherization is to reduce the heating and cooling load, there has to be enough fresh air for the occupants. In addition, if the work is not carefully done, there could be moisture problems that did not exist before weatherization. To make things worse, the tighter the building, the stronger the interaction becomes between components of a home.

In previous times, contractors purposefully left houses leaky to provide good indoor air quality in the fall and spring. Figure 3-2 demonstrates how older homes leak more than newer homes. However, this also made them drafty and expensive to cool in summer and heat in winter. The first energy crisis in the early 1970s inspired many to insulate without considering moisture or indoor air quality. In some homes, the result was rot, mold, and odors.

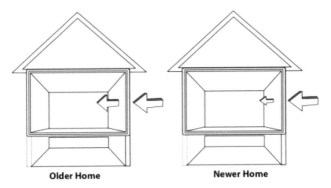

**Older Home**　　　　**Newer Home**

**Figure 3-2. An older home typically has more air leakage than a newer home.**

Newer building codes promote and recognize tighter home construction; however, except for some of the most advanced codes, homes have become tighter without considering some of the potential negative side effects.

## COMPONENTS OF THE HOUSE

Weatherization is concerned with the building envelope, the thermal boundary and the air barrier. These terms are defined as follows:

**The Building Envelope**: The building envelope is the outer structure or shell of the home. It is what separates the interior of a building from the exterior.

**The Thermal Boundary**: The thermal boundary is essentially the insulation in the walls, attic, crawlspace, etc. of a home. Because it is made of insulation, the thermal boundary can be easily identified. Common insulation materials include cellulose, fiberglass, and vermiculite. Insulation acts as a resistor to heat flow or conductive heat loss through the building envelope, and limits heat flow between the inside and the outside of the house. Even small areas of missing insulation can make a big difference in the thermal boundary. Voids in insulation of only 7% can reduce the effective thermal resistance (R-value) by almost half. As an example, if there is a roughly 1000 ft² attic space above a home that is insulated to R-38, the affective R-value falls to 19 when only 70 ft² of insulation is removed.

**The Air Barrier**: The air barrier, also known as the pressure boundary, limits airflow between the inside and the outside of the home, and thereby reduces the loss of heat and moisture in that air. The air barrier is typically the interior drywall or lath and plaster in a home. The optimal air barrier should encompass the living space and nothing else. In other words, the air barrier should not encompass unconditioned areas of the home. The air barrier cannot always be determined through visual inspection. The more elaborate the home, the more difficult it is to have a continuous air barrier, especially around the inside and outside corners of the walls. As a general rule, the inside wall of a home should be chosen as the air barrier in cold weather areas and the outside of the exterior walls should be chosen as the air barrier in warm weather areas. One of the best ways to determine the condition and location of air barriers is by using a blower door fan and pressure testing, as will be discussed in a later chapter.

The air barrier should be in contact with and continuous with the thermal barrier. If not, there is a risk that air will pass through the insulation, carrying moisture and heat with it. One way to make sure that the thermal and air barriers are in contact with each other is to use materials that serve as both. Such materials include foam board, spray foam, and properly installed dense pack cellulose.

In Figures 3-3 and 3-4 are examples of where the air barrier and thermal barrier are aligned or continuous and where the air barrier and thermal barrier are mismatched or not aligned.

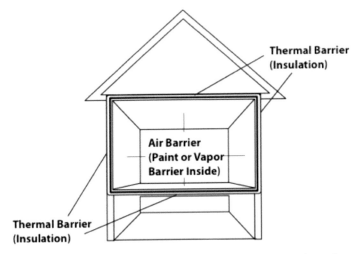

**Figure 3-3. A Home With Thermal and Air Barrier Aligned**
The air barrier (shown on the inside of this building) and the thermal barrier (shown on the outside) are aligned because they are right next to each other. The dark line represents the air barrier while the lighter line represents the thermal barrier or insulation.

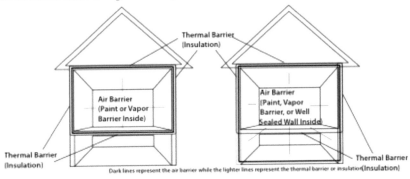

Matching                                          Not Matching
**Figure 3-4. Thermal and Air Barrier**
With the air barrier shown on the inside and the thermal barrier shown on the outside of these buildings, it is not difficult to see that the crawlspace on the home at the right has the best air barrier around it even though the floor is insulated and represents the thermal barrier. This mismatch could be due to openings in the floor where plumbing penetrations, etc., exist, excluding the use of the floor as an air barrier. Use ZPD testing to determine which parts represent the better air barrier.

One of the most challenging air barrier issues comes with "Cape Cod" style homes as shown in Figure 3-5. In these 1-1/2 story homes, it is not uncommon for there to be relatively free communication of air from the attic spaces (A and E) into the vaulted ceilings and knee walls of the upper floor (D and C respectively), and through them, the floor of the upper floor (B). In a cold weather area in the winter, the upper story B floor often feels cold. This space is between the upper and lower floors and should be warm. However, because of air leakage from the attic space all the way down into the floor space of the upper floor, it feels cold.

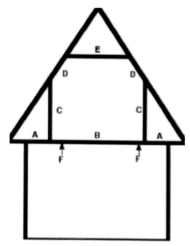

**Figure 3-5.**
**1½ Story Cape Cod Home**

**Fire Barrier:** A fire barrier resists fire. It serves a different purpose than the insulated thermal barrier used in the energy arena even though it may use rock wool insulation as a true fire resister. It should keep the underlying materials from reaching 250°F for at least 15 minutes. A firewall or fire partition on the other hand, is rated for at least 15 minutes and is intended to stay standing during a fire.

## THE EFFECT OF AIR ON THE HOUSE SYSTEM

### Air Flow

Airflow is measured in cubic feet per minute (CFM), or ft³/min. A cubic foot is a little larger than the size of a basketball. Therefore, a flow of 100 CFM would be roughly the equivalent of the air in 100 basketballs flowing in one minute, Figure 3-6.

Air moves from high to low temperature areas and from high to low-pressure areas through holes. For every conditioned (heated or cooled) CFM that goes out of the house, a CFM of untreated outside air enters the home, but typically at a different location. This condition is shown in Figure 3-7. Consider this example: Hot air is less dense

# Air Leakage Comparison

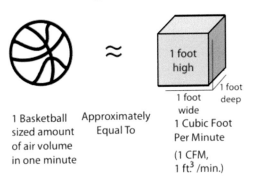

1 Basketball sized amount of air volume in one minute | Approximately Equal To | 1 foot high
1 foot wide
1 foot deep
1 Cubic Foot Per Minute
(1 CFM, 1 ft³ /min.)

**Figure 3-6. Air Leakage Comparison**

than cold air and so it exists at a higher pressure due to its buoyancy. Therefore, in winter, the air wants to find a way to get cooler. It will find holes in the house to get outside to the cold. When hot air escapes a home in cold weather, it takes heat energy with it. Since escaped warmer air must now be replaced in the house, cold air will leak in to replace it. This cold air now needs to be heated until it is equilibrated with the warm air in the house.

These pressure and temperature differences drive air movement into, out of, and throughout the home. The biggest concern is with air movement between the inside and outside of the house. Air movement within the living

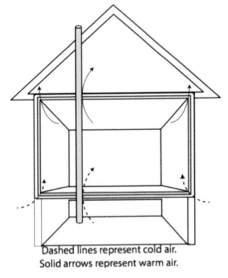

Dashed lines represent cold air.
Solid arrows represent warm air.

**Figure 3-7. Cold Air In, Warm Air Out**
In winter, stack effect creates a vacuum in the lower parts of the home when cold air is drawn in through leaks in the wall around joints, pipes and wires or through cutouts in the floor for P traps, ductwork, vents, etc. In the upper parts of the home, stack effect creates pressure, which pushes warm air out of the house. This warm air is replaced with the cold air from below, which must be heated.

space is less important for energy efficiency. The bigger the temperature or pressure difference between inside and outside, the greater the heat flow and airflow. Retrofit measures that reduce heating needs are often more cost-effective than those that reduce cooling needs. The temperature difference between inside and outside in a colder winter climate (indoor = 70°F, outdoor = 20°F, $\Delta T$ = 50°F) is typically 2 to 3 times more than the difference in even the hottest summer climate (indoor = 70°F, outdoor = 90°F, $\Delta T$ = 20°F).

The temperature difference between inside and outside is known as the Delta T or $\Delta T$. The higher the $\Delta T$, the greater the force of air and heat to enter or escape the building. Thus, the rate of air and heat transfer increases as the $\Delta T$ increases. To use a winter/summer analogy, the rate of natural leakage in the summer of a home might be only 40 CFM, but because of the increased $\Delta T$ in the winter, the rate can increase to 80 CFM, as shown in Figure 3-8.

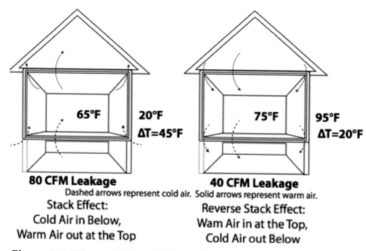

**65°F**   **20°F**   **75°F**   **95°F**
**$\Delta T$=45°F**   **$\Delta T$=20°F**

**80 CFM Leakage**                    **40 CFM Leakage**
Dashed arrows represent cold air. Solid arrows represent warm air.
**Stack Effect:**                    **Reverse Stack Effect:**
**Cold Air in Below,**                **Wam Air in at the Top,**
**Warm Air out at the Top**           **Cold Air out Below**

**Figure 3-8. Winter Makes A House Leakier Than in The Summer**
In winter, the negative pressures below and the positive pressure above cause cold air to be drawn into the lower parts and warm air to be pushed out of the upper parts of a home. Temperature differences between inside and outside tend to be greater in winter than in summer, resulting in greater air leakage in winter.

**Pressure Differences**

The greater the $\Delta T$, the greater the pressure differences. Pressure can act on all sides of a home like the pressure inside a balloon. If the house is pressurized, the air on the inside is denser than the air on the outside. Some of the forces that cause pressure differences in homes include wind, heat, and fans. A driving wind can create a positive pressure on the windward side of a home while creating a negative pressure on the other sides of the home. The protected or leeward side of the home will have a negative pressure that sucks air out of the home. The greater the number of stories, the more a home can be affected by the higher winds that occur farther above the ground. At the same time, nearby buildings, fences, trees, terrain (including berms) can protect or shield a home from the wind and its effects, as shown in Figure 3-9.

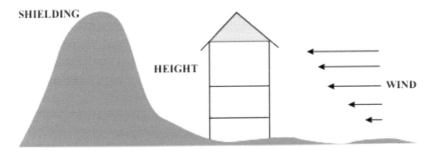

**Figure 3-9. A home may be exposed directly to the wind or protected, as in this diagram, with a berm.**

Bathroom, kitchen and other exhaust fans, as well as combustion appliances can create negative pressure in the home that leads to increased infiltration. Some combustion appliances can create a negative pressure in the home because they take oxygen-rich combustion air from inside the home and exhaust the resultant oxygen-poor combusted air out the flue, as demonstrated in Figure 3-10.

Duct leaks can cause even greater negative and positive pressures than small exhaust fans in different parts of the home. If the duct leak is in a return air duct, then the negative pressure inside the ductwork draws air in from around the duct (for example, from the crawlspace air). If the duct leak is in a supply air duct, then the positive pressure inside the ductwork will push air out of the duct into the space around the duct, as in Figure 3-11. Therefore, even though a hole is a set size, flow through the hole will be greater when pressure differences are greater. Thus, given the same size hole in a duct, expect a higher flow when the pressure differences are greater. Since furnace fans are much larger than bathroom or other exhaust fans, these effects can be even more dramatic.

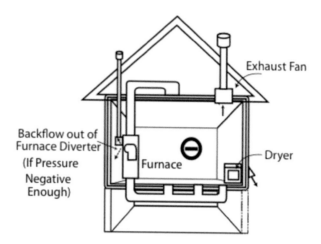

**Figure 3-10. Exhaust fans, dryers, and even combustion appliances can create negative pressures in homes with respect to outside.**

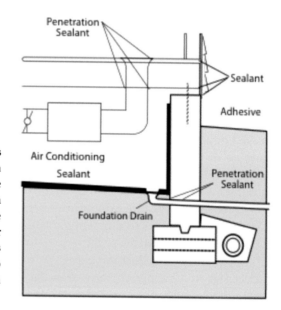

**Figure 3-11. While ducts and walls should both be sealed, the leakage caused by the ducts in this home can have the potential for greater leakage than the leaks through the wall due to the higher pressures in ducts.**

In addition, flow will take the path of least resistance—where the flow is the easiest—not where you want it to flow. Thus, the "makeup air" needed to replace air that exhaust fans etc. push out of the house is provided by air being drawn into the home through leaks in the building shell or, in a worst case scenario, being pulled down from inside the chimney or a combustion appliance flue when there are not enough leaks available. Pulling bad air back down into the home is a health and safety issue, as shown in Figure 3-12.

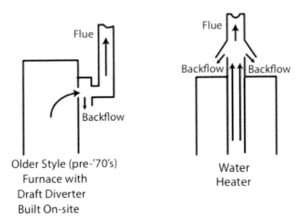

Figure 3-12. Combustion Appliance Backflow is a Health and Safety Hazard

**Air Leakage**

Air leakage can be classified into two categories: direct and indirect leakage. Direct leakage occurs where there are direct openings to the outdoors such as around doors and windows. Other examples of direct leakage include dryer vents and other places there are penetrations in the building envelope. With indirect leakage, the air enters one spot, then moves through some building cavities, and exits at a different location. This leakage typically travels through an "intermediate zone" that is either "inside" or "outside" the building. Indirect leakage occurs more commonly in older homes where there are no top plates on interior walls. Another good example of indirect air leakage is a porch roof that meets the side of the house because siding is often very leaky or not present at these junctures. Cold air is then allowed to flow from the porch attic into the floor and wall cavities and soffits inside the home. As a result, these walls and floors feel cold in winter. This is illustrated in Figure 3-13.

Indirect air leakage can occur through bypasses and gaps in the insulation. This can include the small holes for pipes, wires, or chimney chases that pass into the attic from the level below. Other problems can include openings in floors under showers and tubs to accommodate pipes and P traps, poorly sealed ductwork, poor air isolation between

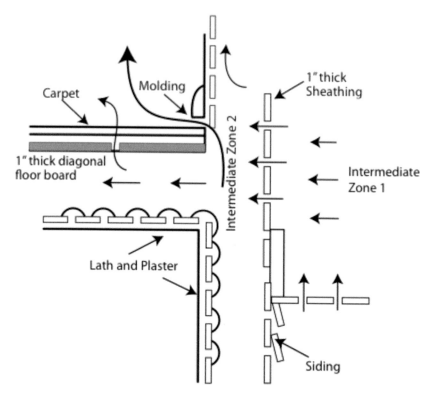

**Figure 3-13. An example of how air in a balloon frame home can easily make its way from the leaky outside porch ceiling, through the outside wall sheathing, and even into the interior floor spaces.**

the garage and the house, dropped soffits, and fireplaces without glass doors. Quality testing can track these leaks.

Because air leakage requires both a hole and a pressure difference across the hole, there would be no need to worry about holes if there were no pressure differences. A bigger hole or a higher-pressure difference will create a greater volume of air leakage. To reduce airflow, reduce the size of the hole or lower the pressure difference. Often if the size of the hole is reduced, the pressure difference is also lowered as a result. This is shown in Figure 3-14.

**Ventilation**

While infiltration and exfiltration are defined as air leakage coming in or going out respectively, ventilation is defined as controlled air

Smaller holes =↓ leak          Hole sizes bigger = ↑ leak
**Figure 3-14. Effect of Hole Size**

leakage. A good example of infiltration would be cold air coming in under a door in the winter. An example of exfiltration would be warm air traveling up through recessed can lights into the attic in winter. Natural or passive ventilation is ventilation that occurs without the use of fans or other mechanical equipment. Mechanical or intentional ventilation is the ventilation that results from the use of fans or other mechanical equipment, see Table 3-1. An example of mechanical ventilation would be bathroom or kitchen hood fans.

**Table 3-1. Natural and Mechanical Ventilation**

| Natural Ventilation | Mechanical/Intentional Ventilation |
|---|---|
| Window Leaks Door Leaks Indirect Leaks | Bathroom Fan Kitchen Fan |

**Stack Effect**

Stack effect is a driving force that creates pressure differences in the home, as shown in Figure 3-14. Rising warm air escapes out the top of the house and creates a suction pulling outside air in near the bottom of the house. The positive pressure at the top of the home causes exfiltration to the attic or to the exterior. Remember that every CFM that escapes from the home has to be replaced with a CFM from somewhere else. The corresponding negative pressure that typically occurs in the lower portion of the building leads to air infiltration there. A good example of a useful version of the stack effect would be the proper drafting of a chimney. A simple flue stack allows hot air from a furnace to be drawn out of the house because hot air wants to rise.

To better understand stack effect, think of a room with a high ceiling. The air will be quite a bit warmer near the ceiling than it will be down by the floor. Imagine an even higher building or home with floors interspersed between the lowest level and the highest ceiling in the home. There would be even bigger differentials between the temperature at the highest level and the temperature at the lowest level. As long as there is communication between the floors in a home with more than one level, such as through an open stairway, the house tends to be more like a room with a very high ceiling.

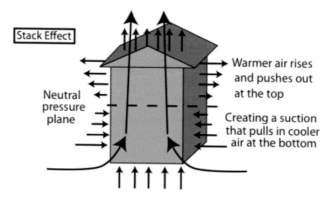

**Figure 3-15. Driving Forces of Stack Effect**

These differences in temperature affect the pressure in the home. For instance, since warm air wants to rise, the flow of air is from the lower floors to the upper floors. At the same time, this flow creates different pressures at different levels in the home. The air escaping from the lower floors to the upper floors tends to create a negative pressure on the lower floors. The warmer air trying to congregate on the upper floor creates a positive pressure on the upper floors. Thus, there may be somewhat of a vacuum drawing the air from the outside into the home on the lower floors, while the upper floors will be pressurized and pushing air from the inside to the outside of the home. In the summer, if the house is being air conditioned, the process can be reversed and is therefore referred to as "reverse stack effect" because the cooler air inside the house wants to drop to the lowest levels, creating the greatest pressure in the bottom of the house.

Air is typically drawn into the house at the lower levels and forced out at the higher levels. There is a point somewhere in between where the pressure is neutral and no air is going in or out. As it turns out, at the middle levels of the home, the pressure between indoor and outdoor is more neutral. It is theorized that at some point a "neutral pressure plane" exists in the midsection of the home depending upon how leaky the crawlspace floor is relative to the attic ceiling. The crawlspace floor and the attic ceiling tend to be places where more air is infiltrated and exfiltrated, respectively. These two areas are typically targeted for air sealing.

## THE EFFECT OF WATER ON THE HOUSE SYSTEM

### Moisture and the Weather—The Water Cycle

The water cycle is an explanation of how water cycles in our weather pattern from collection (lakes, the ocean, rivers, and other bodies of water), to evaporation, to condensation, to precipitation (rain, snow, sleet, etc.). Inside homes, the water cycle is typically reduced to evaporation and condensation. Water movement in the home is typically referred to as moisture transport—regardless of the water state.

### Moisture and Relative Humidity

Condensation is water vapor that has changed to liquid water. Typically, this occurs when air is cooled down. The higher the humidity, the greater the risk of condensation when the air is cooled. Evaporation is the opposite of condensation and involves the transformation of liquid water into water vapor. Absolute humidity tells what the water vapor to air volume ratio is in a given sample of air. The more moisture in the air, the higher the absolute humidity. This is sometimes referred to as the vapor pressure.

Relative humidity is the amount of water vapor contained in a specific volume of air relative to the maximum amount of water vapor that the volume of air is capable of holding. This is expressed as a percentage or ratio. Thus, if the humidity or water vapor in the air is only half of what the air could maximally hold, the relative humidity is 50%.

$$\text{Relative Humidity} = \frac{\text{Weight of water vapor in air}}{\text{Maximum weight of water vapor that air can hold}}$$

Relative humidity is affected by temperature. The higher the temperature of the air, the more moisture it can hold. Thus, cold airs holds less moisture than warm air. Cold winter air typically contains little moisture and by definition has a low relative humidity. If cold air is brought indoors and heated, the relative humidity of the air drops even more. On the other hand, if warm moist inside air leaks into a cold wall to get outdoors, it can condense inside the wall's insulation, reducing the insulation's effectiveness. If the indoor humidity is below 40% in the winter, then the risk of condensation forming inside the wall is reduced.

Indoor relative humidity is a function of the moisture generated in the home from cooking, showers, etc. and how much the home exchanges air with the outside. Figure 3-16 shows various sources of indoor moisture. Outdoor relative humidity is a function of rainfall, wind, cloudiness, and other weather and environmental factors.

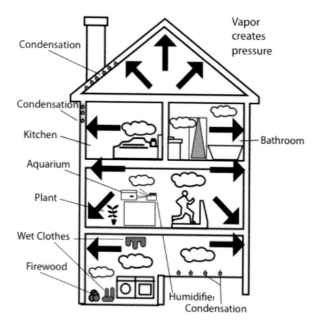

**Figure 3-16. Sources of Indoor Moisture**

If the relative humidity is 100%, the air is saturated with as much moisture as it can possibly hold at that temperature, and condensation begins to occur on all exposed surfaces. This is also referred to as the dew point. If no hard surfaces are exposed to the 100% relative humidity air, then there will be no condensation. The only reason raindrops fall from clouds that are at 100% relative humidity is that the moisture attaches to a speck of dust (a hard surface) that then grows to become the raindrop.

As a rule, people typically do not handle extremes of humidity well. For most people 15% to 75% relative humidity is tolerable. In addition, our tolerance to the upper relative humidity levels drops as we become more active. In other words, we seem to be able to tolerate 75% relative humidity when we are sitting, but not when we are running. Conditions above 75% relative humidity can interfere with our bodies' natural cooling system. Thus, to be comfortable, the indoor relative humidity should be less than about 60% in the summer. Below 15% relative humidity, physical problems such as the failure of furniture glue can occur and medical issues, such as a dry respiratory tract, can begin to arise.

It is interesting to note that keeping relative humidity below 50% eliminates and/or reduces many common household pollutant problems. A relative humidity between 20% and 50% helps avoid breakdown of immune systems, dust mite invasions, mold growth, and viruses. Medical studies show that most people are the most comfortable and healthiest when the relative humidity is between 30% and 50%, absent a medical condition of the occupant.

### Where Does Moisture in the Home Come From?

There are many seemingly innocent sources of water vapor or humidity in homes. For instance, up to 40 quarts of water vapor per day can come from building materials during the first year after a home is built. In addition, water is one of the products of combustion. Thus, an unvented gas range contributes far more moisture load in the home than showering and bathing. Each burner on a gas range produces between one and two pints of water vapor per hour just from burning natural gas. Likewise, unvented kerosene or gas-fired space heaters add water vapor to the air at the rate of 7.5 pints per gallon of fuel burned. Backdrafting combustion appliances can create similar problems. One quart of green firewood produces 3.5 pints of water each day. If clothes are hung dry inside the home, each wash load has about 5 pints of water that need to be dried out. Washing the floor of a room sends about 4 pints of water into the air per room. Virtually all the water to grow indoor plants eventually evaporates into the room. An aquarium can represent the equivalent of 12 potted plants. Cooking in pots without lids produces about 1/6 of a pint of water per person per meal. Humans and pets create about 1 pint of water from sweat, etc. for every 50 pounds of body weight per day. In fact, under the right circumstances, in one day, a family of four can give off up to 2.5 gallons of moisture by exhaling and sweating. If dishes are hand washed and rinsed, about 1/3 pint per person per meal of water goes into the air. Even faucet aerators create about 0.02 pints of moisture per minute. A shower can push a quart of water into the air. Table 3-2 summarizes various sources of moisture.

**Table 3-2. Moisture Sources and Amounts**

| Moisture Source | Amount in pints |
|---|---|
| Kerosene heater | 7.5/gallon of fuel |
| Green firewood, qt | 3.5 |
| Clothes hung dry, load | 5 |
| Floor washing, room | 4 |
| Dishes, hand washed | 1/3 per person |
| Cooking in open pots | 1/6 per person per meal |
| Shower | 0.5 per 5 minutes |

Basements or crawlspaces that have standing water can allow moisture to migrate its way upstairs. The primary cause of moisture problems in homes is cracked or leaky foundations. In addition, the water table (the level of water in the ground) can be an issue around the foundation. Drainage issues can often be solved by making sure gutters and down drains are free of leaks and leaves so they can take the water far away from the house. It can also be helpful to cap the ground next to the foundation with impermeable clay or concrete. Water table issues could be addressed by using sump pumps in the basement as long as the water from the pump is delivered far enough away, and preferably downhill, from the home. It is very important that a complete drainage plane surrounding the house be in place by appropriate lapping of shingles, building paper, siding, grading around the house, flashing of windows, doors and other penetrations, and, if necessary, a French drain system around footings, etc. to appropriately drain water away and to help avoid moisture intrusion. These various items are shown in Figure 3-17.

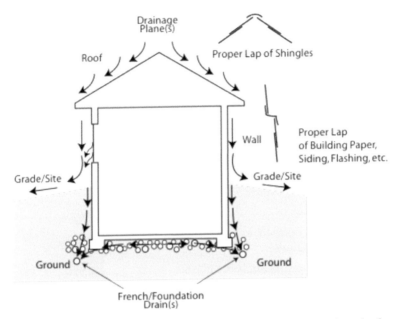

**Figure 3-17. This shows each of the different levels of protecting the house through drainage planes whether the plane is the roof, the walls, or the basement foundation. Examples of lapping and flashing are also shown that are critical in protecting the house from water intrusion from outside.**

Water gets into the house through four basic transport mechanisms or methods: bulk moisture, capillary action, diffusion, and air transport—see the example in Figure 3-18. Bulk moisture is defined as large amounts of water intrusion into the home from wind-driven rain or subsurface water. Examples of this would be a leaky or ice dammed roof, or water in the soil getting into the basement or crawlspace. Capillary action is defined as the movement of liquid water across a material as a function of the surface tension of water and the porosity of the material. Capillary action carries water up a tree and wicks moisture up through concrete.

Capillary action can be so strong, that, unbelievably, water can wick up through a one-inch column of concrete over a mile into the sky. A good example of this would be with a thin layer of water on the floor of a bathroom, but the water being drawn up along the drywall on the wall and wetting it, despite no direct contact of water on the drywall above the water level.

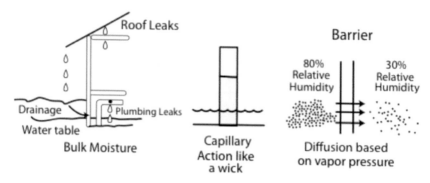

**Figure 3-18. Types of Moisture**

Diffusion is defined as the movement of water vapor through a material as a function of the driving force across the material and the porosity of the material. The driving force needs only be a difference in the humidity or vapor pressure of the air on each side of the material, as shown in Figure 3-19.

Thus, if the humidity outside is 80% and the humidity indoors is 30%, moisture will be driven into the home through the wall if there is no vapor barrier in the wall to prevent it. If a vapor barrier exists but there are gaps in it, moisture will pass through these gaps or holes. Because water vapor is smaller and lighter than other molecules in air, it will diffuse through wall materials faster and more easily than air itself.

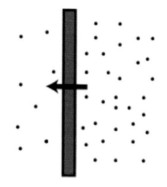

**Figure 3-19. Water vapor moves through permeable barriers from the area of high concentration (high humidity) to the area of lower concentration (low humidity).**

Air transport is defined as the movement of moisture vapor by convective air currents and is affected by wind, stack effect, and mechanical equipment, such as fans. The permeance rating of a material is the rate at which vapor diffuses through it. Everything is permeable to some degree, even 6-mil poly sheeting. A material that has a "perm" rating of 1 or less qualifies as a vapor barrier to block the flow of water vapor. Table 3-3 shows perm ratings of various common materials.

Table 3-3. Perm Ratings of Some Common Materials

| Material | Perm Rating (Perms) |
|---|---|
| Gypsum board, unpainted | 50 |
| 4 in. unfaced mineral wool | 30 |
| Plaster on metal lath, ¾ in. | 15 |
| Plaster on wood lath, ¾ in | 11 |
| Gypsum board, latex paint | 2-3 |
| Concrete block, 8 in thick | 2.4 |
| Insulation facing, Kraft | 1.0 |
| Brick, 4 " thick | 0.8 |
| Vapor retarder latex paint | 0.45 |
| 0.006 " Polyethylene sheet | 0.06 |
| Aluminum foil, 0.001 in. thick | 0.01 |

Moisture can also get into the home from outside sources including abutting dormers, low slope roof additions, roof leaks, ground slopes that guide water towards the home, and even clogged or poorly installed rain gutters. Any type of mechanical ventilation including clothes dryers, whole house fans, central vacuums, bathroom and kitchen fans, and even central heating and cooling systems, can create such a negative pressure in the home that moist exterior air is pulled into wall and ceiling assemblies. Many homeowners purposefully humidify their homes for health or comfort reasons. This clearly will elevate the level of moisture in the home and result in condensation and potential mold.

## What Does Moisture Do?

When warm air leaks through the air barrier, it brings moisture with it, which can cause damage when it condenses on cooler surfaces. A very good example of warm moist air condensing on cold surfaces that many of us are familiar with comes from the "ice tea" effect. This is where, on a hot summer day, virtually without exception, you will find that droplets of water (condensate) forms on the outside surface of a glass of ice tea. This is from warm moist air circulating along the outside surface of the glass and when it is cooled enough by the cold tea glass, it deposits some of the moisture onto that glass surface. A good example in the building arena is warm air on the upper floor that escapes into a cold attic and the moisture in the air condenses on the bottom side of the roof decking on the most northerly walls (Figure 3-20). This can cause mildew or mold and may lead the occupants to believe there is a roof leak because so much condensation builds up and runs down into the house.

**Figure 3-20. Frost can build up in attics when air leaks into the attic from inside the home. This can make the 2 x 4's look a little as if they have been in a freezer that has not been defrosted frequently enough.**

Far too often, kitchens and bathrooms are not vented to the exterior, but rather to the attic or crawl space, causing further moisture problems. Exhaust fans are often underpowered, or the fan ductwork is restricted or leaks, thus leaving more moisture in the house. Moisture and condensation can lead to rot or corrosion of building materials. This is one of the ways air and moisture affect building durability.

Humid air from a crawlspace or basement without a ground vapor retarder can find its way through chimney chases all the way to the attic and damage wood all along the way, including the roof and even shingles. This water infiltration also can cause rot of wood or corrosion/rust of metal structural components and reduce the effective R-value of insulation. Moisture can also lead to mold, dust mites, termites and carpenter ants. It can also deteriorate mortar and warp wood. Glue can also fail when there is too much moisture. Likewise, leaky return air ductwork in the basement will not only draw cold air from the basement or crawlspace into the home, but also basement or crawlspace moisture and other pollutants into the home.

Minor to moderate condensation problems commonly occur during warm, humid seasons. Condensation appears as a result of the warm humid air coming in contact with some of the cool surfaces found in a basement or crawl space such as the concrete floor, the foundation walls, or metal ducts providing air conditioning. This condensation can saturate the building materials and cause mold. In essence, condensation and mold will form on a surface if the surface temperature is below the dew point or mold threshold (the temperature below which the relative humidity becomes 100%).

Mold tends to grow very well in the 70 to 80% relative humidity range. Any temperature above freezing can potentially allow mold to grow. Mold can grow on rafters, sheathing, or the underside of flooring and can lead to structural issues.

High moisture loads can cause severe paint peeling on the exterior and interior when the moisture works its way through the wall and forces the paint off the wall. Moisture can also warp tile.


COMFORT AND CLIMATE

**Environmental Factors**

Most people share a common range of comfort of between 68°F and 85°F air temperature and a relative humidity of between 15% and 75%. Other factors can also affect comfort. Since air movement increases the rate of heat transfer, our internal evaporator cooling system works better if air is moving over us. This makes us feel cooler at the same temperature. For instance, people typically feel cooler on a hot day if there is some breeze and the relative humidity is low, than they will feel if the air is humid and still.

**Comfort Factors**

Mean radiant temperature of objects around us affects comfort. The transfer of radiant heat occurs without any medium—no air is required. Perhaps even more importantly, the temperature of the air around us does not affect mean radiant temperature either. If the surface temperature of the body is higher than the temperature of surrounding surfaces, then net heat is radiated to the surroundings, regardless of the air temperature. Because a window has a colder surface temperature than a wall in winter, you feel colder when you step in front of a window. The reverse also happens and explains why, on a cold day, a wood stove feels so much better. In fact, if you were to pick a day in winter and a day in summer when the air temperature in the house is the same and you used an infrared camera to show the surface temperature of the walls, you would see the wall temperature in winter is colder and you would feel colder on the winter day because of radiant heat transfer, Figure 3-21.

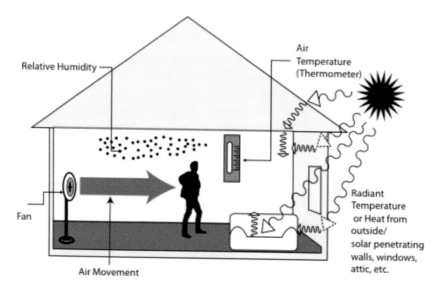

**Figure 3-21. Radiant heat transfer**

You lose more net radiation to a colder wall then to a warmer wall. Corner rooms are often more uncomfortable because you are exposed to the radiant surface temperature of exterior walls and windows on two walls rather than one.

The passive house system places a special emphasis on comfort related to the radiant surface temperatures of the walls around us by requiring significant insulation in the walls and the elimination of thermal breaks (see Appendix Q).

**Personal Factors**

Our bodies attempt to maintain a core temperature of approximately 98°F by dumping heat if our surroundings are too warm, or by burning fat if our surroundings are too cold. At the same time, because we never stop burning fat, we are typically not comfortable unless we are in the process of losing at least some heat to our surroundings. Another factor that affects comfort is our activity level. The more active we are, the warmer our bodies will be. For example, a more sedentary person would typically be more comfortable at a higher humidity and higher temperature than an active person would. Our bodies' built-in metabolism also plays a large role. We are in thermal equilibrium when we are neither warm enough to sweat nor cold enough to shiver, regardless of the conditions.

Conditions we are accustomed to can also affect comfort. A Florida resident might feel colder at 70°F while someone visiting him from North Dakota in the winter might suggest to him that it is very warm. Another personal factor is the insulation value of our clothing.

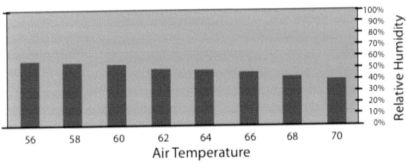

Figure 3-22. Temperature affects the relative humidity of air with a constant amount of water in it.

Many people begin to become uncomfortable when the relative humidity is greater than 60%. The relationship between temperature and relative humidity is shown in Figure 3-22. Individuals often find themselves more comfortable at higher temperatures and lower humidity than at lower temperatures and higher humidity. The key to comfort is a 62°F dew point. The best range for comfort is between 68 and 76°F and between 20% and 40% relative humidity, as shown in Table 3-4.

Table 3-4. Comfort Conditions

| Comfort Conditions For Humans | |
|---|---|
| Best Temperature | 68 to 76°F |
| Best Relative Humidity | 20 to 40% |

All of these thermal comfort factors interact to determine whether we feel comfortable or not. Most of us are familiar with the air temperature factor. If we are surrounded by air that is at a significantly different temperature than our comfort zone, we are going to be uncomfortable.

## POLLUTANTS

### Sources of Pollutants in the House

The concentration of pollutants determines the indoor air quality (IAQ). The possible sources of pollutants include stored household chemicals (including easily vaporized volatile organic compounds or VOCs), unvented space heaters, mold/mildew, vehicles and other equipment in attached garages, plumbing leaks, carbon monoxide and carbon dioxide, animals in the home, and wet crawlspaces/basements caused by poor landscaping, exposed dirt floors, standing water, open sump pumps, and malfunctioning or no gutters. The occupants' exposure to pollutants can depend on other factors as well, including the concentration of the pollutant and the natural air change rate in the home. Low levels of a pollutant in a very tight home are more harmful to the occupants than low levels of pollutants in a leaky home or one with a high natural air change rate.

Another concept sometimes mentioned in this arena is indoor environmental quality or IEQ. IEQ involves all that indoor air quality does and more. It also includes other health, safety, and comfort issues such as aesthetics, ergonomics, acoustics, potable water surveillance, lighting, and electromagnetic frequency levels.

### Dealing with Pollutants

There are three primary methods of dealing with pollutants in a home: elimination, encapsulation, and dilution, see Figure 3-23. The priority is to first attempt elimination, then encapsulation if elimination is inadequate, and finally dilution to deal with pollutants if elimination and encapsulation both have not been successful.

Elimination involves removing the source of the pollution itself. Some examples include removing an unvented space heater from the home, properly collecting and disposing of old paint cans or other old chemicals at a local waste facility, and installing or repairing gutters and down spouts so they properly divert rain runoff away from a basement.

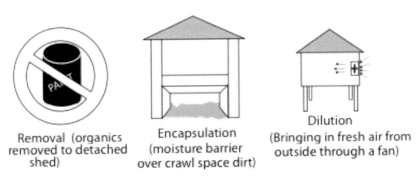

Removal (organics removed to detached shed)

Encapsulation (moisture barrier over crawl space dirt)

Dilution (Bringing in fresh air from outside through a fan)

**Figure 3-23. Ways to Deal With Pollution in a Home**

Encapsulation involves containing the pollutant so that it does not affect air quality. Some examples include putting a poly groundcover over a dirt crawlspace to help contain moisture from the ground, air sealing the ceiling of a tuck-under garage and putting auto-closer hinges on the door to the house from the garage to prevent CO from getting into the living space, and putting a concrete floor over an exposed dirt crawlspace to help contain radon gas.

Dilution involves providing adequate ventilation to reduce the concentration of pollutants to an acceptable level. Some examples of this include running the kitchen range hood to help exhaust combustion byproducts from an unvented gas range while cooking, or opening windows while using an unvented space heater to prevent pollutant buildup. For dilution, the question is "how much outside air is needed to attain good indoor air quality and how do you get the outside air?" The amount of outside air needed has been established by ASHRAE (American Society of Heating Refrigeration and Air Conditioning Engineers). According to ASHRAE standards, outside air can be provided by mechanical equipment, such as a fan, or by natural air change. Because of wind and stack effect, some exterior air moves through all buildings. The amount of air provided through natural conditions depends on a number of factors including building height, building exposure and location, inside to outside temperature difference, wind strength, and the location, size, and type of holes in the building envelope. Blower doors can be used to see how much natural air leakage occurs in buildings. This will help determine if mechanical ventilation is needed to satisfy the standards for bringing in outside air.

## WHOLE HOUSE FANS

One way of circulating air is a whole house fan, as shown in Figures 3-24 and 3-25. However, these fans deserve special consideration. A whole house fan is typically an airplane propeller-type fan placed in a hall or centrally located ceiling area that is used to draw air out of the house and into a ventilated attic. Louvers usually open as the fan begins to operate.

**Figure 3-24. A Whole House Fan From the Attic Side**

The windows in each room are opened slightly to allow the fan to pull air from outside the house and through the attic. Whole house fans are common in the northeastern United States but less common elsewhere in the U.S. They are mostly effective when there is moderate summer weather so that the combination of air temperature and the "breeze" it generates in the home combine to have a cooling effect on the skin without having to use any energy for cooling the air - only the electricity to run the fan is used. Since they pull a great deal of air through the house, if they are operated without the windows being opened wide enough, they can create a very strong negative pressure - strong enough to backdraft even some of the better drafting combustion appliances and even those with powered flue fans. This could occur anytime of the year the fan is used and therefore, the option that seems to best protect the client is to tell them to consider having it disabled or removed because if the windows are not opened wide enough, water heaters can backflow carbon monoxide into the house creating a hazardous situation. (This would also reduce the possible additional exposure to radon that a negative pressure in the house might create over time).

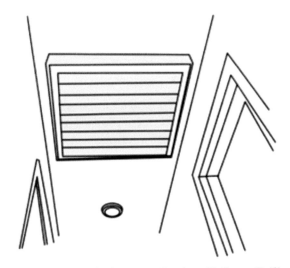

**Figure 3-25. A Whole House Fan in a Hallway Ceiling**

COMFORT, SAFETY, AND EFFICIENCY

In summary, qualities that can help make for a comfortable, safe, and energy-efficient home include the following:

1)   A completely insulated thermal boundary or thermal envelope.

2)   An intact and continuous air barrier that is well sealed. Make sure that in cold-weather climates, heated air is not escaping to the outdoors, and in warm weather climates, cool air is not escaping to outdoors. Also, be concerned about moisture. The air barrier plays a major role in the movement of heat and moisture in the building.

3)   Continuity or "match up" of air and thermal boundaries such that they are in contact with each other; thus, the phrase "seal tight, insulate right."

4)   Properly sized and efficient equipment for conditioning the living space and heating water. Bigger is not always better. Proper sizing

is best evaluated and conducted by an appropriately trained and, if necessary, licensed heating and air-conditioning professional.

5)    A balanced and well-designed air, hot water, or steam distribution system.

6)    Good indoor air quality. This is a key health and safety concern.

An interesting question is "What is more important, good air sealing or good levels of insulation?" Good air sealing is the answer. See Appendix H for results of advanced calculations. The results are obtained with concepts you will learn as you continue to study this book.

# Chapter 4

# The Auditor's Tools and How to Use Them

## TYPICAL TOOLS

One of the first tools an auditor needs to become familiar with is the audit field form. Typically, each weatherization program or independent auditor has a form that auditors use to work their way through the audit process. Building data is collected relating to square footage, age, framing style, insulation location, R-value, moisture, exhaust equipment, approximate CFM, attic ventilation, and blower door test results. Heating system data is typically collected on the form including age, condition, Btu input/output, efficiency, carbon monoxide testing, and combustion appliance zone/closet testing. Depressurization tightness limit results, and ductwork or delivery system evaluation are also typically recorded on the form. The form usually lists the major appliances along with their approximate runtime (if available), refrigerator evaluation, and an analysis of the need for changing from incandescent bulbs. Client and fuel use information are also included.

Typical equipment used by an auditor includes a blower door, a manometer or pressure gauge, and a ladder. A number of assorted hand tools are also helpful including screwdrivers, adjustable wrench, razor knife, cutting pliers, 25-foot tape measure, digital camera, flashlight, pen, paper, and a non-conductive plastic crochet hook for fishing insulation samples through small openings. It is good to place smaller tools in an organized tool case for easier carrying and organization.

There are also some helpful specialized tools that can be used including: a rechargeable cordless drill motor with an assortment of drivers, a hammer, a long reach flat blade screwdriver, a reamer to enlarge any smoke pipe hole when testing heating appliances' efficiency, pry

bars and nail pullers, "nippers" (Figure
4-1) with the face ground flat to be used
to pull nails from cement board siding,
drill index, both long reach and locking
pliers, and a square and star bit driver for
mobile home siding.

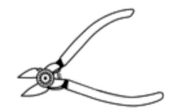

In addition, there are tools that can
help see things in hard to reach locations.

**Figure 4-1. Nippers**

These tools would include: a digital cam-
era that can take pictures anyplace a hand will fit, a borescope (Figure
4-2) that uses a flexible tube for viewing inside cavities, including inside
walls to check for insulation level, and an infrared camera.

### Moisture and Relative Humidity Measurement

There are several ways
to evaluate for moisture in a
home. It is best to take measurements more than once and in
more than one location in the
home. Use your eyes to notice
mold or condensation. A moisture meter is used to evaluate
the amount of water in wood
and other wood-based building
materials. Some moisture meters have sharp pins to puncture
the wood for an accurate read-

**Figure 4-2. Borescope.** A borescope is
used for viewing into areas that are difficult to see, such as inside of walls, etc.

ing, while others rely on electronics only so the wood can be checked
without damage. Even using a pocketknife that easily penetrates wood
saturated with moisture can be a useful tool to identify moisture issues
if the wood is not finish material.

A moisture meter (Figure 4-3) determines the moisture level between two sets of probes on the meter. A longer set of probes helps measure surfaces behind insulation without removing the insulation. The
shorter probes are used to determine the moisture between two points
on an exposed hard surface such as wood. Some moisture meters come
in an electromagnetic form that does not require poking pins into the
material. An electromagnetic moisture meter simply requires touching

the device on the surface of the suspect material.

To determine relative humidity, a sling psychrometer can be used, although a simple indoor/outdoor thermometer (Figure 4-4) that provides relative humidity can be accurate enough for your purposes and is a lot less fragile than a sling psychrometer. A sling psychrometer is basically two thermometers side-by-side, one has a cotton wick that has been wetted around the bulb of the thermometer (wet bulb), and the other is open to air (dry bulb). By spinning the sling psychrometer in the air long enough, the temperature stabilizes. When the wet bulb temperature stabilizes, you have found the wet bulb temperature. The dry bulb temperature is like the regular air temperature taken with a thermometer while the wet bulb temperature is what the temperature would be if, say, you wet your finger and held it to the wind. You can take these two temperatures and place them on a psychometric chart to determine dew point and relative humidity (see Appendix E).

**Electrical Testing**

For purposes of circuit testing and tracing, a live wire tester (Figure 4-5) can help determine if a wire is "hot." This simple device has a light that turns red whenever it is placed near a live wire. This is especially useful if you want to check if a wire has been abandoned, such as knob and

**Figure 4-3. Moisture Meter.** An electromagnetic version might look much the same but without the prongs.

**Figure 4-4. Thermometer with Relative Humidity Reading**

tube wiring. This tool is also called a non-contact voltage tester, test light, or electrical sniffer.

A circuit tracer (Figure 4-6) is a two-part tool used to determine which fuse or breaker controls a circuit. Plug one part into an outlet on the circuit being checked and move the other part over the circuit breakers or fuses in the panel box to determine which breaker or fuse services that outlet. Some tracers can tell the percent of overload on the circuit. This information can be used to determine if it is safe to place insulation around the wire serving the outlet.

A wattmeter (Figure 4-7) and recorder can measure and record the watts consumed by an appliance. This is very useful for baseload analysis. Some of these instruments can extrapolate future use from a two-hour runtime sample.

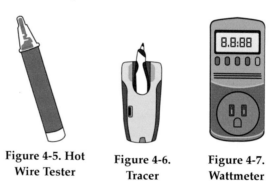

**Figure 4-5. Hot     Figure 4-6.       Figure 4-7.**
**Wire Tester         Tracer            Wattmeter**

### Combustion Testing and Analyzer

To test combustion appliances special equipment is used including a draft gauge and a pressure gauge or manometer for measuring the pressure inside the house, the pressure inside flues, and the flow in a blower door. Other equipment needed includes a long lighter, an inspection mirror for checking in confined spaces, gas leak detection (Figure 4-8) bubble solutions, a combustion analyzer, a digital probe thermometer for checking fan operating temperatures and testing temperature rise, an inspection mirror, pipe wrenches, an electronic gas leak detector, a smoke tester for oil burners to measure the amount of smoke produced in the combustion process, and a smoke pump and smoke test filter paper. It is probably best to keep combustion appliance test equipment together in one case if possible. Also, include in the kit drill bits for drilling test holes in flues and high temperature sealant to seal off these holes.

A combustion analyzer (Figure 4-9) can help determine the safety and efficiency of the combustion process in an appliance. The combustion analyzer can measure the oxygen and carbon dioxide content along with the flue gas temperature. It can also calculate the combustion or steady-state efficiency of the appliance. True steady-state efficiency is represented as the percentage of the total heat generated by the appliance that is captured by air, steam, or water to help heat the house.

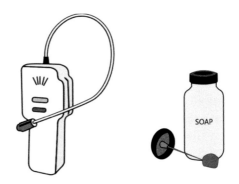

Figure 4-8. Gas Leak Detection—Electronic Gas Detector (left) and Soap Bubble solution (right). Usually the soap is only used to confirm a leak and avoid a false positive reading by the detector when volatile pipe sealant has been used.

**Smoke**

There are a number of techniques using smoke including the so-called Wizard stick (Figure 4-10) to check for leaks around the house. (Note that the chemical-based smoke generators can create havoc for nearby electronic equipment.) The smoke is drawn into the cracks of the wall, etc. when the house is pressurized. If there is a negative pressure in the house, smoke can be used to check for leaks from the attic side. In the attic, the smoke will stream to openings or leaks that are connected to the living space below.

Figure 4-9.
Combustion Analyzer

BLOWER DOORS

A blower door with a modern digital manometer or pressure gauge uses negative pressure and flow readings to give an approximate leakage rate or hole size of the home's envelope. Old magnehelic gauges were the predecessors to the modern day digital manometers and required considerable operator expertise to arrive at accurate results. Manometers/gauges give readings for CFM much as a speedometer tells how fast a car is going. There are two common varieties of more modern manometers or gauges. The DG-700 is a product of "The Energy Conservatory" (EC). Another commonly used gauge is the DM-2, available from "Retrotec." Retrotec gauges (Figure 4-11) can be used on equipment from other manufacturers as well as Retrotec equipment, but gauges from other manufacturers do not allow full use of other equipment.

Blower doors can also be used to create a positive pressure so that smoke, feel or an infrared camera can be used to help pinpoint leaks in a home. A blower door test before air sealing will tell how

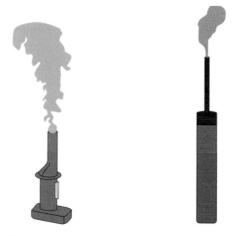

**Figure 4-10. The "Wizard Stick" smoke generator on the left uses a virtually harmless chemical to generate smoke. The smoke tube on the right does not require batteries or refilling with chemicals, but it can produce a smoke caustic enough to deteriorate electronic equipment.**

much air sealing can be safely completed and will give a baseline to compare how well the retrofit helped seal the leaks in the home. A blower door test can also help identify the air barrier.

Figure 4-12 shows two commonly used blower door systems. There are a number of components that are part of a blower door system

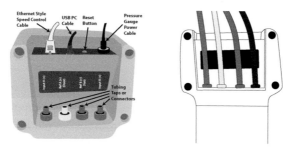

**Figure 4-11. The backside of a Retrotec manometer (pressure gauge) where the connectors are color coded to better assure proper placement of hose connections.**

including a fan, a speed controller, a pressure gauge/manometer, the hoses that need to be connected to the gauge, an adjustable frame with a special-made cloth that sits in the doorway. The cloth is placed across and around the metal frame and then the frame is adjusted so that it fits tightly in the door. The cloth has a round elastic hole in it near the bottom where the fan is tightly installed. This arrangement allows the fan to be used on the home without creating significant leaks around the temporary door.

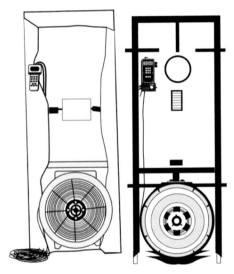

**Figure 4-12. Two Commonly Used Blower Door Systems**

The fan serves to pump air and create a negative or positive pressure in the home compared with the outside. A speed controller allows for control of the speed of the fan to create the pressure you want without having to use a gauge to continue to control it. Once you have found the right speed to maintain the pressure you want, the speed controller will maintain that speed.

The pressure gauge or manometer measures the pressure in two spaces, or a pressure and a flow, at the same time. For instance, if you want to set a home for -50 Pascals with respect to outdoors, there needs to be a hose that can tell what the pressure is outside and a hose or open tap on the gauge that can tell what the pressure is inside. Otherwise, there is no way of determining how negative the house is with respect to outside.

**Pressure Gauges**

The pressure gauges or manometers are basically designed for providing two readings at a time: one on Channel A and one on Channel B. Pressure sensors exist for each channel inside the gauge. "Taps" or hose connectors on the gauges allow pressure readings to be made from as far away as a football field length by connecting hoses to the gauges. Dirt, dust, insects, water, etc. can plug up the hoses and result in some interesting, but highly erroneous pressure measurements. As a result, many technicians store their hoses in oversized, air tight, plastic bags when not in use to assure that they do not develop problems while they are being stored.

Each channel on the gauge measures a difference in pressure from one area to another. Thus, you need two taps for each channel to complete a comparison. Each channel has a "reference" and "input" tap. The pressure read through the input tap is compared to the pressure at the reference tap. Thus, the phrase "with reference to" or WRT is often used to indicate what the pressure on the input tap is compared with when the gauge is sensing a pressure difference. For instance, say you have connected one end of a hose to the Channel A input tap with the other open end going to the garage. At the same time, you have connected another hose to the Channel A reference or WRT tap with the other open end going into the house. This will give a comparison of the pressure difference between the house and garage. If you turn on a blower door so that the house is at -50 Pascals and obtain a reading of 45 Pascals on Channel A with this setup, you are seeing the pressure change from the house to the garage by 45 Pascals. Thus, the pressure in the garage WRT the house is 45 Pascals. Remembering that the gauge's reference tap represents the "with reference to" aspect can help you keep a consistent orientation to how you are taking pressure measurements.

Another important note—if you are standing in an area that is to be a part of the pressure comparison, you need not connect a hose to the appropriate tap. This is true whether you are standing in the input area (in the example above, the garage) or the WRT (in the example above, the house) area. If you are standing with the gauge in your hand in the area that is to be the input area, you need not connect a hose to the input tap on the gauge—you can leave the input tap "open to air." Likewise, if you are standing with the gauge in your hand in the area that is to be the reference or WRT area, you need not connect a hose to the reference tap on the gauge—you can leave the reference tap "open to air" as shown in

Figure 4-13. The following diagram shows the set up with the reference tap on Channel A open to air (except on the blower door gauge on the upper left where the outdoor hose is connected to the reference tap—the reason it reads -50 Pa WRT outside):

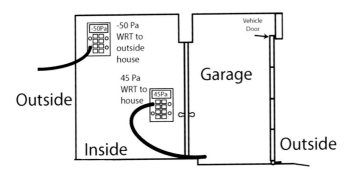

**Figure 4-13. EC WRT House with Reference Tap Open (no hose)**

In addition, you can change a positive to a negative and vice versa if, after measuring, you want to record the numbers as if the garage were the WRT or reference and the house were the input. Thus, you would be able to say that the house with respect to (WRT) the garage is -45 Pascals, even without taking any additional measurements.

Channel A on gauges is locked into only reading pressures. Channel B, on the other hand, can be changed by the technician to read pressure (when trying to get two pressure readings from two sets of areas at a time), flow (as when you are using a blower door fan or a duct blower fan), or flow extrapolated to a certain pressure (such as "CFM@50 Pa" as when the house is so leaky the blower door cannot create enough flow to reach the required pressure and the gauge extrapolates what the flow would be at a higher pressure). On various different gauges, Channel B can also be set to show a variety of other readings (Modes) and units including Effective Leakage Area, Equivalent Leakage Area, air handler flow, etc.

**Converting Flow To Pressure**: You can get a flow rate on a fan using only a pressure sensor in the gauge. This is done using the principle that air flowing perpendicularly across an opening in a hose will create a vacuum in the hose. This vacuum or negative pressure is transferred through the hose to the pressure gauge where the pressure sensors detect the negative pressure. An electronic calculator in the gauge converts that negative pressure to a flow rate, which can be read on the gauge display. The faster the airflow caused by the fan, the higher the negative pressure. The ends of hoses are designed and carefully manufactured into each fan to create this negative pressure on the hose(s) connected from the fan taps to the gauge taps, Figure 4-14.

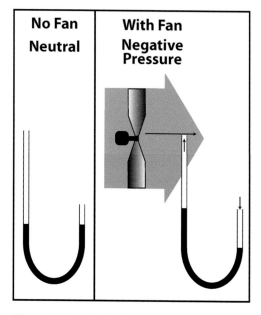

Each different type of fan may need a different equation for the electronic calculator to convert the negative pressure to a flow rate for the fan. With some exceptions, most gauges are designed to be used only with the fans made by the manu-

**Figure 4-14.** Air flowing across the end of a hose creates a negative pressure as shown with this simple water manometer. This is the same principle that allows a pressure gauge to convert a negative pressure from a blower door into a flow rate.

facturer who also manufactured the gauge. An exception to this is the Retrotec gauge that allows the use of other manufacturer's fans with its gauge.

Also, by the way, we typically consider pressure testing that involves taking pressure measurements in the home either with or without operating any equipment that is part of the home such as bathroom fans, kitchen fans, dryers, etc. as operational testing. If you are using special equipment while taking pressure measurements, such as blower door fans, duct testing fans, etc., you typically refer to these pressure measurements as being under imposed conditions.

**Proper Gauge Setup**: All gauges must be properly set up in advance of a test to indicate what device (blower door fan, duct blower fan, flow meter, etc.) is being used. In addition, the gauge must be set up for other parameters for the gauge to read properly. This includes setting the gauge for proper "mode" (for example, pressure on channel

A and flow on channel B) and for the type of narrowing rings or flow restrictors (the configuration or Config setting) that are on the device at the time of the test. If the technician does not set up the gauge for the proper settings to match the device, mode, and configuration that is being used, gross errors in the readings will occur.

**Baselines**: Some tests require that a "baseline" pressure measurement be integrated into the gauge calculations so the gauge reads the proper results. For example, if wind or stack effect creates more of a positive or negative pressure outside relative to the pressure inside the house, you do not want the gauge to take into account that artificial pressure difference when providing a readout of the flow on a blower door. In that case, the pressure gauges have a way of eliminating those pressure differences to avoid unnecessary errors in the calculation of flow and pressure differences. The process for conducting baseline corrections on gauges is described throughout this book in the individual procedures for each test where it is required.

**Time Averaging**: Sometimes pressure readings are unstable as a result of the wind outside, etc. The gauges have a time-averaging function that averages out the instability to give a more reliable number at a given point in time.

**Other Functions**: Many gauges have other useful settings such as a "hold" button (grabs the numbers that are momentarily on a display for easier recording on paper), an "autozero" button that helps reduce drift in the readings, a "set pressure" or cruise control function (keeps the pressure constant even if the flow must change), etc.

Figures 4-15 and 4-16 are some examples of gauges used. Notice the "config" symbols represent, for instance, going from the largest flow ring to the smallest.

## Blower Door Theory

There are natural driving forces that create pressure differences between the inside and outside of a home. These forces include stack effect, wind effect, and mechanical effects related to combustion and ventilation. Wind and temperature changes that affect these naturally occurring pressure differences are generally not considered repeatable. In fact, under natural conditions, the pressure differences are far too small to be measured reliably. Thus, early in the modern history of energy auditing it was discovered that artificially creating a higher exaggerated vacuum in homes makes measurement more reliable.

## Figure 4-15. Energy Conservatory Gauge

Bold Items represent the button name or what actually shows on this gauge display (location of info on display in parenthesis)

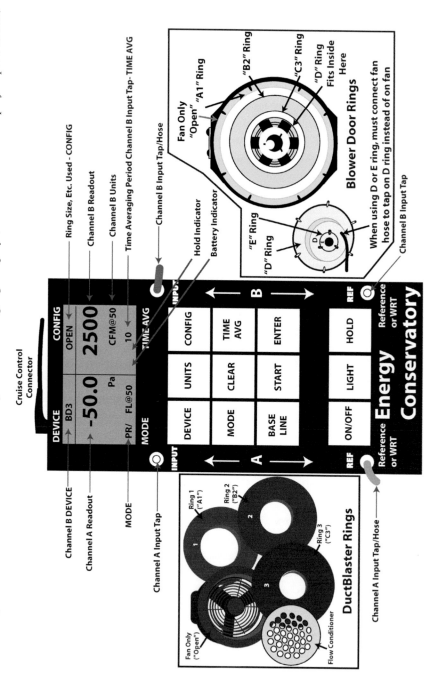

<u>ON/OFF</u> - Turns the gauge on or off

<u>MODE</u> - To select what the gauge will read for output on the two channels: A and B (Channel A results shown on middle left while Channel B results shown on middle right)

Channel A/Channel B displays: Pressure/Pressure - **PR/PR**; Pressure/Flow – **PR/FL**; Pressure/Flow@25 – **PR/FL@25**; Pressure/Flow@50 – **PR/FL@50**Pressure/Air Handler – **PR/AH**; Pressure/Velocity – **PR/V**

<u>DEVICE</u> - To indicate which equipment is connected to gauge on Channel B (upper left corner of display)
Model 3 Blower Door fan - **BD 3**; Model 3 (220V) - **BD 3 220**; Model 4 (220V) - **BD 4**; Series A Duct Blaster® - **DB A**; Series B Duct Blaster® - **DB B**; Exhaust Fan Flow Meter - **EXH**; TrueFlow® Air Flow Meter - **TF**

<u>CONFIG</u> - To indicate (upper right corner of display) which ring, door position, or plate has been installed from large to small.
No rings installed – **OPEN**; For BD's: **A1, B2, C3, D, E**; For DB's: **A1, B2, C3**; For EXH: **A1, B2, C3**; For TF: **14, 20**

<u>BASELINE</u> - To correct for the pre-existing inside/outside pressure differential
Shows "**BASELINE**" on Channel A results spot while establishing the baseline and then "**ADJ**" just below Channel A results after the baseline is established

<u>START</u> - To begin the baseline or True Flow functions

<u>ENTER</u> - Accepts and enters a baseline reading

<u>CLEAR</u> - Clears the Baseline numbers and function

<u>TIME AVG</u> - To obtain a better measurement over a longer period of time when the gauge readings are less stable (the time frame for averaging shows in the lower right of the display)
1 sec - **1**; 5 sec - **5**; 10 sec - **10**; Keeps averaging until button is pressed again - **LONG**, Lower right

<u>UNITS</u> - Used to change the units displayed

<u>LIGHT</u> - To turn the backlight on the display on or off, ,

<u>HOLD</u> - To hold the readings showing on the display at the moment the button is pushed or to return to the non-hold mode when pushed again.

## Figure 4-16. Retrotec Gauge

Bold Items represent the button name or what actually shows on the gauge display (location of info on display in parenthesis)

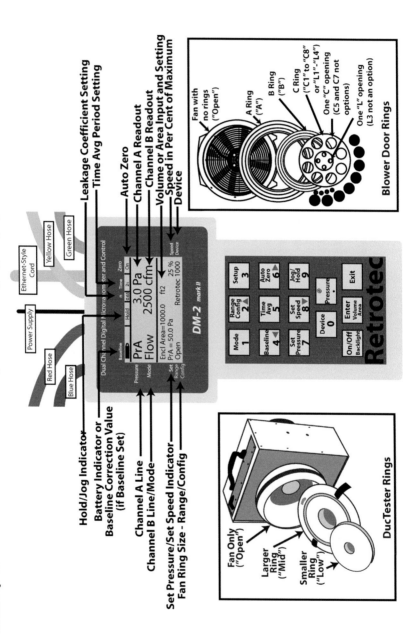

Older Model Info available on gauge but not shown here. Rings under "Range/Config" settings: Rings are labeled except C1 – C8 are how many caps are on the C ring. L1-L4 is how many mini-caps on the C ring.

**ON/OFF** – If tapped, turns on the gauge, once on, if held down momentarily, turns the gauge off. If just tapped while on, without being held down, turns backlight on/off

**MODE** - To select what the gauge will read for output on Channel B (Channel A results are shown next to the "Pressure" text while Channel B results are shown next to the "Mode" text to the left of the panel)

Channel A always shows a pressure (next to the "Pressure" line as "PrA") while Channel B displays (all settings but PrB allow for interpolating at an inputted pressure in Pa "@Pressure" as "@XPa"): Pressure - as **PrB**: Flow - **Flow**: Flow@25 or 50 Pa – **FL@25** or **50Pa**: Equivalent Leakage Area - **EqLA**: Effective Leakage Area – **EfLA**; Air Changes per hour– **Air Chg**: **Flow/Area; EfLA/Area; EqLA/Area; Hole Flow.**

**DEVICE** - To indicate which equipment is connected to the gauge on Channel B (lower right corner of display)

Retrotec fans: **1000**: Q46 & Q56 **2000**: Q4E, Q5E and QMG – **3000**: Q4E, Q5E and QMG with green tap no fan – **3000SR**
Model 200 & DU200 Duct Tester
Minneapolis ™ (Mn) fans: **Model 3 (120V); Model 3 (240V); Model 4 (240V);** Series B Duct Blaster® - **DuctBlaster™ B**

Infiltec fan: **Infiltec E3**

**RANGE/CONFIG** - To indicate (lower left corner of display) which ring has been installed (from large to small ring); **Open** = No rings

Fans: 1000 – **Open, A, B, C8, C6, C4, C2, C1, L4, L2, L1**: 2000/3000/3000SR – **Open, A, B, C8, C6, C4, C3, C2, C1, L4, L2, L1**: DU200 – **Open, Mid, Low**
Mn Models 3/4 – **Open, A, B, C, D, E.** DuctBlaster™B – **Open, Ring 1, Ring 2, Ring 3.** Infiltec E3 – **Open, 7 Holes, 4 Holes, 3 Holes, 2 Holes, 1 Hole**

**BASELINE** - To correct for the pre-existing inside/outside pressure differential
Shows the baseline correction in the upper left corner of display once the baseline is established

**TIME AVG** - To obtain a better measurement over a longer period of time when the gauge readings are less stable (the time frame shows below the "Time" text on the panel near the upper right corner of the display): **Off, 1s, 2s, 4s, 8s, 10s, 20s, 1m, 2m**

**ENTER** – Selects a menu item or inputs a value such as a baseline reading

**EXIT** – Clears the Baseline numbers and function and backs out of whatever other functions you are in

**SETUP** – Used to set up the gauge for units, significant figures, language selection, etc.

**AUTO ZERO** – Keeps the readings from drifting by automatically zeroing them every 8 seconds.

**JOG/HOLD** - To hold the readings showing on the display at the moment the button is pushed or to return to the non-hold mode when pushed again. The JOG function allows you to use arrows on the keypad to modify pressures or speeds you are setting.

**SET PRESSURE** - Used to automatically adjust the speed of the fan so that it delivers a specific pressure continuously.

**SET SPEED** - Used to automatically keep the speed of the fan constant regardless of how the pressure changes.

**@ PRESSURE** - Used to have the gauge estimate what the flow or other Mode would be at a given pressure, even if it is not at that pressure.

Note: If you dial *9 on this gauge you can make international calls.

So, blower doors are used to create an artificially high-pressure difference between the inside and outside of the home to determine how leaky a home is. Blower door leakage numbers can then be used to calculate an estimated leakage in the home under natural conditions. A blower door test is used for at least two very important reasons: determining the air leakage rate of the home at -50 Pascals and determining the location of air leaks in the home.

A blower door test measures the relative pressure difference between one space and another. This reading is called a pressure measurement "with reference to" abbreviated as WRT. For instance, you should not say the attic is at 50 Pascals, you should say it is at 50 Pascals with reference to the inside of the house. With the classic blower door test, a negative pressure or vacuum of 50 Pascals is obtained in the home WRT the pressure outside. As part of the test, air leakage is measured under these depressurized blower door conditions in cubic feet per minute (CFM) at -50 Pascals. This condition is designated by the term "CFM 50." Note since vacuum or depressurization is the standard in testing homes, the negative symbol is left out. This is the amount of airflow through openings and leaks when the home is depressurized to -50 Pascals.

For larger or especially leaky buildings, more than one fan may be required. In that case, it is best to use a software program that allows the results of all fans to be calculated at once (see Appendix O). In addition, conditions such as the outside temperature can affect the blower door results. For instance, a home that has 4,000 CFM50 when the air outside is at 65°F, will show only 3430 CFM50 when the outside air is at -10°F (Karg ZipTest Pro3™).

In an energy audit, you can also test for pressure under natural or normal operating conditions without a blower door in use. An example of this is Combustion Appliance Zone or CAZ testing, which will be discussed later.

## Effective Leakage Area (ELA)

The larger the holes in the home, the larger the amount of air allowed into the home with a negative pressure inside the home. Sometimes clients can better understand when the overall leakiness of a home is expressed in terms of the "effective leakage area." In fact, the flow of air in CFM50 can be converted to the equivalent size of a hole in the house if all the small holes were added together and placed into one

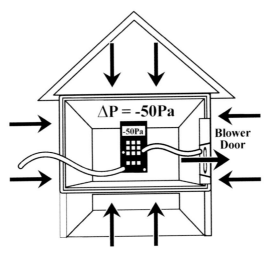

Figure 4-17. This shows that air leaving the house through the blower door, creating a -50 Pascal air pressure (vacuum) in the house with respect to (WRT) outside.

opening. Industry standards indicate dividing the CFM50 by a correction factor of 10 to come up with the approximate total square inches of holes in the home represented by the leakage. For example, if the CFM is 5000, then 5000 CFM50/10 = 500 in.² (the hole would be approximately 500 square inches). Since the square root of 500 is approximately 22, the hole would be roughly equivalent to about a 22" x 22" opening. This would be the ELA at 50 Pascal pressure.

**CFM50 and CFM Natural Calculations**

As mentioned previously, CFM50 refers to the level of air leakage measured when testing the home with the blower door at -50 Pascals. Since -50 Pascal is not the normal pressure in a home's natural condition, there is a calculation involving an "N" factor to find out how much air leakage there is under natural conditions. Natural conditions are defined as a 4 Pa difference in pressure between inside and outside.

The N factor is based on the climate zone of the home, the building height, and how well the home is shielded from the wind by trees, berms, other homes, etc., as shown in Figure 4-18. The climate zone number is determined based on the typical local outdoor temperature-based stack effect, combined with local wind conditions.

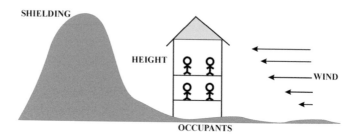

Figure 4-18. Other N factor considerations—Trees are also an example of shielding.

The N factor is found by checking the map (Figure 4-19 and Table 4-1) which were developed by Max Sherman of the Lawrence Berkeley National Laboratory (LBNL). These zones give an indication of the wind factors in a given area.

Divide the CFM50 by the N factor to obtain an estimate of the natural leakage rate in the home. The N factor ranges from 9.8 to 29.4, but commonly averages around 20. If a home has a high N factor, say because it is only a one story house surrounded by a heavily wooded area, then the blower door is creating even more exaggerated conditions than would exist under natural conditions. If a lower N factor applies, the blower door reading is closer to the level of the natural leakiness of the home. 80-90% of homes should be placed in the normal category. In other words, it should be uncommon to designate homes as well shielded or exposed.

Figure 4-19. Zones for Calculating N Factor

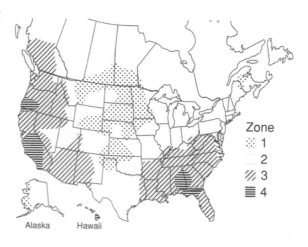

## Table 4-1. N-Factor Table

| Zone | # of Stories> | 1 | 1.5 | 2 | 3 |
|------|---------------|------|------|------|------|
| 1 | Well-shielded | 18.6 | 16.7 | 14.9 | 13.0 |
| | Normal | 15.5 | 14.0 | 12.4 | 10.9 |
| | Exposed | 14.0 | 12.6 | 11.2 | 9.8 |
| 2 | Well-shielded | 22.2 | 20.0 | 17.8 | 15.5 |
| | Normal | 18.5 | 16.7 | 14.8 | 13.0 |
| | Exposed | 16.7 | 15.0 | 13.3 | 11.7 |
| 3 | Well-shielded | 25.8 | 23.2 | 20.6 | 18.1 |
| | Normal | 21.5 | 19.4 | 17.2 | 15.1 |
| | Exposed | 19.4 | 17.4 | 15.5 | 13.5 |
| 4 | Well-shielded | 29.4 | 26.5 | 23.5 | 20.6 |
| | Normal | 24.5 | 22.1 | 19.6 | 17.2 |
| | Exposed | 22.1 | 19.8 | 17.6 | 15.4 |

Note that the lower the N factor, the higher the CFM natural because it is divided by the N factor. Thus, the more exposed, the more stories, and the higher the climate zone number, the larger the CFM natural.

As an example, say there is a reading of 4000 CFM50 in the home and an N factor of 20:

$$4000 \text{ CFM50}/20 = 200 \text{ CFM natural}$$

Thus, under natural conditions, this house would have a natural air leakage rate of 200 CFM.

Unfortunately, one of the assumptions with using the N factor is that there is a random assortment of hole shapes and sizes randomly distributed throughout the building envelope. When a home is weatherized, there is a tendency to concentrate on sealing holes in the lower and upper portions of the building. This takes the randomness out of the situation, which is something that the N factor does not take into account.

## Air Changes per Hour

You may know the CFM50 in a house, but how do you know it is acceptable? How can you compare it to other homes? For instance, if the blower door test reveals 4000 CFM50, that would be very leaky for a smaller home, but it could be extremely tight for a much larger house. To make the comparison, use the concept of "air changes per hour" or ACH. ACH50 refers to how often the air in a home is refreshed when the pressure difference between the inside and outside of the house is –50 Pascals. ACH50 takes into account the size of the home. It is calculated by multiplying the CFM50 by 60 minutes/hour and then dividing by the volume of the home. Figure 4-20 shows how volume is determined.

ACH = CFM50 x 60min/hr x 1/V

Correspondingly:
CFM50 = ACH/60xV

A typical newly constructed platform framed house will have an ACH50 of 5 to 10, while an older balloon-framed home typically has an ACH50 of 11 to 15. Some homes even have an ACH50

**Figure 4-20. Home volume is the height from the floor to the ceiling multiplied by the length and the width of the home.**

of up to 30, making them extremely leaky homes. This means that if you run the blower door for a full hour, the air in the house would change 30 times. If the N factor is assumed to be 20, divide the 30 ACH50 by the N factor of 20 and get an ACH natural of 1.5. While any home with an ACH50 of 11 or more might be considered leaky, any home with an ACH natural of more than .5 has to be considered a relatively leaky home.

## MINIMUM VENTILATION REQUIREMENT (MVR)

The minimum ventilation requirement (MVR) is the rate that ventilation must enter a home from outside when the home's windows etc. are closed to provide for adequate healthy indoor air quality at a minimum level. This MVR is the lowest CFM level allowed when a home is sealed. MVR is based on CFM50, typically with a correction for already existing exhaust fans. To see how MVR is calculated, see Appendix D.

## Air Sealing Target

In the weatherization program, some local programs have flexibility to set their own target as long as it stays at or above the MVR. For instance, some local programs set their air sealing targets somewhere between the MVR and the pre-retrofit blower door reading. Other programs may require that a reduction in CFM50 of 40% be achieved for the leakiest homes, 20% for the homes in the middle range of leakiness, and a 0% reduction in CFM50 for the tightest homes.

## Using the Blower Door CFM50

Even if the blower door reading at CFM50 is not used for any other calculations, it can give an idea of how successfully the house was sealed if a blower door test is conducted before and after the house is retrofitted. A blower door test will also tell when it is cost-effective to spend time finding and sealing leaks in a home. If the blower door test reveals that the home is already fairly tight and does not leak very much, time could be better spent improving the baseload energy numbers, the insulation, and providing appropriate mechanical measures to the home. Thus, on a tight home, you may not spend very much, if any, time looking for or sealing leaks. On the other hand, if you find, through the blower door test, that a home is extremely leaky, then it could be very cost-effective to spend time finding and sealing leaks on the home.

## Trouble Shooting

If the home is relatively tight, you may have to use "low flow rings" (Figure 4-21) to help correct for the home's tightness. The sensors on the fan require a minimum air speed passing by them to accurately read the flow. If, as on a tight home, it does not take very much airflow through the fan to reach the -50 Pascals level, the air speed will be too slow for the fan to be accurately read. That is why "rings" that look like flat discs are installed to restrict the fan opening so that even at low flow rates air passes by the sensors at an adequate speed to be read accurately. This way you can still create a -50 Pascal pressure in a tight home and get adequate air speed for the fan to read accurately.

There are special adapter rings that can be used to close off the fan opening on a blower door. These adapter rings allow some of the tightest homes to be tested by the same fan that tests leaky homes. By narrowing the opening to the fan, the lower flow of air can still provide enough speed to allow the fan system to determine the flow rate.

On the other hand, if it is an extremely leaky home, even if the fan is at its highest speed, it sometimes still cannot create -50 Pascals pressure in the home. This is sometimes referred to as the "cannot reach 50" situation. Fortunately, most modern fans and gauges are able to automatically calculate what the CFM50 would be had the fan been able to create a -50 Pascals pressure in the home. Older fans or gauges may require the use of tables to calculate the correction for this.

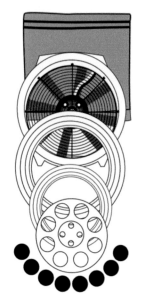

**Figure 4-21. "Low Flow" Rings**

If you get unusual readings on the gauge, make sure that the flow sensor openings on the center portion of the fan are not blocked. Also, check to see that the hose on the outside of the house is not in the air stream of the fan blowing the air outdoors. Make sure this hose to the outdoors is placed along the edge of the house at least 5 feet away from the blower door. Also, make sure that any of the hoses inside the house do not become drawn into the fan in the process of conducting the blower door test.

As air flows across the sensor hub on the fan, it causes air pressure in the sensor opening. In some of the older fans, these sensor openings were caulked. Caulked seals on older fans can fail. If you happen to be using one of these older blower doors, you can test this seal by placing your fingers over the sensor holes while creating suction on the connector or tap at the top of the fan. If there is no other way to create suction, you can even suck on the tap and see if you can get your tongue to stick to the tap. If your tongue sticks then the caulking has not failed.

It should be mentioned that if you only do a blower door test at one pressure level, such as -50 Pascals, you are only conducting a "single point blower door test." This means there is only going to be a blower door CFM reading at one pressure: -50 Pascals. This type of test should be distinguished from the so-called "multi-point test" that involves taking a blower door CFM reading at various pressure levels. For instance, for the benefit of better accuracy, when commercial buildings are tested with blower door systems, it is generally considered good practice to measure the CFM at multiple depressurizations such as -25, -35, -45, -55, -65, and -75 Pascals. This would be called a six-point multi-point blower door depressurization test. The test could also include similar pressurizations of the house.

## HOME SETUP AND PREPARATION FOR BLOWER DOOR

The following table gives a step-by-step procedure of how to prepare the house for blower door testing regardless of the type of equipment being used. The two subsequent tables explain the specific procedures to use for two brands of blower doors, EC and Retrotec. Select the appropriate table that applies for the type of blower door you are using. Figure 4-22 shows the blower door setup. When you are done with the testing, always remember to return the house and its appliances to their original conditions.

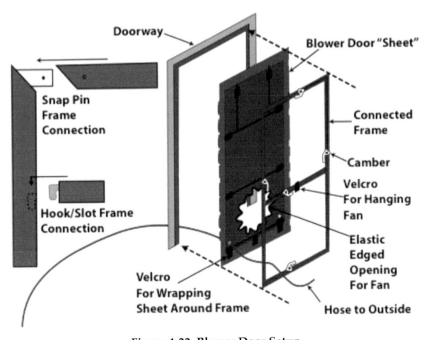

**Figure 4-22. Blower Door Setup**

**Table 4-2. Step-by-Step Blower Door Preparation and Set-up for Using a Blower Door for Either Depressurization or Pressurization**

1. If you notice any vermiculite or asbestos insulation in the attic, mold, etc. inside of the house during your inspection, do not proceed.

   If this is a new house being tested under a specific energy efficient home protocol, follow that protocol instead.

   If the house is in an area where the home interior could be exposed to freezing temperatures during the next year if unheated and the home is new or has been vacant for a while, obtain permission from the owner to refill the P traps, floor drains, and toilets with antifreeze if allowed in your locality.

   Notify the homeowner that no one can open or close doors, turn on furnaces or air conditioner systems, fans, etc. during the test.

2. Close any vents to the outside, whether powered by a fan or not, including fireplace or wood stove dampers and doors.

3. Open all HVAC registers/diffusers/grilles and interior doors, including closet doors. Open the basement door if you see evidence of basement storage, washer, dryer, etc. Only open an attic doorway or hatch if it appears the attic space is insulated from the outside (it is conditioned space), such as finding insulation on the attic side of the roof deck or between the roof rafters, etc. If there is an attached garage attic and it looks like it is insulated from the house (the attic is unconditioned space with insulation on the "floor" of the attic space), open that attic hatch and the vehicle door. If the crawl or attic spaces are unconditioned, open crawl space and attic vents, as applicable. If the house has T-bar type suspended ceilings, or fluorescent light lenses/diffusers, with the owner's permission, remove one ceiling tile in each room they exist in, unless they have insulation laying on top of the ceiling tiles. In that case, watch the ceiling closely as you slowly depressurize the home to be sure you are not pulling down/destroying the ceiling tile. If it appears they are being affected, you have to use the "@ pressure" button (Retrotec) or "can't reach 50" factors/"Pressure flow@50" mode (EC)—press the "Mode" button until '@50" shows up on the screen.

4. Close exterior doors and latch all windows (including skylight windows), storm windows, crawlspaces, hatches and attic hatches, unless specifically instructed otherwise. Seal off and cover any window air

conditioner units and cat/dog doors. Close doors to rooms/closets with furnaces, boilers, water heaters, etc. only if they have open-air access/makeup/supply from the outside.

5. Turn off all HVAC system electrical and the water heater to pilot. Turn off all fans—bathroom & kitchen fans, clothes dryers, central vacuums, broiler hoods, whole house fans, swamp coolers, window air conditioner units, ceiling fans, solar air panel fans, attic/crawlspace powered ventilators, HRV/ERV or other fresh air ventilation systems, etc. Leave your car keys on top of the water heater as a reminder to turn these appliances back on—you will not be able to start your car if you do not return to the water heater to turn it and the other appliances back to their original setting.

6. Make sure all fireplace and wood stove fires have been out for at least 24 hours and that all ashes have been cleaned out. If the ashes have not been removed or you are unable to remove the ashes, completely cover them with wet newspapers. Close all closeable fireplace and wood stove doors.

7. If there is an attached space with combustion appliances that freely communicates with the unit you are testing (such as a townhouse that has a common, unseparated attic or source of outside air, etc. with the unit), be sure that these other appliances in the other unit are also off to avoid the same risk there.

8. Make sure, during the test, all pets, children, small objects, loose papers, etc. are kept away from around the inside or outside of the door where you are placing the blower door.

9. Set-up of Blower Door—Select an exterior door, other than a sliding door, where to place the blower door. The clearer the area directly outside and especially inside of the door, the better the blower air will be able to flow unhindered. If you cannot create at least 5 feet of open clearance in front of the fan on the inside (if depressurizing—if pressurizing, the important clearance is on the outside), you can try to use one of the fan rings to reduce the flow through the fan and still get an accurate reading.

10. Open the door and the storm door, if applicable, of the doorway you have chosen, as much as possible and prop them so they stay open. Connect the door frame pieces together at the snapping connectors to form

a rectangle and put the middle cross bar in place using the hook/slot connection.

11. Fit the adjustable metal blower doorframe in the doorway by first loosening the three knobs on all of the lower, middle and upper horizontal frame pieces, wedging the frame inside of the doorjambs, and pushing the side frames away from each other against the side doorjambs. While holding the frame in place, tighten these same three knobs.

12. Loosen the knobs on the two side frames, and with your foot over the lower frame, pull the top frame up to the top of the door frame and tighten the two side knobs while holding the frame in place.

13. Now that you have "fitted" the frame for the door, remove it, lay the fabric or cloth door sheet on the floor and set the frame over the top of the fabric with the knobs on the frame facing up.

14. Giving plenty of extra length, attach the long Velcro strip over the top of the frame.

15. Pull the bottom of the nylon sheet/panel under the doorframe and attach the two Velcro strips at the bottom so they hold the sheet in place.

16. Readjust the top Velcro strip so it tightens the sheet from top to bottom.

17. Pull both sides of the sheet around the sides of the frame and Velcro them to hold both of them in place.

18. Fit the frame with the sheet Velcroed to it into the doorframe. Fit the cross bar into the frame just above the circular fan opening and tighten the knob to it while pushing the two side frames into the door jamb on each side. If you tightened everything properly in the first place, you need only turn the cam levers for each of the five knobs so that the frame with sheet is tightened into the doorframe against a doorstop. If the doorframe does not fit tightly into the door, disengage the cams, readjust the frame so it fits tighter in the door, and then re-engage the cams.

19. Run at least 5 feet of tubing (red hose for Retrotec or green hose for Energy Conservatory) through the hole in the sheet and along the side of the house on the outside so that it does not rest in the path of the blower door air stream. Protect the hose from getting water or dirt in it.

*EC Pressurization*: If you are pressurizing the building rather than depressurizing it, be sure to run another set of tubing out through the other access hole in the door sheet so that one end of the hose will rest just outside the door sheet and just next to, but not in the stream of, the blower door fan.

20. Install the blower door fan, with all the rings and cover plate installed, facing the house side (or facing outside if pressurizing with a Retrotec fan) with the handle up, into the round elastic opening in the blower door sheet. This works easiest if you tip the fan towards you while placing it next to the opening and then tipping it over the bottom of the elastic opening. You should then pull the rest of the elastic opening around the top of the fan to fully enclose the opening around the fan casing.

*EC fans*: Use the "Flow" switch on the fan to select the proper arrow for flow direction. To depressurize, switch the button towards the arrow pointing towards the outside of the house. To pressurize, switch the button towards the arrow pointing towards the inside of the house.

*Retrotec fans*: Only if *pressurizing* the house, as indicated above, face the rings and cover plate towards the outside of the house when placing the fan in the door.

21. Pull the fan so it sits perfectly vertical and approximately 2 inches off the bottom frame by strapping the Velcro strap hanging off the middle cross-bar through the fan handle and back up on itself.

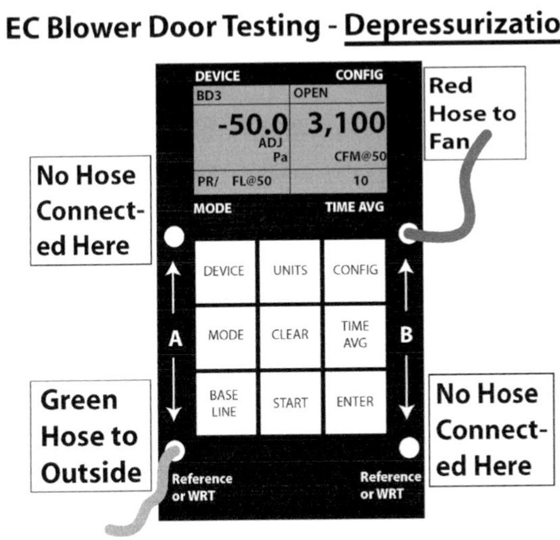

Figure 4-23. Depressurization

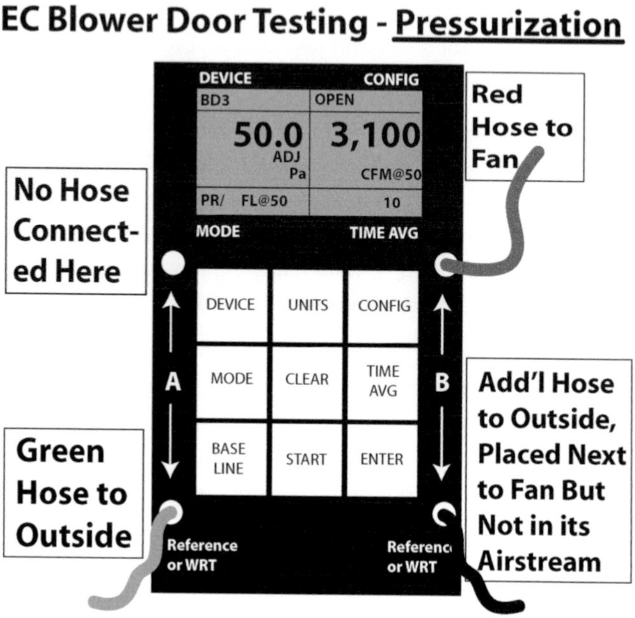

# EC Blower Door Testing - <u>Pressurization</u>

**No Hose Connected Here**

**Green Hose to Outside**

**Red Hose to Fan**

**Add'l Hose to Outside, Placed Next to Fan But Not in its Airstream**

**Figure 4-24. Pressurization**

**Table 4-3. Setting up the Energy Conservatory (EC)/Minneapolis™ Blower Door with DG-700 Gauge (Gauge) for Either Depressurization or Pressurization**

1. **Before Starting: Be sure to conduct all of the steps in the "Step-by-Step Blower Door Preparation and Set-up" procedure above.**
   Make sure your gauge batteries are fully charged or that you have adequate outlets and extension cords to connect the pressure gauge to its recharger/transformer.

   Attach the hoses to the gauge—red tubing coming from the tap on the fan to the Channel B input tap (upper right tap) and green tube(from outside) to the Channel A reference tap (lower left tap). *Pressurization*: If you are pressurizing using an EC fan, be sure to connect the other end of the extra hose you placed just outside the door sheet and just to the side of the fan (NOT the Green tube). Connect it to the Channel B reference tap on the lower right of the EC gauge.

2. Attach the gauge mounting board and fan speed controller (with the speed control dial on it) onto the blower door (there is a special short

door frame attachment that the mounting board can be hung on using a hook/slot attachment).

3. Attach the gauge with Velcro on the mounting board.

4. Insert the female plug from the speed controller into the receptacle on the fan.

5. Make sure the fan speed controller dial is turned as counterclockwise as possible. Plug the regular power cord plug from the speed controller into a wall outlet.

6. Check the fan direction switch on the top of the fan so that the arrow shows the fan will blow air to the outside (depressurization) or towards the inside, in the rare event you are pressurizing the home.

7. If your fan speed controller and gauge have "cruise control" capability, install the low voltage cruise control connector between the gauge (on its top) and the speed controller (on its side). Otherwise, ignore this step.

8. Turn on the gauge—Press "On/Off."

9. Press the "Mode" button twice to put the gauge in the PR/FL@50 mode.

10. Check that the device showing in the upper left corner of the display matches the blower door you are using. It should say "BD3" if you expect to use the Model 3 blower door. If it does not indicate the blower door model, press "Device" until "BD3" shows up on the display.

11. Push "Baseline," push "Start," and wait for 20 seconds or until the reading stabilizes, and then push "Enter." The Channel A display should now show an "ADJ" icon in it to show that the reading will be adjusted for baseline.

12. Remove the fan covers.

13. Press "Config" until the proper flow ring you are using shows up on the upper right of the display. If you are not using a ring, the config should show "OPEN."

14. Press the "Time Avg" button until you see "5" displayed in the lower right corner (set it for "10" if it is a windy day).

15. Turn on the blower door fan to -25 Pa: Turn the fan controller clockwise. Continue to increase the fan speed until the pressure shown on the left side of the display (Channel A) shows roughly -25 Pa. You need not at-

tempt to keep the pressure at exactly -25.

16. Check around the house to make sure no problems are arising.

17. If everything seems okay in the house, turn on the blower door fan to -50 Pa:

    *Manual Control*—(if you only have one gauge and need to use it to do Duct Pressure Pan, ZPD, etc. tests) Turn the fan controller clockwise. Continue to increase the fan speed until the pressure shown on the left side of the display (Channel A) shows between -45 and -55 Pa. You need not attempt to keep the pressure at exactly -50 as the gauge will calculate a flow reading assuming the pressure were -50. If the flow is fluctuating more than you would like, press the "Time Avg" button until you have selected the 10-second averaging period.

    *Cruise Control*: Turn the speed controller dial to the "just on" position and press the on/off switch to "on" if it has one. Press the "Enter" button and the cruise control will be triggered. The left side of the display (Channel A) should display the number 50. Press the "Start" button and the blower door fan will increase speed until the pressure is approximately -50.

18. Read the flow rate that now shows up on the right side of the display (Channel B). *If it shows a series of hyphens—*("-----") instead, turn off the fan with the controller and step up to larger flow rings one at a time until there is adequate flow, the hyphens disappear, and you get a flow number in its place. Turn off the fan to change flow rings.

    *If it shows "LO"* instead of a number, turn off the fan with the controller and step down to smaller flow rings one at a time until the "LO" disappears, and you get a flow number in its place. Turn off the fan to change flow rings.

    *Be sure to press "Config"* until the ring displayed in the upper right corner of the gauge display matches the ring on the fan. If you are using the cruise control, press "Clear" before you press the "Config" button and then follow the instructions in the previous step to restart the fan.

    Some suggest that a ring should be added even if the CFM50 is below 2400 whether it shows "Too Low" or not. Restart the fan by following the instructions starting at "Set Pressure."

19. This is a good time to check for air leaks as you are going around the house to do ZPD testing, etc. (see gauges issues below) Locations of leaks: _____

_____

_____

*Two Gauges Available*: Keep this gauge on the blower door. You can use the other gauge to go around the house to conduct pressure pan, ZPD, etc. testing.

*One Gauge Available*: If you have only one gauge, use the manual speed control dial to set the fan rpm's so the gauge reads -50 Pascals. Once you have established the -50 Pascals using your gauge, record the flow and pressure in the next step and disconnect the tubes going to the blower door. If you are conducting ZPD testing indoors, keep the hose to outside connected to your gauge if it is long enough to reach the zones you will be testing.

20. Read and record the CFM number that shows up in the right side of the middle of the gauge (Channel B). _____CFM Compare this recorded number with the MVR or BAS range that you have already calculated for this home. MVR/BAS Range: _____ to _____CFM

*BPI-Require* 100% continuous of the BAS mechanical ventilation because the house CFM is *lower than the low end* of the BAS CFM range? Y — N

*Recommend* mechanical ventilation that makes up for the differential in CFM and BAS CFM because the house CFM falls *within* the BAS CFM range? Y — N

*Weatherization*—check with your supervisor—you may be using 62.2-2007

21. Turn off the fan by holding down the "On/Off" button or dial down the manual speed control dial counterclockwise until it stops if you are using the manual speed control dial. Return the house to normal conditions by returning the combustion appliances and thermostats to their original settings, unseal window air conditioner units, open fireplace dampers, etc. Be sure to check that pilot lights have not blown out from the test. If they have, relight them or have someone else relight them safely. Watch out for build-up of unburned gas.

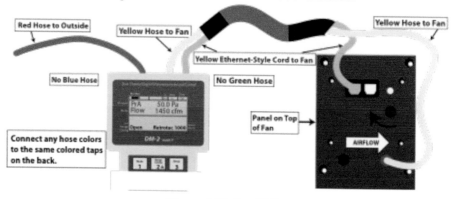

Figure 4-25. Fan 1000

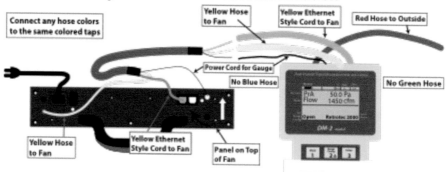

Figure 4-26. Q46 and Q56

**Table 4-4. Setting up the Retrotec Blower Door—the Q46 and the 1000 Fans Using the DM-2 Retrotec Gauge (Manometer) for either Depressurization or Pressurization**

1. **Before Starting: Be sure to conduct all the steps in the "Step-by-Step Blower Door Preparation and Set-up" procedure above.**

   Make sure your gauge batteries are fully charged or that you have adequate outlets and extension cords to connect the pressure gauge to its recharger/transformer.

   Attach the hoses to the gauge from the fan and the red outside hose. Plug the colored hoses into the appropriately colored nipples on the DM-2 gauge—only yellow-to-yellow, red-to-red.

2. Connect the black low voltage (smaller) cord (if the battery is low on the gauge) and the yellow ethernet-style cable on the umbilical cord to the DM-2. Connect the other end of these cords to the fan.

3. Turn on the gauge by pressing and holding down the "On/Off" button.

   Once the gauge is on, press "Auto Zero" button until "On" shows in the upper right corner of the display.

   If the number below the "n" value on the right side of the top line of the display does not show "0.65," press the "Setup" button repeatedly until you have selected the "n" value option. Then press "Enter," type "0.65" (use the numbers & period button—same as the "@ Pressure" button) in as the n exponent, press "Enter" again and then "Setup."

4. Connect the ethernet-style data cable, the DM-2 low voltage power connector, and the yellow hose on the other end of the umbilical into the yellow fan connections. Press "Exit" if the status light on the fan turns red. This should make it turn steady green.

5. Plug the black power cord into the fan connection for it, and the wall plug on the other end into a wall outlet.

6. Press "On/Off" and then "Exit." If you do not see "Retrotec 2000" in the display's lower right hand corner, hit "Mode" once and then "Device" until it shows up.

7. If you do not see "Open" in the display's lower left corner, hit "Range/Config" until it shows up there.

8. If the display upper right corner does not show "On," hit "Auto Zero" once.

9. If you do not see "4s" just below the "Time" setting on the top line of the display, push "Time Avg" until it shows up. If it is windy, set it for 10 seconds instead.

10. Push "Baseline" and wait for 20 seconds or until the reading stabilizes, and then push "Enter"

11. Press "Mode" until "Flow" appears on the display's left edge at the "Mode" mark.

12. Remove the fan covers.

13. Press "Set Pressure" input "25" using the number buttons and then press "Enter" to bring the pressure to 25.

14. Check around the house to make sure no problems are arising.

15. If everything seems okay around the house, you need to set the blower door to stay at -50 Pascals if possible. Push "Set Pressure," "50," and "Enter" to bring the pressure up to 50 Pascals.

16. If the display reads "Too Low," press "Exit" to stop the fan, add the largest ring if there is no ring on it or use the next smaller size ring if there is a ring on it, press "Range/Config" until the size matching the ring is in the display's lower left corner. Some suggest that a ring should be added even if the CFM50 is below 2400 whether it shows "Too Low" or not. Restart the fan by following the instructions starting at "Set Pressure."

17. If the pressure reading is not matching the pressure you set it at—say, you can only get up to 40 Pa but you need to reach 50 Pa—press "Exit" to stop the fan, remove a fan ring and push "Range/Config" until the remaining ring size shows up on the display's lower left corner. Restart the fan by following the instructions starting at "Set Pressure." If you still cannot reach the target pressure, press "@ Pressure" to have the gauge calculate what the flow would be at the target pressure even though you are not able to actually reach it with the fan configurations.

18. This is a good time to check for air leaks as you are going around the house to do ZPD testing, etc. (see gauges issues below) Locations of leaks:

_____

_____

_____

*Two Gauges Available*: Keep this gauge on the blower door. You can use the other gauge to go around the house to conduct pressure pan, ZPD, etc. testing.

*One Gauge Available*: If you have only one gauge, setup and use the manual speed control dial to set the fan rpm's so the gauge reads -50 Pascals. Once you have established the -50 Pascals using your gauge, record the flow and pressure in the next step and disconnect the tubes going to the blower door. If you are conducting ZPD testing indoors, keep the hose to outside connected to your gauge if it is long enough to reach the zones you will be testing.

19. Read and record the CFM or CFM50 number that shows up in the middle of the gauge. _____ CFM

    Compare this recorded number with the MVR or BAS range that you have already calculated for this home. MVR/BAS Range: _____ to _____CFM

    *BPI-Require* 100% continuous of the BAS mechanical ventilation because the house CFM is lower than the low end of the BAS CFM range? Y N

    *Recommend* mechanical ventilation that makes up for the differential in CFM and BAS CFM because the house CFM falls within the BAS CFM range? Y N

    *Weatherization*-check with your supervisor—you may be using 62.2-2007

20. Turn off the fan by holding down the "On/Off" button or dial down the manual speed control dial counterclockwise until it stops if you are using the manual speed control dial. Return the house to normal conditions by returning the combustion appliances and thermostats to their original settings, unseal window air conditioner units, open fireplace dampers, etc. Be sure to check that pilot lights have not blown out from the test. If they have, relight them or have someone else relight them safely. Watch out for buildup of unburned gas.

# Retrotec Blower Door Fans <u>Q4E, Q5E and QMG</u> Set-Up Depressurization and Pressurization

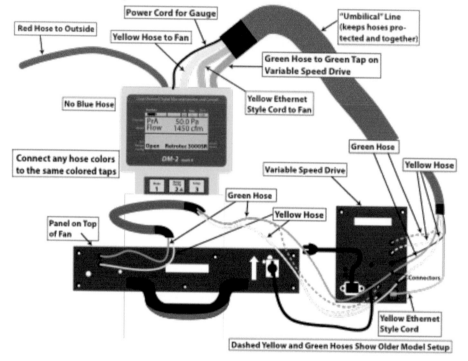

Figure 4-27. Q4E, Q5E, and QMG

**Table 4-5. Setting up the Retrotec Blower Door Q4E, 5E, or QMG Using the Retrotec DM-2 Gauge for Either Depressurization or Pressurization**

1. Before Starting: Be sure to conduct all of the steps in the "Step-by-Step Blower Door Preparation and Set-up" procedure above.

   Make sure your gauge batteries are fully charged or that you have adequate outlets and extension cords to connect the pressure gauge to its recharger/transformer.

   Attach the hoses to the gauge from the fan and the outside hose. Plug the colored hoses into the appropriately colored nipples on the DM-2 gauge—only yellow-to-yellow, red-to-red, green-to-green.

2. Connect the low voltage (smaller) power cord (if the battery is low on the gauge) and the yellow ethernet-style cable on the variable speed drive (VSD) umbilical cord to the DM-2.

3. Turn on the gauge by pressing and holding down the "On/Off" button.

   Once the gauge is on, press "Auto Zero" button until "On" shows in the upper right corner of the display.

   If the number below the "n" value on the right side of the top line of the display does not show "0.65," press the "Setup" button repeatedly until you have selected the "n" value option. Then press "Enter," type "0.65" (use the numbers & period button—same as the "@ Pressure" button) in as the n exponent, press "Enter" again and then "Setup."

4. Connect the DM-2 low voltage power connector, and the ethernet-style data cable on the other end of the VSD umbilical into the VSD on the right side in that order from top to bottom. Then connect the yellow hose, and the green hose on the same end of the VSD umbilical to the yellow and green hoses, respectively, on the fan umbilical using two male-to-male hose connectors.

5. Older models: Connect the yellow hose, and the green hose on the other end of the VSD umbilical into the VSD on the right side in that order from top to bottom. Then plug the fan umbilical cord's yellow and green hoses into the VSD right next to where the fan power cord plugs in at the bottom with yellow below the green.

6. Plug the fan umbilical cord's fan side hoses into their respective colored connectors on the fan and the power cord into the fan power connection (push it in and turn it clockwise to get it to lock into place).

7. Connect the plug end of the power cord for the VSD into a wall outlet and the other female end into the variable speed drive.

8. Press "On/Off" and then "Exit" on the gauge. If you do not see Retrotec "3000SR" in the display's lower right hand corner, hit "Mode" once and then press "Device" until it shows up.

9. If you do not see "Open" in the display's lower left corner, hit "Range/Config" until it shows up there.

10. If the upper right corner of the display does not show "On," hit "Auto Zero" once.

11. If you do not see "4s" just below the "Time" setting on the top line of the display, push "Time Avg" until it shows up. If it is windy, set it for 10 seconds instead.

12. Push "Baseline" and wait 20 seconds or until the reading stabilizes, and then push "Enter."

13. Press "Mode" until "Flow" appears on the display's left edge at the "Mode" mark.

14. Remove the fan covers.

15. Press "Set Pressure" and input "25" using the number buttons and then press "Enter" to bring the pressure up to 25.

16. Check around the house to make sure no problems are arising.

17. If everything seems okay around the house, push "Set Pressure," "50," and "Enter" to bring the pressure up to 50 Pascals

18. If the display reads "Too Low," press "Exit" to stop the fan, add the largest ring if there is no ring on it or use the next smaller size ring if there is a ring on it, press "Range/Config" until the size matching the ring is in the display's lower left corner. Some suggest that a ring should be added even if the CFM50 is below 2400 whether it shows "Too Low" or not. Restart the fan by following the instructions starting at "Set Pressure."

19. If the pressure reading is not matching the pressure you set it at—for example, you can only get to 40 Pa but you need to reach 50 Pa—press "Exit" to stop the fan, remove a fan ring and push "Range/Config" until the remaining ring size shows up on the display's lower left corner. Restart the fan by following the instructions starting at "Set Pressure." If you still cannot reach the target pressure, press "@ Pressure" to have the gauge calculate what the flow would be at the target pressure even though you are not able to actually reach it with the fan configurations.

20. This is a good time to check for air leaks as you are going around the house to do ZPD testing, etc. (see gauges issues below) Locations of leaks: _____
    _____
    _____

*Two Gauges Available*: Keep this gauge on the blower door. You can use the other gauge to go around the house to conduct pressure pan, ZPD, etc. testing.

*One Gauge Available*: If you have only one gauge, set up and use the manual speed control dial to set the fan rpm's so the gauge reads -50 Pascals. Once you have established the -50 Pascals using your gauge, record the flow and pressure in the next step and disconnect the tubes going to the blower door. If you are conducting ZPD testing indoors, keep the hose to outside connected to your gauge if it is long enough to reach the zones you will be testing.

21. Read and record the CFM or CFM50 number that shows up in the middle of the gauge. _____ CFM Compare this recorded number with the MVR or BAS range that you have already calculated for this home. MVR/BAS Range: _____ to _____CFM

    *BPI-Require* 100% continuous of the BAS mechanical ventilation because the house CFM is *lower than the low end* of the BAS CFM range? Y N

    *Recommend* mechanical ventilation that makes up for the differential in CFM and BAS CFM because the house CFM *falls within* the BAS CFM range? Y N

    *Weatherization*-check with your supervisor—you may be using 62.2-2007

22. Turn off the fan by holding down the "On/Off" button or dial down the manual speed control dial counterclockwise until it stops if you are using the manual speed control dial. Return the house to normal conditions by returning the combustion appliances and thermostats to their original settings, unseal window air conditioner units, open fireplace dampers, etc. Be sure to check that pilot lights have not blown out from the test. If so, relight them or have someone else relight them safely. Watch for build-up of unburned gas.

## MEASURING EXHAUST FAN FLOW

Exhaust fan flow is used to help calculate the MVR. You can deduct exhaust fan flows from the MVR (see MVR section in Appendix D). To measure exhaust fan flow, use an exhaust fan flow meter that is really nothing more than a calibrated pressure pan as shown in Figure 4-28. In fact, the flow meter will function as both a flow meter and a pressure pan if you completely close the adjustable opening on the flow meter. A gauge is also needed to use this device. It only works for measuring the intake on an exhaust fan; it cannot be used to measure any fan's discharge. Set the flow meter pan completely over the fan intake when the fan is on. You only need to adjust the flow meter's opening until you get an appropriate reading on the gauge (see test procedures below).

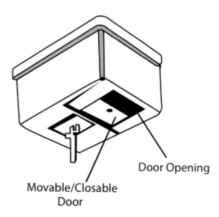

Door Opening

Movable/Closable
Door

**Figure 4-28. Exhaust Air Flow Meter**

Because you may not know whether you need to utilize fans to get the MVR, you may decide to wait until it is conspicuous that you need fan flow before spending time conducting this test. Thus, you may find that this test will only need to be done by retrofitters and then only if they discover a need for fan flow to help with MVR. Some auditors, if they do not find a need to actually measure flow in exhaust fans, will conduct only the tissue paper test to make sure it seems to be minimally working—seeing if a single layer of tissue paper is able to stay up against an exhaust fan grille with the exhaust fan on.

## EC Exhaust Fan Flow Testing

| DEVICE | | CONFIG |
|--------|--|--------|
| EXH | | A1 |

**Hose to Flow Meter**

| 0.0 | 65 |
|-----|----|
| Pa | CFM |

| PR/FL | 10 |
|-------|-----|

**No Hose Connected Here**

| MODE | | TIME AVG |
|------|--|----------|

| DEVICE | UNITS | CONFIG |
|--------|-------|--------|

**Exhaust Fan Flow Meter Box**

A

| MODE | CLEAR | TIME AVG |
|------|-------|----------|

B

| BASE LINE | START | ENTER |
|-----------|-------|-------|

**No Hose Connected Here**

Reference or WRT

Reference or WRT

**No Hose Connected Here**

**Figure 4-29. EC Exhaust Fan Flow Testing**

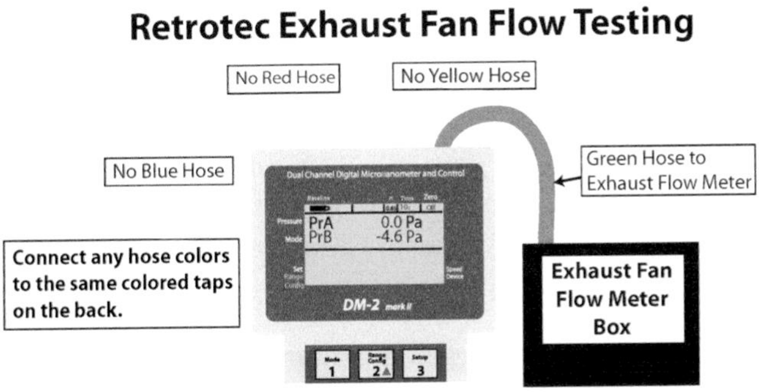

## Retrotec Exhaust Fan Flow Testing

Figure 4-30. Retrotec Exhaust Fan Flow Testing

**Table 4-6. Exhaust Fan Flow Testing Procedure Using the EC Exhaust Fan Flow Meter with either a Retrotec or EC Gauge**

1.  Make sure your gauge batteries are fully charged or that you have adequate outlets and extension cords to connect the pressure gauge to its recharger/transformer.

    Connect some tubing on one end to the Exhaust Fan Flow Meter box tap/connector.

2.  *EC Gauge:* Connect the other end of the tubing to the Channel B Input tap (upper right tap).

    *Retrotec Gauge:* Connect the other end of the tubing to the green tap on the back of the gauge.

3.  If it is not already set in the E1 position, set the small adjustable door opening on the flow meter to the E1 setting by gently and slowly pushing down on the metal snap pin button until the door slides so that the snap pin ends up in the "E1" setting of the door.

4.  Screw in a painter's roller pole extension into the end of the flow meter handle and attach the handle to the flow meter by way of the Velcro surfaces.

5.  *EC Gauge:* Press the "On/Off" button. Press the "Mode" button once to set the gauge in the "PR/FL" mode as shown in the lower left corner of the display. Then press the "Device" button repeatedly until the "EXH" appears in the upper left corner of the display. Finally, press the "Con-

fig" button repeatedly until "A1" shows in the upper right corner of the display.

*Retrotec* Gauge: Switch on the gauge by pressing and holding down the "On/Off" button and then the "Exit" button immediately thereafter. Once the gauge is on, press "Auto Zero" button until "On" shows in the upper right corner of the display. Then Press the "Mode" button repeatedly until PrA/PrB appears in the left edge of the display.

6. Turn on the exhaust fan you are testing. Create an airtight seal against the wall or ceiling around the fan as you place the flow meter completely over the fan grille and housing. Watch the pressure or flow reading on the gauge to see if it is relatively stable. If it is not stable:

*EC*: Press the "Time Avg" Button until "10" shows up in the lower right corner of the display.

*Retrotec*: Set the time average to 10 seconds by pressing the "Time Avg" button until "10 s" shows up just below the "Time" title near the upper right corner of the display.

7. **EC**: Read the fan flow on the Channel B display.

**Retrotec**: Read the pressure for "PrB," find that pressure on the chart below, and move to the right on the chart from there to find the equivalent flow at that pressure under the "E1" column (for example, a pressure of 2.2 Pascals represents a flow of 65 CFM for the E1 setting).

**Both EC and Retrotec:**

*If the flow is above the minimum of 44 CFM and below the maximum of 124 CFM (the range where the E1 box setting is accurate)*: Jump to the last step of this procedure and record the door setting, the flow rate, and the location of the fan you have tested. You are done.

*If the flow is above the maximum of 124 CFM*: Consider using a cardboard box with a specifically known larger size opening to test the fan flow (see Retrotec's description of the cardboard box procedure in its residential manual.)

*If the flow is below the minimum of 44 CFM*: Change the position of the adjustable flow meter door to E2 and make the following changes on the pressure gauge:

EC: Press the "Config" button until "B2" shows up in the upper right corner of the display. Retest.

Retrotec: Retest, read the pressure for "PrB," find that pressure on the chart below, and move to the right on the chart to find the equivalent flow at that pressure under the "E2" column (for example, a pressure of 2.2 Pascals represents a flow of 31 CFM for the E2 setting).

Both EC and Retrotec: If you have a flow above 21 CFM, record the results below. You are done.

*If you cannot obtain a flow above 21 CFM*: Change the position of the adjustable flow meter door to E3 and make the following changes on the pressure gauge:

EC: Press the "Config" button until "C3" shows up in the upper right corner of the display. Retest.

Retrotec: Retest, read the pressure for "PrB," find that pressure on the chart below, and move to the right on the chart to find the equivalent flow at that pressure under the "E3" column (for example, a pressure of 2.2 Pascals represents a flow of 15 CFM for the E3 setting).

Both EC and Retrotec: If you have a flow above 10 CFM, record the results below. You are done.

If you cannot obtain a flow of 10 CFM or more, the flow meter does not have a small enough adjustable door to accurately test that small of an exhaust fan flow. You can record the flow as "smaller than 10 CFM" or consider using a cardboard box with a specifically known smaller size opening to test the fan flow (see Retrotec's description of the cardboard box procedure in its residential manual.) Otherwise, the fan may need repair.

8. Door Setting: E1 E2 E3 (circle one)

Flow: _____ CFM

Location of fan: _____

**Table 4-7. Exhaust Fan Flow Pressure-to-Flow Conversion Chart**

| Meter Pressure (Pa) | Flow (CFM) | | | Meter Pressure (Pa) | Flow (CFM) | | |
|---|---|---|---|---|---|---|---|
|  | E1 | E2 | E3 |  | E1 | E2 | E3 |
| 1.0 | 44 | 21 | 10 | 4.6 | 94 | 44 | 22 |
| 1.2 | 48 | 23 | 11 | 4.8 | 96 | 45 | 22 |
| 1.4 | 52 | 25 | 12 | 5.0 | 98 | 46 | 23 |
| 1.6 | 55 | 28 | 13 | 5.2 | 100 | 47 | 23 |
| 1.8 | 59 | 28 | 14 | 5.4 | 102 | 48 | 23 |
| 2.0 | 62 | 29 | 14 | 5.6 | 103 | 49 | 24 |
| 2.2 | 65 | 31 | 15 | 5.8 | 105 | 50 | 24 |
| 2.4 | 68 | 32 | 16 | 6.0 | 107 | 51 | 25 |
| 2.6 | 71 | 33 | 16 | 6.2 | 109 | 52 | 25 |
| 2.8 | 73 | 35 | 17 | 6.4 | 111 | 52 | 25 |
| 3.0 | 76 | 36 | 17 | 6.6 | 112 | 53 | 26 |
| 3.2 | 78 | 37 | 18 | 6.8 | 114 | 54 | 26 |
| 3.4 | 81 | 38 | 19 | 7.0 | 116 | 55 | 27 |
| 3.6 | 83 | 39 | 19 | 7.2 | 117 | 56 | 27 |
| 3.8 | 85 | 40 | 20 | 7.4 | 119 | 56 | 27 |
| 4.0 | 87 | 41 | 20 | 7.6 | 121 | 57 | 28 |
| 4.2 | 90 | 42 | 21 | 7.8 | 122 | 58 | 28 |
| 4.4 | 92 | 43 | 21 | 8.0 | 124 | 59 | 28 |

## DUCT TESTS

### Duct Blower or Fan

Another tool is the duct blower, a miniature version of a blower door fan, see Figure 4-31. A duct blower can be used to check how badly ductwork is leaking in a more quantitative way as CFM than with a pressure pan test. It is generally considered more accurate in determining duct leakages than a pressure test, but is much more costly. In addition, it is more commonly used to evaluate new construction ductwork because better sealing can be achieved when all the ductwork is exposed when a home is under construction (see New Construction chapter). The ducts can be either pressurized or depressurized (exposed to a vacuum) to a 25 Pascal level when doing a duct blower test. This is referred to as the CFM25 total or CFM25_total. While the traditional test involves pressurization of the ducts, some believe that depressurization is more accurate because it will not push out the taping/masking over the registers or return grilles when used.

If you combine the duct blower with a blower door test, you can distinguish the duct leakage to the indoors from the duct leakage to the outdoors or CFM25out or CFM_out, a very important difference. Duct leakage to the outdoors or unconditioned spaces represents a more significant loss of energy than duct leakage to the indoors or conditioned spaces.

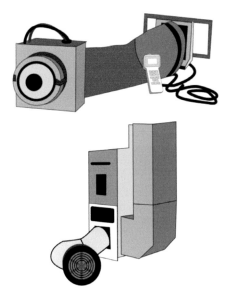

**Figure 4-31. Two Types of Duct Blowers**
Upper: Attached to a return air duct
Lower: Attached to a fan compartment of a furnace.

The following table is a step-by-step process for preparing the house and ducts for testing regardless of the type of duct blower being used. The two subsequent tables explain the specific procedures to use for two types of duct blowers. Select the appropriate table that applies to the type of duct blower you are using. When you are done with the testing, return the house and its appliances to their original conditions.

### Table 4-8. General Step-By-Step Duct Blower Set-up

1.   If you notice any vermiculite or asbestos insulation in the attic or mold, etc. inside the house during your inspection, do not proceed.

If this is a new house being tested under a specific energy efficient home protocol, follow that protocol instead.

If the house is in an area where the interior of the house could be exposed to freezing temperatures during the next year if unheated, *and* the home is new or has been vacant for awhile, obtain permission from the owner to refill the P traps, floor drains, and toilets with antifreeze if allowed in your locality.

Notify the homeowner that no one can open or close doors, turn on furnaces or air conditioner systems, fans, etc. during the test.

2.   Close any vents to the outside, whether powered by a fan or not, including fireplace or wood stove dampers and doors.

3.   Open all HVAC registers/diffusers/grills and interior doors, including closet doors. Open the basement door if you see evidence of storage, washer, dryer, etc. in the basement. Only open an attic doorway or hatch if it appears the attic space is insulated from the outside (it is conditioned space), such as finding insulation on the attic side of the roof deck or between the roof rafters, etc. Open the vehicle door to the garage. Also, open an attic hatch located in the garage only if the attic looks like it is insulated from the house (the attic is unconditioned space). If the crawl or attic spaces are unconditioned, open crawlspace and attic vents, as applicable.

*Outside Only Tests.* If the house has T-bar type suspended ceilings, or fluorescent light lenses/diffusers, remove one ceiling tile in each room where they exist (with the owner's permission), unless there is insulation laying on top of the ceiling tiles. In that case, watch the ceiling closely as you slowly depressurize the home to be sure you are not pulling down/destroying the ceiling tile.

4.   Close exterior doors and latch all windows (including skylight windows), storm windows, crawlspaces, hatches and attic hatches, unless specifically instructed otherwise. Seal off and cover any window air conditioner units and cat/dog doors. Close doors to rooms/closets with furnaces, boilers, water heaters, etc. only if they have open-air access/

makeup/supply from the outside.

5. Turn off all HVAC system electrical and the water heater to pilot. Turn off all fans- bathroom & kitchen fans, clothes dryers, central vacuums, broiler hoods, whole house fans, swamp coolers, window air conditioner units, ceiling fans, solar air panel fans, attic/crawlspace powered ventilators, HRV/ERV or other fresh air ventilation systems, etc. Leave your car keys on top of the water heater as a reminder to turn these appliances back on—you will not be able to start your car if you have not returned to the water heater to turn it and the other appliances back to their original setting.

6. Make sure all fireplace and wood stove fires have been out for at least 24 hours and that all ashes have been cleaned out. If the ashes have not been removed or you are unable to remove the ashes, completely cover them with wet newspapers.

7. If there is an attached space with combustion appliances that freely communicates with the unit you are testing, (such as a townhouse that has a common, unseparated attic or common source of outside air, etc. with the unit), be sure that these other appliances in the other unit are also off to avoid the same risk there.

8. Make sure, during the test, all pets, children, small objects, loose papers, etc. are kept away from around the inside or outside of the door where you are placing the duct blower.

9. *Set-up of Duct Blower*—Select the air handler for the furnace (preferred) or the return grille closest to the furnace/air handler, to connect the duct blower. Note that once the duct blower has been set up, it is best to avoid any unnecessary turns in the duct blower flex duct hose that connects the blower to the air handler or return grille.

10. Place masking tape or an appropriate sticky mask over all the registers, both supply and return, to seal them off. Be sure the sealing tape will not pull paint off the registers or walls, or leave a residue.

11. Remove all filters in the ductwork, unless the filter is placed directly behind a return grille you are masking. Dirty air conditioner coils can also act like a dirty filter when conducting duct leakage testing. If possible, be sure that the air conditioner coil is clean before starting any duct testing.

12. *Total Duct Leakage Test Only*: Open a large window in the home that is not facing the wind.

**Table 4-9. Setting up the Energy Conservatory (EC)/Minneapolis™ Duct**

## EC DuctBlaster Pressurization Testing

(Gauge would show 0.0 Pa if we were testing for outside leakage only
by adding a blower door fan set to 25 Pascals in an exterior door - see text)

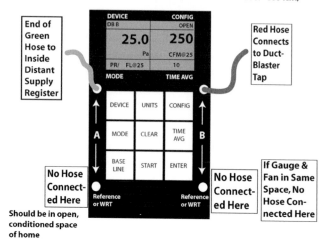

Figure 4-32. Pressurization Testing

## EC DuctBlaster Depressurization Testing

(Gauge would show 0.0 Pa if we were testing for outside leakage only
by adding a blower door fan set to -25 Pascals in an exterior door - see text)

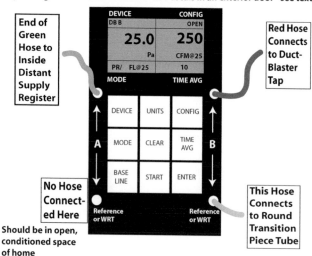

Figure 4-33. Depressurization Testing

## Blaster with DG-700 Gauge—Depressurization or Pressurization for either Total Duct Leakage or Duct Leakage to the Outside Only

1.    **Before Starting: Be sure to conduct all the steps in the "General Step-By-Step Duct Blower Set-up" procedure above.**

Make sure your gauge batteries are fully charged or that you have adequate outlets and extension cords to connect the pressure gauge to its recharger/transformer.

Attach the hoses to the gauge—red tubing coming from the tap on the fan to the Channel B input tap (upper right tap) and green tubing (inserted into the closest supply register) to the Channel A input tap (upper left tap).

*Outside Only Test*: Set up a blower door consistent with the instructions for blower door testing but with the settings for depressurization at -25 Pascals or pressurization at 25 Pascals. If possible, use cruise control (see below) with the blower door to provide a more accurate test. You must have two gauges to do a Duct Blaster test and use cruise control with the blower door. Set up the blower door for the same pressurization/depressurization as you plan to use with the Duct Blaster. Do not set them up for different set-ups (DO NOT set up one for depressurization and the other for pressurization).

2.    Attach or place the gauge mounting board and fan speed controller (with the speed control dial on it) onto or near the Duct Blaster

3.    Attach the flex duct to the fan intake side (air going in, the side without the fan motor on it) for depressurization, or fan exhaust side (air going out, the side with the fan motor on it) for pressurization. Match the edges of the round-edged transition piece on the end of the flex duct to the round fan housing on the appropriate side of the fan and then slip the hard black rubber connecting trim strip over the outer edges of both to "U" clamp them together. In addition, with depressurization, both a flow ring (perhaps start with the largest ring) and the multi-hole foam flow conditioner must be placed between the fan and the flex duct.

4. Insert the female plug from the speed controller in the receptacle on the fan.

5. Make sure the fan speed controller dial is turned as counterclockwise as possible. Plug the regular power cord plug from the speed controller into a wall outlet.

6. Check the fan direction so the fan will pull air out of the ducts into the house (depressurization) or into the ducts out of the house (pressurization).

7. If the fan speed controller and gauge have "cruise control" capability, install the low voltage cruise control connector between the gauge (on its top) and the speed controller (on its side). Otherwise, ignore this step.

8. Turn on the gauge—Press "On/Off."

9. *Total Duct Leakage Test*: Press the "Mode" button three times to put the gauge in the PR/FL@25 mode.

   *Outside Duct Leakage Test*: Press the "Mode" button only once to set it up for the PR/FL mode.

10. Check that the device showing in the upper left corner of the display matches the Duct Blaster you are using. It should say "DB B" if you are using the Series B Duct Blaster. If it does not indicate the correct Duct Blaster model, press "Device" until the proper Duct Blaster model shows up on the display. Press "Config" until the proper flow ring you are using shows up on the upper right of the display.

11. *Outside Duct Leakage Test Only*: Start the blower door setup to run at -25 Pa (depressurization) or 25 Pa (pressurization).

12. *Outside Duct Leakage Test Only*: Be sure to set up both the blower door and Duct Blaster for either depressurization or pressurization so one is not set up for depressurization and the other for pressurization.

13. **Manual Control**

    *Total Duct Leakage*—Turn the fan controller clockwise. Continue to increase the fan speed until the pressure shown on the left side of the display (Channel A) shows between about -20 and -30 Pa. You need not attempt to keep the pressure at exactly -25 if you are depressurizing (25 Pa if pressurizing) as the gauge will calculate a flow reading assuming the pressure were -25 if you are depressurizing (25 Pa if pressurizing). If the flow is fluctuating more than you would like, press the "Time Avg" button until you have selected the 5 or 10 second averaging period.

*Outside Duct Leakage Only*: Same as "manual" instructions above except turn the fan speed up until the pressure reading shows "0." Then re-check the blower door building pressure to make sure it is still showing the appropriate pressure (-25 Pa for depressurization or 25 Pa for pressurization). Once you have completed that process, recheck/reset the Duct Blaster pressure so it is at 0 Pa.

**Cruise Control**

*Total Duct Leakage*: Turn the speed controller dial to the "just on" position and press the "On/Off" switch to "ON" if it has one. Press the "Enter" button and the cruise control will be triggered. The left side of the display (Channel A) should display the number 25. Press the "Start" button and the Duct Blaster fan will increase speed until the pressure is approximately **-25** if you are **depressurizing** or 25 Pa if **pressurizing**.

*Outside Duct Leakage Only*: Same cruise control instructions above except, after pressing "Enter," press "Config" **twice** if you are **pressurizing**, **three** times if you are **depressurizing**, to show/set the target pressure at "0." Then press "Start."

14. Read the flow rate that now shows up on the right side of the display (Channel B). *If it shows a series of hyphens*—("-----") instead, turn off the fan with the controller and use the next larger flow ring one at a time until there is adequate flow, the hyphens disappear, and you get a flow number in its place.

   *If it shows "LO"* instead of a number, turn off the fan with the controller and add flow rings one at a time until the "LO" disappears, and you get a flow number in its place.

   *Be sure to press "Config"* until the ring displayed in the upper right corner of the gauge matches the ring on the fan. If you are using the cruise control, press "Clear" before you press the "Config" button and then follow the instructions in the previous step to restart the fan.

15. *Pressure Target Not Reached*: If you find that your pressure reading is not matching the pressure you want—say you can only get up to 15 Pa but you need to reach 25 Pa for a "Total" leakage test or you need to reach 0 for an "Outside only" test—use the speed control knob to stop the fan. Check for disconnected or improperly connected hoses, register seals pulling loose or disconnected ducts. If you find these problems, correct them and retest.

    If you still cannot reach the target pressure:

    *Total Duct Leakage Test*: Find the pressure you were able to reach on the ™ on the correction chart below, and calculate the true duct leakage rate using the multipliers in the chart:

    _____ CFM x _____ = _____ CFM (corrected)

    *Outside Only Leakage Test*: Lower the pressure for the blower door until the DuctBlaster™ gauge shows "0" Pascals. Find the pressure on the *blower door* on the correction chart below, and calculate the true duct leakage rate using the multipliers in the chart:

    _____ CFM x _____ = _____ CFM (corrected)

    *All Tests*: Record your results in the next step.

16. Read and record the CFM number that shows up in the right side of the middle of the gauge.

    _____CFM

17. *Manual Control*: Turn off the fan by turning the speed controller counterclockwise to "Off" or by pressing the "On/Off" switch to the "Off" position.

    *Cruise Control*: Turn the fan off by pressing the "Clear" button.

18. Return the house to normal conditions by returning the combustion appliances to their original settings, unseal window air conditioner units, open fireplace dampers, etc. Be sure to check that pilot lights have not blown out from the test. If they have, relight them or have someone else relight them safely. Watch out for build up of unburned gas.

**Table 4-10. Duct Leakage Testing Correction Chart**

| Duct Pressure (Pa) | CRP Factor | Duct Pressure (Pa) | CRP Factor |
|---|---|---|---|
| 24 | 1.02 | 14 | 1.42 |
| 23 | 1.05 | 13 | 1.48 |
| 22 | 1.08 | 12 | 1.55 |
| 21 | 1.11 | 11 | 1.64 |
| 20 | 1.14 | 10 | 1.73 |
| 19 | 1.18 | 9 | 1.85 |
| 18 | 1.22 | 8 | 1.98 |
| 17 | 1.26 | 7 | 2.15 |
| 16 | 1.31 | 6 | 2.35 |
| 15 | 1.36 | 5 | 2.63 |

**Retrotec DucTester 200 Set-Up**

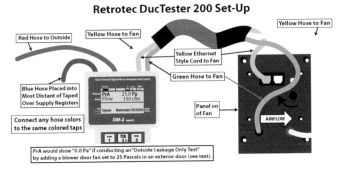

**Figure 4-34. Retrotec DucTester 200**

**Table 4-11. Retrotec DucTester™ 200 Step-By-Step Procedure—Depressurization or Pressurization for either Total Duct Leakage or Duct Leakage to the Outside Only Using the Retrotec DM-2 Gauge**

1. **Before Starting: Be sure to conduct all the steps in the "General Step-By-Step Duct Blower Set-up" procedure above.**

   Make sure your gauge batteries are fully charged or that you have adequate outlets and extension cords to connect the pressure gauge to its recharger/transformer.

   *Outside Only Test*: Set up a blower door consistent with the instructions with "Set Pressure" setting but with the settings for depressurization at -25 Pascals or pressurization at 25 Pascals.

2.  Connect one end of the blue, yellow, and green tubes to their correctly matched color ports/nipples on the manometer/gauge.

3.  Place the other end of the blue hose (preferably with a metal static pressure hose attached to the end of it) through and into the closest taped-over supply register or the supply register nearest where you suspect the greatest duct leakage is occurring.

4.  Plug the proper end of the power cord into the fan and the other end with wall plug into a wall outlet. You should now see a green light on the fan above the fan switch.

5.  If there is one, press the switch on at the fan next to the power cable connection and adjust the knob so you can check if the fan will operate. Then turn back the speed control knob to stop the fan.

6.  Connect the other ends of the green and yellow tubes to the fan.

7.  Place the midrange plate (the larger flow ring for this fan) onto the fan.

8.  *Depressurization*: Attach the flex duct over the inlet of the fan (arrow pointing away from the flex duct) by tightening the Velcro strap.

    *Pressurization*: Have the arrow pointing towards the flex duct and then tighten the flex duct Velcro strap over the inlet of the fan.

9.  Connect the hard plastic flex duct flange to the air handler compartment opening (or a return grille that you have chosen that is closest to the air handler) using masking tape or other appropriate sticky mask material. Be sure to seal over any unused portion of the air handler or return grille outside of the flange. Connecting to the air handler may require cutting out cardboard pieces and taping them to each other and/or the edges of the air handler compartment. By using the flange as a sort of "cookie-cutter" to mark where to cut a hole in the cardboard and then cutting the rest of the cardboard to just cover the air handler opening, you may be able to use just one piece of cardboard.

10. Tightly attach the other end of the flex duct to the flange using the Velcro strap.

11. Position the fan so that you minimize the bends in the flex duct.

12. Switch on the gauge (pressing and holding down the "On/Off" button and then "Exit" immediately thereafter).

    Once the gauge is on, press "Auto Zero" button until "On" shows in the upper right corner of the display.

If the number below the "n" value on the right side of the top line of the display does not show "0.60," press the "Setup" button repeatedly until you have selected the "n" value option. Then press "Enter," type "0.60" (use the numbers & period button—same as the "@ Pressure" button) in as the n exponent, press "Enter" again and then "Setup."

Turn on the fan via the fan switch if there is one.

13. If you do not see "Retrotec DU200" or the name of the other duct blower you are using in the display's lower right hand corner, hit "Mode" once and then "Device" until it shows up.

14. If the upper right corner of the display does not show "On," hit "Auto Zero" once.

15. If you do not show "4s" just below the "Time" setting on the top line of the display, push "Time Avg" until it shows up. If it is windy, set it higher than 4 seconds, until it gives a more stable reading.

16. If you do not see "Mid" in the display's lower left corner, hit "Range/ Config" until it shows up there.

17. Press "Mode" until "Flow" appears on the display's left edge at the "Mode" mark.

18. *Outside Duct Leakage Test Only*: Start the blower door setup to run at -25 Pa (depressurization) or 25 Pa (pressurization). Be sure to set both the blower door and DuctTester for either depressurization or pressurization. *DO NOT* set up one for depressurization and the other for pressurization.

19. *Total Duct Leakage Test*: Adjust the fan speed knob until the "PrA" pressure row on the gauge reads "-25" Pa in the case of a depressurization test or "25" Pa in the case of a pressurization test. If you are able to get the proper pressure without a "Too Low" reading, skip to the last step to record the flow.

    *Outside Duct Leakage Only Test*: Same as "Total" test above except adjust the fan speed knob until the pressure row reads "0" Pa.

20. *"Too Low" CFM Flow Reading*: If the Mode/Flow row displays "Too Low," use the speed control knob to stop the fan. Replace the "Mid" flow ring with the "Low" flow ring. Press "Range/Config" until the "Low" size matching the ring is in the display's lower left corner. Adjust the fan

speed knob until you reach the appropriate pressure and record your results in the last step.

21. *Pressure Target Not Reached*: If you find that your pressure reading is not matching the pressure you want –say you can only get up to 15 Pa but you need to reach 25 Pa for a "Total" leakage test or you need to reach 0 for an "Outside only" test—use the speed control knob to stop the fan. Remove the "Mid" flow ring. Press "Range/Config" until the word "Open" appears in the display's lower left corner. Adjust the fan speed knob until you reach the appropriate pressure and record your results in the last step.

If you still cannot reach the target pressure: Check for disconnected or improperly connected hoses, register seals pulling loose or disconnected ducts. If you find any of these problems, retest after correcting the problem.

If you still cannot reach the target pressure:

*Total Duct Leakage Test*: Find the pressure you were able to reach on the DucTester™ on the correction chart above, and calculate your true duct leakage rate using the multipliers in the chart:

_____ CFM x _____ = _____ CFM (corrected)

Record your results in the next step.

*Outside Only Leakage Test*: Lower the pressure for the blower door until the DucTester™ gauge shows "0" Pascals. Find the pressure on your blower door on the correction chart above, and calculate your true duct leakage rate using the multipliers in the chart:

_____ CFM x _____ = _____ CFM (corrected)

Record your results in the next step.

22. Read and record the CFM or CFM@25 number that shows up in the middle of the gauge. _____ CFM

23. Return the house to normal conditions by returning the combustion appliances to their original settings, unseal window air conditioner units, open fireplace dampers, etc. Be sure to check that pilot lights have not blown out from the test. If they have, relight them or have someone else relight them safely. Watch out for build up of unburned gas.

To visualize where the leaks are in ductwork in unconditioned space, draw fog from a fog-generating machine into the duct blower fan and watch the fog bleed out where there are leaks in the ductwork, for example, in the crawlspace or attic. Also, you could use the simpler method (without a duct blower) of turning on the fan supplying the ductwork and checking for leaks by scanning the supply duct by feeling with your hands for air and using a small smoke stick to see if the return ducts are drawing the air in the joints, as shown in Figure 4-35.

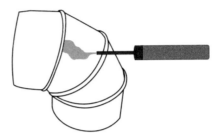

**Figure 4-35. Using smoke at duct joints can prove effective in identifying where sealing work should be directed. In this diagram, there is a leak at a joint in this elbow.**

To the extent that you can visualize openings, cracks or leaks, you can even use a mirror to look into ducts from the vents while turning on a light on the other side of the duct in the crawlspace or attic to see where light is shining through. Once you locate a leak, use a flashlight to look into the vent to better note it.

**Dominant Duct Leakage Testing**

A dominant duct leakage test determines whether the supply ductwork or the return ductwork is more leaky. This test does not tell how leaky the supply or return ductwork is, but only which is more leaky.

All furnaces should have both supply and return air ductwork. If all the ductwork is in conditioned spaces, with none in the attic or basement for example, the pressure inside the home should be relatively similar to the pressure outside the home. However, if some of the ductwork lies in unconditioned spaces (for instance with return ductwork passing through the attic and supply ductwork passing through the crawlspace or basement) then on a cold day, warm air can be lost from the supply ducts to the basement, or cold air can be drawn into the return ducts in the attic, etc. This occurs because supply ducts are pressurized while return ducts have a vacuum in them. If the holes in the return ducts are larger than the holes in the supply ducts, there will be some net air added to the flow in the form of cold air from the attic; for instance, entering into the return ducts. In this case, the air handler

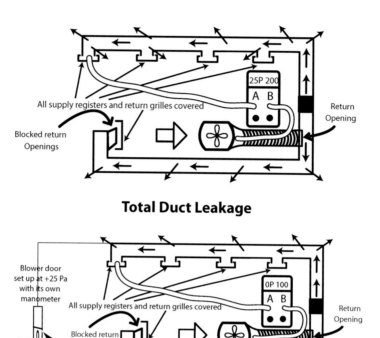

**Total Duct Leakage**

**Duct Leakage to the Outside Only**

**Figure 4-36. Duct Blower Testing**
Total leakage testing done with only a duct blower and duct leakage to outside/
outdoors involving the additional use of a blower door. With the total duct leakage test, there is a 25 Pascal pressure difference in the ductwork WRT the areas
not in the ductwork. There is a 0 Pascal difference in pressure in the ductwork
WRT the house pressure because of the additional use of the blower door in the
duct leakage to outdoors/outside. Note that in these cases the home and ducts
are pressurized, however, many believe that depressurization is the preferred
method with duct blower testing.


has to take in more air and provide that additional air to the interior of
the home to the supply registers. This means there is more air coming
out of the supply registers in the conditioned space than going into the
return grilles from the conditioned space. The effect is to pressurize the
home because the supply ducts are adding more air to the conditioned
space than the return grilles are taking from it. The reverse can happen

if the supply ducts are leakier than the return ducts. Some air in the supply ducts will leak into the unconditioned crawlspace or basement. This means that less air will be delivered to the conditioned space of the home from the supply registers than is going out of the conditioned space into the return grilles. With more flow going out of the house than coming in, there is a negative pressure or slight vacuum in the home.

Whether you have created a positive or negative pressure in the home because of greater return duct or supply duct leakage, the tighter the home, the greater the pressure differential will be under the circumstances with respect to the pressure outdoors. If the pressure is positive that means that the return leaks are more significant than the supply leaks to the outside or unconditioned spaces. If, on the other hand, the pressure is negative, that means the supply leaks are more significant than the return leaks to the outside or unconditioned spaces. If there is roughly a neutral pressure, that would indicate that the supply and return leaks, if they exist, are relatively equal. One good example of a dominant return duct leak would be a large disconnected return duct in a well-vented attic. This would cause the conditioned space on the inside of the home to have a positive pressure with reference to outside.

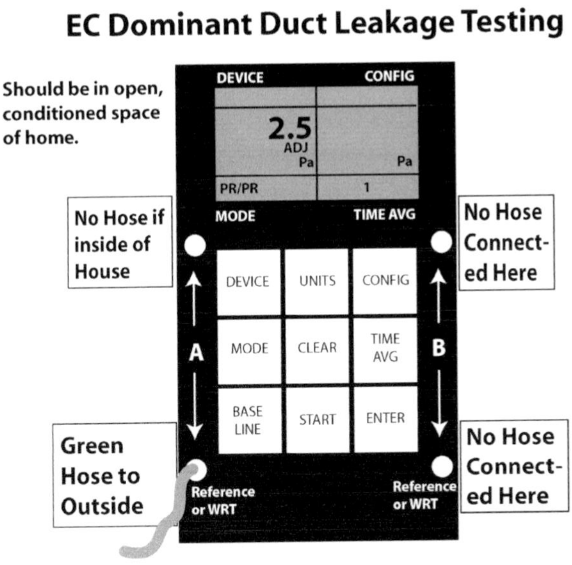

## EC Dominant Duct Leakage Testing

Figure 4-37. EC Dominant Duct Leakage Testing

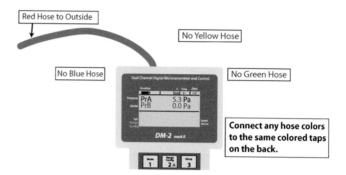

Figure 4-38. Retrotec Dominant Duct Leakage Testing

**Table 4-12. Dominant Duct Leakage Test Procedure (Using either the Retrotec DM-2 or EC DG-700 Gauges)**

1. Make sure your gauge batteries are fully charged or that you have adequate outlets and extension cords to connect the pressure gauge to its recharger/transformer.

   Close all the exterior openings to the house including windows and doors. Open all the interior doors, including closet doors.

2. Run a hose to the outside by attaching a metal hose to the end of the hose and placing the metal hose underneath a closed exterior door to avoid having the door pinch off the hose. Keep the hose inside the house.

3. *EC*—Turn on the gauge—Press "On/Off." If "PR/PR" is not showing in the lower left corner of the display, keep pressing the "Mode" button until it appears. Connect the other (inside) end of the hose to the gauge's Channel A reference tap (lower left tap). Disconnect all other hoses to the gauge. Do a baseline as follows:

   Press the "Baseline" button once.

   Press "Start" once.

   Once the numbers on the gauge seem to level off without as much change, press the "Enter" button once.

   *Retrotec*—Turn on the gauge by pressing and holding down the "On/Off" button. Once the gauge is on, press "Auto Zero" button until "On" shows in the upper right corner of the display. If both "PrA" and "PrB" do not show up on the left edge of the display, repeatedly press the "Mode" button until they do. Connect the other (inside) end of the hose to the gauge's red tap on the back). Disconnect all other hoses to the gauge. Do a baseline:

Press the "Baseline" button once.

Once the average shown near the bottom of the gauge seems to level off without as much change, press the "Enter" button once.

4. Turn on only the air handler of the forced air system and record this pressure difference of the house with reference to outside with the air handler fan on. _____Pa (H/O)

5. Turn off the air handler.

6. If you are concerned you may not be getting accurate results, repeat this process as many times as needed.

---

### Delta Q Test

To get an indication of the actual amount of leakage in each supply and return ductwork system, combine turning on the furnace (as in the dominant duct leakage test) with running the blower door at various intervals of negative and positive pressure settings (for example, 11 readings spaced at 10 Pascals intervals between -50 and +50 Pascals). This is a particularly convenient test since it does not require the use of a duct blower. The delta Q test compares the difference in flow (at a given pressure inside the house) through the blower door before and after turning on the furnace fan, "Delta Q." This gives a separate and specific CFM of leakage for the supply and return ductwork located outside the conditioned space. Because of the complexity of the calculations involved, this method requires the use of software. Software also allows preset control of the fan so this test can be done more quickly.

### Subtraction Method

The subtraction method uses a blower door test first with and then without the heat registers and return grilles tightly sealed. The flow through the blower door is recorded when the registers and grilles are open and when they are sealed. The difference in flow between the two measurements is treated as the leakage rate of the entire ductwork system, both supply and return. This method is usually considered one of the most unreliable and inaccurate.

### Other Similar Duct Leakage Tests

Other similar duct leakage tests exist, but they are used less frequently. These include Supply/Return Split, Nulling Pressure Test (NPT), House Pressure Test (HPT), Irvine Quality Plus (IQ+), and Tracer Gas. These are rarely used for a variety of different reasons, including time involved, lack of familiarity, lack of common availability of necessary items (equipment, training, supplies, etc.), risk of operator error, inaccuracy, etc.

### Pressure Pan Test

A pressure pan test combines the use of a blower door (see Blower Door Section) with the use of a "pan" that covers the supply and return grilles to temporarily seal them off. It looks similar to the exhaust fan flow meter but without a door opening on the pan, see Figure 4-39. The pan has a hose tap/nipple/port in it so you can measure the pressure in the pan to find out how leaky the ducts are near the heat register being tested.

The leakier the ducts are to the outside or unconditioned spaces, the greater the pressure differential reading. In fact, the goal is to keep pressure pan readings to 1.0 Pascal or less.

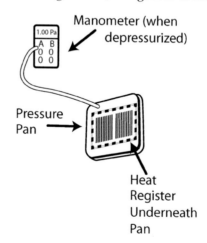

Many auditors and retrofitters view pressure pan test as one of the most cost effective ways to evaluate existing home ductwork to quickly direct attention to the most significant duct leaks. An additional benefit is that this test concentrates on leakage to the outside. To set up the blower door for this test, depressurize the house to -50

**Figure 4-39. Pressure Pan**

Exhaust air flow meters and pressure pans can be used upside down when it is necessary to test ceiling registers. Exhaust flow meters can be used as pressure pans to measure the amount of air being drawn into a fan that exhausts to the outside, but cannot be used to determine register flow, etc. where the air is being pushed out of the covered opening rather than being pulled into the opening.

Pascals (or similarly pressurize the house). Then go around the house, test each register (both supply and return), and record the readings. Here is a specific procedure for pressure pan testing:

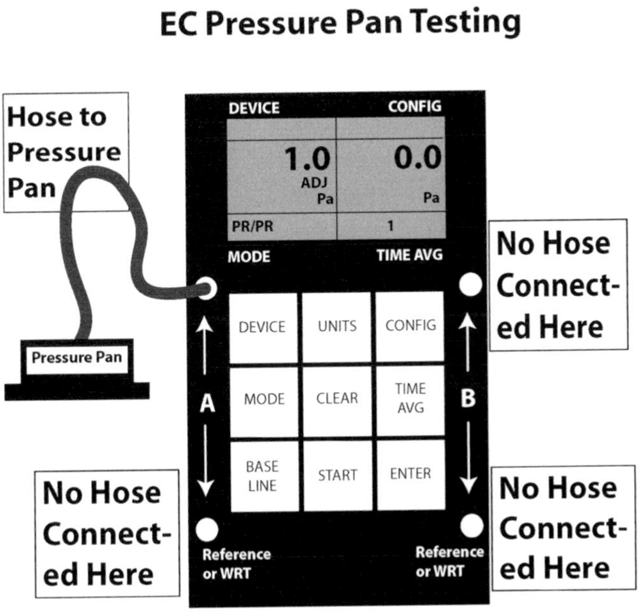

## EC Pressure Pan Testing

Figure 4-40. EC Pressure Pan Testing

## Retrotec Pressure Pan Testing

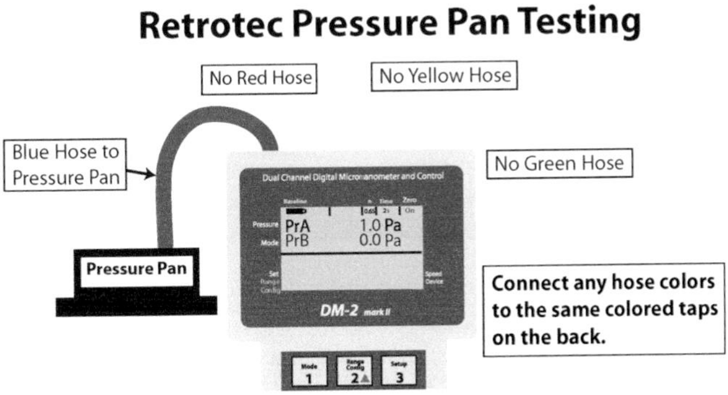

Figure 4-41. Retrotec Pressure Pan Testing

**Table 4-13. Pressure Pan Testing Procedure Using the EC DG-700 or Retrotec DM-2 Gauges**

1. Make sure your gauge batteries are fully charged or that you have adequate outlets and extension cords to connect the pressure gauge to its recharger/transformer.

   Seal all supply and return grilles located in a garage or other similarly semi-conditioned space (these registers should be permanently sealed as a safety measure as part of the retrofit after notifying the owner anyway).

2. Set up a blower door consistent with the instructions for blower door testing with the settings for depressurization at -50 Pascals or pressurization at 50 Pascals. If you have two gauges available, use the "Set Pressure" setting (Retrotec) or cruise control (EC) with the blower door to help provide a more accurate test. If you do not have two gauges, use the manual speed control and the gauge to set the blower door speed so it provides -50 Pascals in the home. Then disconnect the gauge from the fan system to use it for pressure pan testing

3. Start the blower door system.

4. Check the pressure between the house and unconditioned spaces where ductwork is located, such as crawlspaces or attics. If you are not reading at least a 45 Pascal difference in pressure, check to make sure that any vents, windows, etc. to the outside from that unconditioned space have been fully opened. If you are unable to achieve at least a 45 Pa difference, the readings related to ducts in the unconditioned space are not likely to be as accurate.

5. Make sure the manometer/gauge is reading down to the 0.1 Pascals level. Connect tubing between the gauge and the pressure pan:

   *EC Gauge*: Connect the tubing from the pressure pan nipple/port to either one of the upper "Input" nipples/ports on the gauge (either Channel A or B). Press the "On/Off" button. If it is not already set for it, set the time average to 1 second by pressing the "Time Avg" button until "1 s" shows up in the lower right corner of the display.

   *Retrotec Gauge*: Connect the tubing from the blue (Input A) nipple/port on the gauge to the pressure pan nipple/port. Disconnect all other hoses to the gauge. Switch on the gauge (pressing and holding down the "On/

Off" button and then "Exit" immediately thereafter). Once the gauge is on, press "Auto Zero" button until "On" shows in the upper right corner of the display. Make sure the "Pressure" setting is at "PrA" and the "Mode" setting is at "PrB" (Press the "Mode" button until these are indicated on the left side of the display). Set the time average to 1 second by pressing the "Time Avg" button until "1 s" shows up just below the "Time" title near the upper right corner of the display.

6. Place the pressure pan over each of the supply registers in the home, one at a time. Make sure the rubber seal on the pressure pan firmly seals against the floor, ceiling, wall etc. where it is placed. Similarly, test each of the return grilles. However, since the pressure pan may not fit over these typically larger registers, you may have to tape or mask over them (see Duct Blower test). You would then place the hose that would normally go to the pressure pan through and below the tape, and into the register to get a measurement on the large or unusually shaped registers.

7. Once each pressure pan reading has stabilized, record the reading and indicate the location of each register in the process.

If a register is oddly shaped or too large (return grilles often are), temporarily tape or seal over it, poke a hole for the hose to reach inside the register, and then read and record the result. With the blower door still running, some auditors then proceed to the area where the ductwork is exposed, such as a crawlspace, basement, or attic and use smoke to find leaks around the ductwork from the boot to the trunk line of the register that showed the highest Pascal reading. They then mark the joints that showed leakage on the duct and on their diagram. If the pressure reading at a duct is more than 3 Pascals, then there is probably a significant leak in the ductwork near the register, or there may be a gap and subsequent leak where the sheet metal boot attaches to the floor or ceiling that the boot passes through. See Figure 4-42.

If you notice this gap, include a comment on your report requiring that mastic be put over these gaps before the installer retests each register with the pressure pan. These gaps tend to be especially leaky if the other side of the floor or ceiling is unconditioned space, such as a crawl space or attic that is "outside" or ventilated to the outside. In addition, a single leak in ductwork typically affects the pressure pan reading on more than one register. The data collected during pressure pan testing should at least be listed for each register or shown on a layout diagram showing the location of each register on each floor, as shown in Figure 4-43.

**Figure 4-42. One of the worst locations for air leaks is where the boot of the ductwork is not adequately sealed to the floor, wall, or ceiling material around it, especially when the ductwork is in an unconditioned space.**

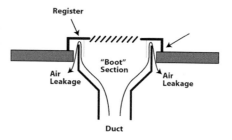

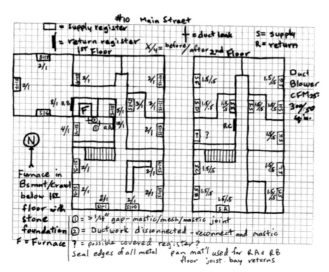

**Figure 4-43. Providing a layout diagram of the location and pre-retrofit pressures of registers tested with a pressure pan can be a great benefit to an installer/retrofitter. Notice the installer has recorded the "after" retrofit pressure pan numbers for each of the registers on this diagram. Also, notice that the initial pressures for the second floor are lower than the initial pressures for the first floor. This could be explained by the ducts for the lower floor being exposed to a basement with a leaky stone foundation ("outside" the air barrier) while the ducts for the second floor are in the floor joist bays in between the first and second floors ("inside" the air barrier). In addition, notice the labeling for the registers can be done with a combination letter and number system ("S1-1" for first story first supply register) shown for the first floor or a straight letter system ("SA" for first supply register) shown for the second floor.**

By far, return air ducts are the most common culprits for leaks in ductwork, especially if they utilize floor joist bay or stud bays capped with sheet metal pan material. Leaks closest to the furnace will leak the most and create the greatest pressure differences in pressure pan testing, so seal them first. Leaks around the edges of supply and return grilles exchanging air with crawlspaces, attics, etc. are also common. Some studies have shown that roughly 75% of duct leakage arises from the combination of leaks nearest the fan and at register.

Because so often the worst pressure pan readings come from return air registers, some auditors recommend that the installer first repair the return air plenum and duct leaks. The installer can then conduct a mid-repair retest at all the registers that did not pass the pressure pan test before determining which supply ducts to repair next, if any, and after repair, retest the ones that still failed. After leaks have been repaired, the installer can retest to evaluate how effective the sealing has been. Remember, all readings should be roughly one Pascal or lower in pressure differential.

Readings over 25 Pascals indicate a major duct disconnect. If you get a high reading, say around 10 Pascals on all vents, it is likely a plenum leak. If only the registers along a section using the same trunk are higher than 3 Pascals, you could have a major leak somewhere along that trunk. If the registers with the greatest leaks (highest Pascal pressure difference) do not appear to use the same trunk and are more randomly spread around, many auditors and installers have had great success sealing the ductwork from the trunk to the register on the top two to four highest Pascal registers and then retesting.

It may be necessary to strip insulation off the ductwork at the joints and seal all the joints. You can also get a general, though less quantitative analysis, by feeling for air or using smoke to identify air coming out of the registers during the blower door depressurization test. Furthermore, the boots on flex duct systems may not have been properly connected and may require proper renovation to reduce leakage.

It should be noted that if you find a pressure difference between the house and an unconditioned space to be less than 45 Pa, consider adjusting where the "cutoff" is for acceptable pressure pan readings. For instance, if you find an attic or crawlspace that was not well ventilated to the outside and the pressure difference was only 25 Pa between the house and the unconditioned space, the new cutoff would be half of the original 1.0 Pa cutoff used for a roughly 50 Pa differential. In that in-

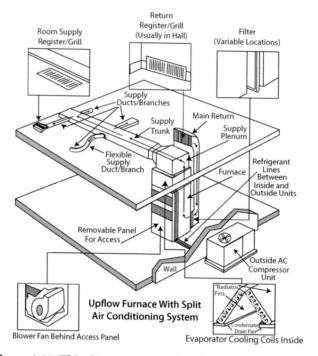

Figure 4-44. This diagram shows the plenum, the trunk, the branches, and the registers, etc. in a hot air furnace distribution system.

Table 4-14. Guidelines on Pressure Pan Testing

| | |
|---|---|
| >25 Pa | Duct disconnected |
| >10 Pa  all registers/vents | Plenum leak |
| > 3 Pa   all registers/vents | Leak in duct to that register |
| ≤1 Pa    one register | No leaks in or near this register |

stance, you might choose a cutoff of 0.5 Pa. Likewise, if you are only able to get the house blower door pressure up to 25 instead of 50 Pa because the house is leaky, the maximum differential between the house and unconditioned spaces could be no larger than 25 Pa and you could halve the cutoff to 0.5. Conversely, if you get a reading of 15 Pa with only a differential of 25 Pa between the house and the unconditioned space, you should probably treat that as a 30 Pa (likely disconnected duct) pressure pan reading.

Some experts have established general guidelines for deciding the next step after conducting pressure pan tests. (Generally speaking, these guidelines are not intended to be used in evaluating new construction.) They are as follows:

1) If the pressure pan readings on all registers is < 1.5 Pa (or equivalent if lower pressure differentials involved)– This is a tight duct system and no further testing or repair is necessary unless more accurate testing is a requirement of the retrofit program.

2) If the pressure pan readings on 3 or more registers is > 2.0 Pa (or equivalent if lower pressure differentials involved)– This is a leaky duct system and while repair is necessary, no further testing need be done on this home unless it is a requirement of the retrofit program.

3) For all other houses not included in the two above categories— This duct system lies in the grey area of duct leakage. Further testing may be needed to determine more accurately how much leakage is going to the outside. Such tests could include a duct blower test measuring leakage in CFM. It could be a program requirement to do more accurate duct testing. Otherwise, to decide whether to seal duct leaks and to what extent, you should consider the cost-effectiveness of the situation, such as accessibility of the ductwork lengths most suspect, concentrated leakage areas of ductwork, cost and extent of repair necessary, etc.

A pressure pan can also be used to check for pressure at interior wall outlets, recessed lighting fixtures, and other more confined areas that might represent bypasses or air leakage sites that are connected directly or indirectly to the outside. However, a more accurate method of testing for pressure in a zone behind a wall ceiling or flow is to use a small non-conducting probe on the end of the tube (see Zone Pressure Diagnostics section).

Pressure pan tests are much faster, and although not as accurate as a duct blower test (see next section) they seem to be accepted as good enough to confirm whether there is a major problem, or after the work has been done, whether the duct sealing was generally adequate. Duct blower testing should be done to test buildings under construction at pre-cover (before drywall or any insulation is installed) because the access then allows for plugging virtually any duct leak.

### "Room to Room" Or Duct Induced Pressure Difference Testing

To determine if there are restrictions in the return air getting back to the furnace, use the "room to room" or duct induced pressure difference test on homes where the return grilles are not in the rooms. Run a hose from the gauge's reference tap to the inside of a room by placing it underneath a closed interior room door while standing in the hall with a gauge. Turn on the forced air system air handler and go from room to room testing the pressure differential between the inside of the room and the hall with the room door closed. If air supplied to the room from the supply registers is not allowed to easily return to the return grilles (typically in the hall), there will be a positive pressure in the room relative to the hall.

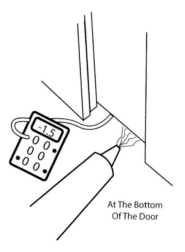

**Figure 4-45. You can also use the room-to-room test to determine whether to open a door during worst-case set up (see Heating Chapter) by testing, either with a gauge (left) or a smoke tester (right). This test checks the pressure in a room with respect to (WRT) the hall. In this case, the room is negative with respect to the hall (such as a bathroom with the fan on) so leave this door open for purposes of establishing the worst-case conditions.**

At The Bottom Of The Door

In essence, the bedroom is being pressurized with reference to the house or hall. Another way to check is to simply use smoke to see if air is being drawn out of, or into the room at the bottom of the door. The faster the smoke moves, the more significant the problem. If there are pressure imbalances over 4 Pascals (some suggest 3 is the cutoff), you need to find an easier way to get supply air to the returns.

Table 4-15 gives a specific procedure for room-to-room pressure testing.

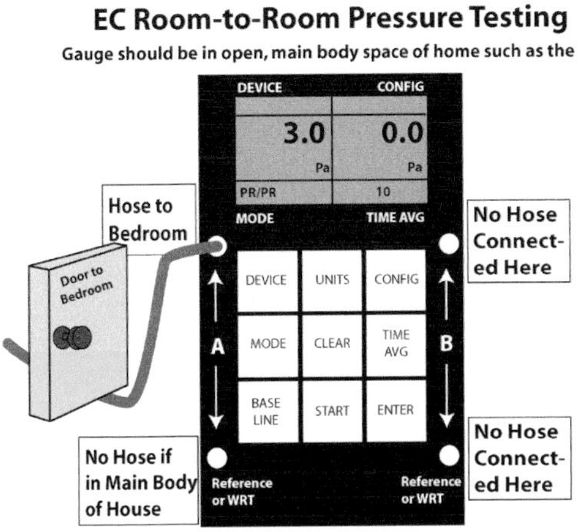

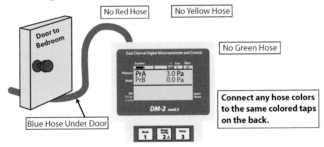

## EC Room-to-Room Pressure Testing

Gauge should be in open, main body space of home such as the hall

Figure 4-46. EC Room to Room

## Retrotec Room-to-Room Pressure Testing

Gauge should be in open, main body space of home such as the hall

Figure 4-47. Retrotec Room to Room Pressure Testing

**Table 4-15. Room-to-Room Pressure Test Procedure (Using Either the Retrotec DM-2 or EC DG-700)**

1. Make sure your gauge batteries are fully charged or that you have adequate outlets and extension cords to connect the pressure gauge to its recharger/transformer.

   Close all the exterior openings to the house including windows and doors. Close all the interior doors including closet doors.

2. *EC*—Turn on the gauge—Press "On/Off." If "PR/PR" is not showing in the lower left corner of the display, keep pressing the "Mode" button

until it appears. Connect the end of a hose to the gauge's Channel A reference tap (lower left tap). Disconnect all other hoses to the gauge.

*Retrotec*—Turn on the gauge by pressing and holding down the "On/ Off" button. Once the gauge is on, press "Auto Zero" button until "On" shows in the upper right corner of the display. If both "PrA" and "PrB" do not show up on the left edge of the display, repeatedly press the "Mode" button until they do. Connect the end of a hose to the gauge's red tap on the back). Disconnect all other hoses to the gauge.

3.  Turn on only the air handler of the forced air system.

4.  While standing in the hall or other main body of the house, reclose the door to each room after laying the other end of the hose on the floor of the room. Record the pressure difference of the room compared to the main body of the house (for example, in the hall) with the air handler fan on.

5   Pressures:

| Room | Result | Room | Result |
|------|--------|------|--------|
|      |        |      |        |
|      |        |      |        |
|      |        |      |        |

6.  Pressure readings greater than 4 Pascals should be retrofitted to a lower pressure.

7.  Turn the air handler fan off.

---

One way to correct pressure imbalances is to cut the bottom of the bedroom doors up high enough (2 to 2½ inches) so the pressure imbalances do not exceed 4 Pascals, see Figure 4-48. Another method is putting in jumper or other ducts (such as transfer ducts) through the wall. (See also the retrofit section about jump ducts, etc.) Yet, another solution is to put grilles in the doors or walls, Figure 4-49.

## ZONE PRESSURE DIAGNOSTICS

Zone pressure diagnostics (ZPD) tests are conducted to better understand where the air barrier is in the home, how effective the air bar-

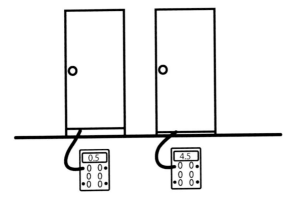

**Figure 4-48. An example of how a door with a greater undercut can relieve the pressure more easily. This is a measurement of the pressure in the room with respect to (WRT) the hall. Notice that the pressure is positive, indicating the room is pressurized WRT the hall, a result of the air supply to the room from the HVAC system.**

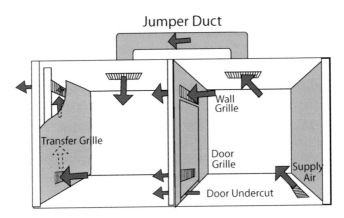

**Figure 4-49. There are a number of ways to relieve pressure in a room caused by inadequate return air access.**

rier is, whether the thermal boundary and air barrier are aligned, and if serious leaks exist between different transition/intermediate zones in the house. A transition or intermediate zone is an area that lies between the living space and the outside. The ZPD tests are most commonly used if there are IAQ issues involved, such as when there is an attached garage (including tuck-under garages). However, they are also commonly used in warm humid areas for crawlspaces and in cold weather areas for attics where IAQ issues can also arise. ZPD testing can tell where to concentrate the efforts for air sealing—the upper floor ceiling, the walls, the lowest level floor, or the foundation wall.

Transition zones in the house could include the following: the attic space, the basement or crawl space, the inside of interior and exterior walls, any attached garage especially tuck under garages, soffits, dropped ceilings, cantilevered or overhanging floor space including integrated balconies or bay windows, attic knee walls, upper "half story" floors, roof areas above the upper floor of Cape Cod homes, plumbing, electrical and heating ductwork/flue penetrations or chase ways through wall plates (top or bottom) or exterior walls, masonry chimney surround chases, stairways to the attic or basement, common walls in multilevel homes or older style balloon frame walls, exterior porch roofs and similarly positioned roof overhangs, cathedral or vaulted ceilings, commercial style false dropped or "T" ceilings with removable ceiling tiles, ceiling soffits, slant walls/roofs found on the upper floor of mansard roofs, vanities or other cabinets with supply or drainage plumbing or other openings to the wall in them, or in-wall medicine cabinets, electrical panel boxes, built-in appliances or similar in-wall arrangements. Figure 4-50 shows some areas of intermediate zones.

ZPD testing is often referred to as series leakage evaluations because you are trying to evaluate the leakage through a series of at least

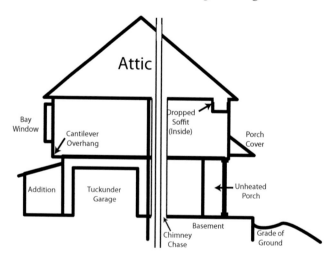

**Figure 4-50. Areas of Intermediate Zones**

two air barriers, for example, attic testing both the ceiling of the living space and the roof. ZPD testing helps determine where the most significant leaks exist and what sealing in certain areas is warranted. An example would be trying to determine whether the top floor ceiling or the roof deck is a better air barrier for an attic zone.

Typically, zones with an opening (door, hatch, vent, etc.) through them (basements, attics, crawlspaces, attached enclosed porches, etc.) are referred to as primary zones while those without openings to them (chases, soffits, cantilevers, outside porch ceilings, interior floor or wall cavities, etc.) are referred to as secondary zones. Some suggest that you need not go to great lengths to do regional ZPD testing if there is little leakage in the house as a whole. These same sources say not to worry about ZPD testing if the house does not leak at a higher rate than the calculated building tightness limit (BTL). Obviously, there are exceptions to this (see "Levels of ZPD Testing" below).

## Equipment

With any ZPD testing, the equipment needed, at a minimum, includes a blower door and a gauge. See the procedures table to learn the set up and hose arrangements for the gauges.

You can perform the basic ZPD tests with one gauge if you use a fan speed control dial (both EC and Retrotec fans) and set it for the correct speed that produces -50 Pascals. Then you can use your one gauge for ZPD testing. Better yet, if you have two gauges, you can use the EC cruise control or the Retrotec "Set Pressure" control to be better assured you are getting -50 Pascals throughout the ZPD testing.

## Basic Tests of Relative Leakiness Only

Basic ZPD testing (sometimes referred to as Level I Zone Testing) tells relatively how well an air barrier is doing its job by measuring the pressure differences between one side of the air barrier and the other. In the terminology of the energy arena, "Is a zone or space on the other side of the air barrier more "inside" the home or "outside" the home?" When there is less pressure differential between the inside and outside of the air barrier than wanted, then the intermediate/transition zone or space is more "inside" the home and the air barrier is not serving its purpose. When there is a good pressure differential between the inside and outside of the air barrier, then an intermediate zone is more "outside" the home and the air barrier is tighter, provides greater resistance to air flow or leakage, and is serving its purpose. You typically do not open or close doors or hatches between zones when conducting basic ZPD tests. Thus, when doing basic ZPD tests, take pressure readings without opening or closing a door or hatch, except to place a hose in the zone.

For ZPD testing, most experts recommend achieving a -50 Pascals using a blower door—see Blower Door Instructions (this -50 is the inside WRT the outside). If you cannot reach -50, try to find any obvious leaks and seal them to get -50 Pascals. To proceed with ZPD testing, follow the detailed ZPD test procedures near the end of this section.

**Door or Hatch Available to the Zone**: Place one end of a hose into the zone by placing it through the door or hatchway and then closing the door or hatch on the hose. If you think you may be pinching the hose in the process, use a metal hose that tightly fits this end of the hose so that the door or hatch is pinched against it rather than the flexible portion of the hose. If you have to leave the hatch or door open to a significant degree that compromises the reading, consider using a smaller non-conducting rigid tube, adapted to fit the hose. Slip the non-conducting hose around the outside edge of an electrical box (after removing the cover) or other penetration through the barrier and into the zone to get a pressure in the zone.

**No Door or Hatch Available to Zone**: If you do not have a door or hatch through the barrier into the zone, consider using a smaller non-conducting rigid tube, adapted to fit the hose. Slip the non-conducting hose around the outside edge of an electrical box (after removing the cover) or other penetration through the barrier and into the zone to get a pressure in the zone. If there are no penetrations through the barrier, with the homeowner's approval, create a small hole with as little damage as possible in an area of the barrier in which it will be less noticeable. Good locations for this include a closet ceiling, floor, or wall adjacent to an attic, crawl space or chimney zone respectively. Consider using the smaller non-conducting rigid tube, here as well. Slip the non-conducting hose through the small hole in the barrier to get a pressure in the zone.

To create a hole to get the hose through to the zone you are checking when there are no penetrations, you may have to use a small dull tipped, non-conducting plastic awl to puncture through drywall. This awl should be only slightly larger than the hose used for pressure testing. If, in a rare instance, you have to check the pressure above a porch covered with wood sheathing, and there are no fixtures, etc. to get the hose into the intermediate zone, avoid using power tools, drills etc. to create a hole in the sheathing for the tube. Instead, consider using a flat-headed off-set screwdriver of the smallest diameter possible for the hose but with a "stop" on it to keep from going too far through the sheathing and hitting wires etc. behind the sheathing, see Figure 4-51. You can set a stop on the screwdriver by wrapping the screwdriver with enough tape to act as a stop. You could also use a ring with a setscrew on it tightened down onto the appropriate point on the shaft of the screwdriver to match the thickness of the sheathing. Spinning the screwdriver in one spot on the sheathing should allow you to get through the sheathing to test the zone on the other side of the sheathing.

This technique helps reduce the risk of electrical shock or damage if you were to hit a wire or pipe, etc. behind the sheathing. In the rare case that you might have to make a hole in the drywall or sheathing, you could be responsible for patching and repainting the hole. Thus, making even a small hole through a building material should be used only as a last resort and only with the homeowner's approval.

Compare the pressure where you are standing with the pressure where the hose has been inserted, such as in the attic, basement, garage, etc. If the pressure reading of the area, such as the attic WRT the inside is 0, that means the pressure in the attic is the same as the pressure in the house which means the ceiling serves as no air barrier at all. If the pressure difference is 50, that means the ceiling is an excellent barrier. If the attic WRT the inside is 35, that means the barrier is not perfect and needs some work. If the basement reading WRT the inside is 45, the attic reading WRT the in-

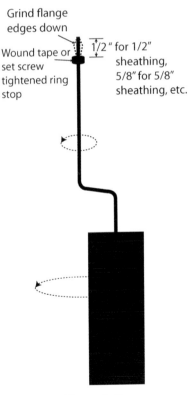

**Figure 4-51.**
**Offset Flat-Headed Screwdriver**

Figure 4-52. In
this scenario,
the home WRT
(with respect
to) outside has
been depres-
surized to -50
Pa (upper left

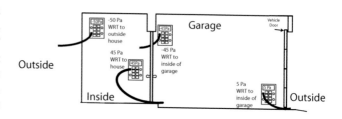

gauge). The pressure inside the gauge WRT the house is +45 Pa (lower middle
gauge). The upper middle gauge shows that the pressure in the house is -45
WRT the inside of the garage (as expected, it is the negative of the pressure
inside the house WRT the garage). The lower right gauge shows the pressure
outside WRT the garage is +5 Pa, meaning there is a great deal of communication or leaking between the garage and the outside, which is typical. As with
most garages, the garage is more "outside" than "inside." Thus, the air barrier
between the garage and the outside is relatively much leakier than the air barrier between the house and the garage.

side is 25, that would indicate that the attic is leakier than the basement,
and work should probably be concentrated there. You can also test the
intermediate zone with the outside while the blower door is running. In
that case, if the difference in pressure between the outside and the zone
is closer to zero, there is little communication between the zone and the
inside. (See "Gauge Outside" instructions in the procedures at the end
of this section.)

Table 4-16 shows the relative size of leaks in the barrier between
the zone and the house, and the zone and the outside given the pressure differential between the house and the zone assuming there is a 50
Pa pressure difference between the house and the outside. In the first
column under "zone pressures" find the measured pressure differential,
say between the attic and the house (the house-zone differential). The
second column shows that number subtracted from 50 to show the difference in pressure between the zone and the outside of the house. Thus,
if you measure a 12 Pa difference between the attic and the house, the
table shows that there is an attic to outside pressure differential of 38
(50-12).

The columns under "relative size of leaks" show the relative leakiness of the zone-to-house air barrier and the zone-to-outside air barrier.
Thus, with a 38 Pascal differential between the attic and the house, the
relative leakiness of the zone-to-house air barrier and the zone-to-out-

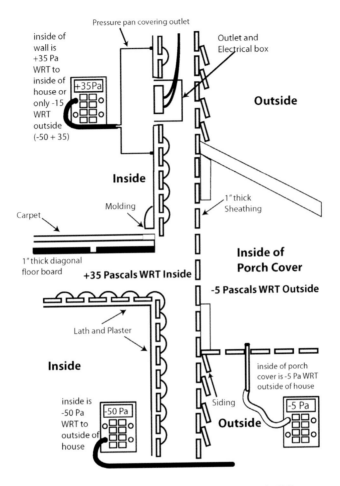

Figure 4-53. This diagram shows four different areas with different pressures: the inside of the home, the inside of the wall, the inside of the porch cover, and the outside of the house. It appears that the inside of the wall is "outside" (less than a 25 Pa difference in pressure between outside and the zone), while the inside of the porch cover expectantly is even more "outside." This means this leaky home must be well sealed to prevent air leakage. In cases like this, filling the wall with dense pack insulation is the best remedy because it seals and insulates at the same time. Incidentally, it is better to use a stiff non-conducting hose connected to the end of the flexible hose from the gauge to get around the edge of an electrical box to check for pressure inside of a wall rather than a pressure pan.

**Table 4-16. Leakiness Ratio (Pressure Ratio Chart)**

| Zone Pressures | | Relative Size Of Leaks | |
|---|---|---|---|
| Zone-to-House | Zone-to-Outside | Zone-to-House | Zone-to-Outside |
| 12 | 38 | 2 | 1 |
| 25 | 25 | 1 | 1 |
| 37 | 13 | 1/2 | 1 |
| 41 | 9 | 1/3 | 1 |
| 45 | 5 | 1/4 | 1 |
| 48 | 2 | 1/8 | 1 |
| 49 | 1 | 1/13 | 1 |

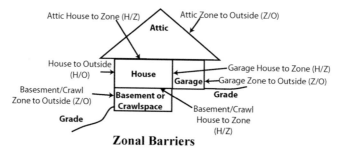

**Zonal Barriers**

**Figure 4-54. Examples of Different Barriers/Walls in ZPD Testing**

side air barrier is 2 to 1. Thus, the zone-to-house barrier is twice as leaky as the zone-to-outside barrier. This can also be viewed as a leakiness ratio. As another example, assume the zone-to-house pressure difference is 37 Pascals and the zone-to-outside pressure difference is 13, then the ratio of relative leakiness is 1/2 to 1. If this were an attic zone, the ceiling between the upper story and the attic would be half as leaky as the attic to outside air barrier (roof). If the zone-to-house pressure difference were 45 Pascals, the ceiling between the house and the attic would only be one fourth or 0.25 times the leakiness of the roof. Notice that the relative size of leaks changes more dramatically when the house-to-zone (H/Z) pressure is over 45 and the zone-to-outside (Z/O) pressure is less than 5. If you mistakenly measure only one Pascal off the true pressure, this could lead to a very different result in the relative leakiness.

A good example of when this may be misleading would be when measuring pressures in a well-vented attic or crawlspace. Since the zone-to-outside "leakiness" is due to larger intentional ventilation

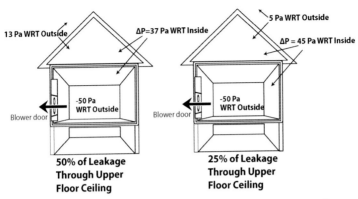

**Figure 4-55. Comparison of How the Percent of Leakage Effects
Pressures in a Home when Pressure Testing
(1/2 to 1 = 50%, ¼ to 1 = 25%**

openings, you could have a very leaky house-to-zone barrier and not know it. Some testers will actually tape over vents in cases like these to get a more accurate relative leakiness of the two barriers, but this is more impractical for attics and garages than it is for crawlspaces. Getting a pressure measurement between the house and attic is a basic test and provides only a relative number. It tells you that the bigger the difference in pressure between the house and the attic, the smaller the equivalent "hole" size in the ceiling between the house and the attic relative to the attic to outside.

In summary, basic ZPD testing helps narrow the evaluation to sometimes very small portions of the air barrier to identify what may be the more serious leaks. Basic ZPD testing also allows you to break down even a complex home into smaller parts to better identify the areas with the most air leakage, and therefore the best choice for air barrier. See infrared section for identifying specific leakage sites.

Interconnectedness Testing

In ZPD testing, if a zone has connections with another neighboring zone, it can interfere with the test, showing better leakage reduction than what is really happening. Interconnectedness testing is a cross between basic and advanced ZPD testing because it involves doing a basic test and opening a door or hatch but without using any charts to calculate leakiness. It helps assure that you are measuring a pressure that accurately portrays the leakiness involved in a zone. With this test, compare the pressures in the neighboring zone both before and after opening a door in the primary zone. If the pressure in the neighboring zone changes significantly upon opening the door or hatch, it is interconnected to the primary zone. If you ignore significant interconnectedness, you are going to overestimate the expectations for sealing potential primary zone barriers or double count the leakage from two barriers. You must find a way to seal off the interconnections between the primary zone and its neighboring zone. Perhaps one of the most common places where this occurs is in Cape Cod knee wall and other multiple space attics. Typically, only use this interconnectedness test when you suspect interconnectedness in a particular situation.

## ADVANCED ZPD TESTS THAT ESTIMATE LEAKAGE FLOW AND HOLE SIZE

To take zone pressure diagnostics and series leakage testing further and determine an estimate of the actual size of the "hole" and leakiness, there are several types of "advanced" zone pressure diagnostic tests. These tests can estimate, albeit crudely, the CFM leaking through an air barrier between the house and a zone, and between the zone and outside. Thus, for example, you could get a leakage rate for the wall between the house and the garage as if the outside walls of the garage were not there. You could also get a leakage rate for the wall between the garage and outside as if the wall the between the house and the garage had been removed. As a result, the total or combined leakage through both barriers can also be estimated although the total leakage CFM is always less than either of the individual barrier's leakages. This is because the air must go through both air barriers to count as combined or total leakage. For instance, ZPD testing may suggest a leakage rate of 11,000 CFM50 from an attic to outside and 2500 CFM50 from inside to the attic, while the total or combined leaking may be only 2000 CFM50. Advanced ZPD testing can be complicated and the values obtained, in some cases, are only estimates, so some auditors and retrofitters are content using primarily only basic ZPD testing to generally locate leaks in the air barrier. They then rely on the blower door test for overall leakiness in the home, and to see how well they sealed the house. They then use advanced ZPD testing only in special situations, such as to evaluate the house/garage, house/attic, house/basement or house/crawlspace barriers (see "Levels of ZPD Testing" below). You should be especially concerned when IAQ issues are involved, such as attached garages, attics in cold weather areas and crawlspaces in warm, humid climates. Advanced ZPD testing can tell you where the leakiest air barriers are in the house, and, if you can seal them in a significant way, how much tighter the home may become. Thus, if you have a whole house blower door reading of 3000 CFM50 and a combined or total leakage rate for air going from the house-to-outside through the garage of 1000 CFM50, that means the house, other than the garage, is leaking at a rate of 2000 CFM50 (3000 CFM50—1000 CFM50 = 2000 CFM50).

There are three different advanced ZPD tests suitable for quick and easy use: Open-a-Door (see Figure 4-56), Open-a-Hatch, and the Vent test. These tests are typically only done on primary zones that have doors or hatches (basements, attics, crawlspaces, attached enclosed porches, etc.). The measurements are considered more reliable if there is a minimum 15 to 25 Pascal difference in house-to-zone pressures at some point during these tests. If you are completing an advanced ZPD test using only simple charts and graphs, then it is considered by some to be a Level II zone test. If you are using software, etc. to calculate the results, these tests are considered by some to be Level III tests. Thus, Level III tests include the Open-a-Door, Open-a-Hatch, and the Vent test when using software rather than just charts and graphs in the process. At the same time, some other advanced ZPD tests have been developed that must use software as well, and these tests, such as the Hybrid Method, are typically based on the Level II tests in some way. The Hybrid Method compares the results from both the Open-a-door and Open-a-hatch methods to choose the most reliable result based on the uncertainty calculated.

o

This test only works if there is a door sized (or larger) opening that can be closed, and then opened, between the house and the zone or between the zone and the outside. You must be able to open at least a full-size door in the air barrier, such as a door to the basement zone, an unheated but enclosed porch zone, or an attic zone. With this test, you must maintain the indoor to outdoor pressure difference at 50 Pascals throughout the entire test using the "Set Pressure" (Retrotec) or the cruise control (EC) function.

Another phrase for this test could be the "Equilibrated House-to-Zone Pressure Test." This method can estimate the total size equivalent of the hole in the boundary between the interior of the home and the zone, as well as the total hole size equivalent between the zone and the exterior. It allows you to establish which barrier is the better air barrier, the house-to-zone barrier (such as the floor above the crawlspace) or the zone-to-outside barrier (such as the crawlspace foundation wall). The Open-a-Door method is generally considered the most accurate of the advanced tests. Several examples of Open-a-Door and Open-a-Hatch calculations are given in Appendix R.

**Open-a-Door Test**

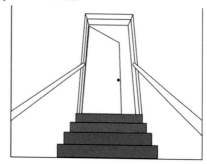

**Figure 4-56. Basement and attic doorways must be used with the "Open-a-door test."**

## Low Tech Air Leakage EfLA/Flow Test
## (Alternative to Advanced ZPD Testing)

A special Open-a-Door/Open-a-Hatch test is a low-tech variety that simplifies the process but also may not be as reliable. It involves the same set up as an Open-a-Door and Open-a-Hatch test by setting up a blower door to -50 Pascals with two gauges—one to remain with the blower door to keep the house at -50 Pascals while you keep the other gauge with you to check house-to-zone (H/Z) pressures, etc. Then after the house-to-zone pressure difference is measured with the door or hatch closed, the door or hatch between the zone and the house is slowly lodged open (lodge the door open with a door stop or lift one edge of the hatch to one side of the opening while leaving the other edge of the hatch against the opening). Open the door or hatch up until the house-to-zone (H/Z) pressure goes down to half of what it started at with the opening closed. For instance, if the H/Z pressure were 32 Pascals, you would open the door or hatch until the pressure dropped to 16 Pascals. You then measure that opening created (in square inches) to find the Effective Leakage Area (EfLA) of the H/Z barrier. Multiply this number by 10 to get the approximate CFM50 for the H/Z barrier.

### Open-a-Hatch or Add-a-Hole Test

A special type of zone pressure diagnostics is called the Add-a-Hole or Open-a-Hatch zone leakage measurement. This test is almost like the Open-a-Door test; however, it is used if you cannot get the pressure differential between the house and the zone to 0 Pascals as expected in the Open-a-Hole test. Examples of this would be attics, basements, or crawlspaces that are only accessible by smaller openings such as hatches or pull-down stairs.

With this test, maintain the indoor to outdoor pressure difference at -50 Pascals throughout the entire test. Unlike the Open-a-Door test, you do not have to establish a 0 Pascal pressure difference between the house and the zone after opening the hatch. The hatch between the house and the zone is closed before starting the test. Once you have started the blower door and established the 50 Pascal pressure, record the blower door flow and measure the difference in pressure between the house and the zone. Then open the hatch to the zone, and measure the new pressure difference between the house and the zone as well as the new blower door flow (@50 Pascals) that changed with opening the hatch. You need at least a 6 Pascal minimum drop in differential pressure between the house and the zone when comparing the pre-opening (hatch closed) with the opening (hatch open) pressure differentials. Thus, another phrase for this test could be the "Non-Equilibrated House-to-Zone Pressure Test" since there cannot be a 0 Pascal pressure differential between the house and zone after opening the hatch. The test involves calculating the difference in blower door flow from before and after opening the door through the air barrier (attic, basement etc.), while maintaining the -50 Pascal house-to-outside blower door pressure. It can be advantageous to have a blower door equipped with a "Set Pressure" or "cruise control" that keeps the pressure constant by increasing the rpm even when the hatch is opened. The flow and pressure data is analyzed using charts or computer programs to calculate how leaky the air barrier is between the house and the zone. With this method, you can also estimate the total size equivalent of the hole in the boundary between the interior of the home and the zone, as well as estimate the total hole size equivalent between the zone and the exterior. In the end, this test also allows you to establish which barrier is the better air barrier, the house-to-zone barrier (for example, floor above the crawlspace) or the zone-to-outside barrier (for example, the crawlspace foundation wall). The Open-a-Hatch method is considered less accurate than the Open-a-Door test. Examples of Add-a-Hole calculations are given in Appendix R.

**Vent Area Test**

    Open-a-Door and Open-a-Hatch tests can tell the leakage between the house and the attic or other zones. The vent area ZPD test involves calculating the size of vents in the zone (attic, basement, crawlspace, etc.). From this you can calculate the approximate leakage in the air barrier between the house and the zone. The vent area test is the least reliable of the advanced ZPD tests. (See "net free area" section for more information on vent area)

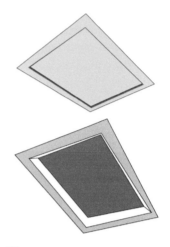

Figure 4-57. By measuring the change in pressure that results from opening a hole in the barrier between conditioned and unconditioned spaces, you can crudely estimate the leakage through the barrier.

LEVELS OF AND STANDARDS FOR ZPD TESTING

    As the level of ZPD testing goes from Level I to Level III, the time and difficulty of the ZPD testing also goes up, but the amount of information and the accuracy of the testing is improved. Thus, when you go from Level I (Basic ZPD Testing) to Level II (Open-a-Door, Open-a-Hatch), you go from only a relative leakiness of the house/zone and zone/outside barriers, to a rough approximation of actual leakage at CFM50 and effective leakage area. Going from Level I to II, only a small amount of time is added if there is no need to cut a hole into the attic space, etc. to create a door or hatch. When going from Level II to Level III (Level II with software), correction factors are automatically integrated into some software to reduce the risk of errors, but it takes more time and much more equipment (software, calculator or computer). As a rule, do Level I (Basic) ZPD testing on all primary and secondary zones in all houses and evaluate for relative leakiness using the leakiness or pressure ratio charts. Typically only do interconnectedness testing where interconnected zones are suspected, as in Cape Cod knee wall or other multiple space

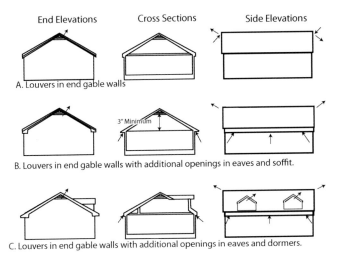

Figure 4-58. Ventilation Areas For Gable Roofs

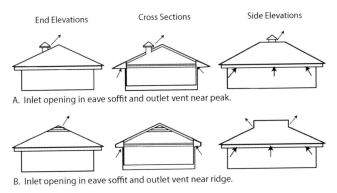

Figure 4-59. Ventilation Areas for Hip Roofs

attics. Typically only do Level II ZPD testing (only open-a-door or hatch tests) on garages that are attached to the house, on crawlspaces that have mold or moisture issues, and, if in a colder climate, on attics, particularly if the house is a relatively "tight" home with the potential for high moisture. (See "Summary of Work Flow" section for how this fits into the sequence of work).

Unless there is a significant, but very unusual problem that cannot be solved, there is no need to use Level III ZPD testing in most

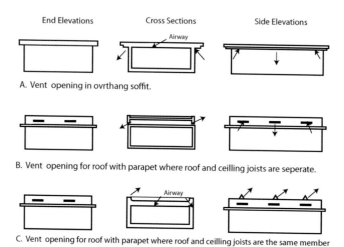

End Elevations          Cross Sections          Side Elevations

A. Vent opening in ovrthang soffit.

B. Vent opening for roof with parapet where roof and ceiling joists are seperate.

C. Vent opening for roof with parapet where roof and ceiling joists are the same member

**Figure 4-60. Ventilation Areas For Flat Roofs**

cases. Multiple interconnected zones in a building might be an example of where you might need to use Level III testing if you are unable to access the zones to close off any conspicuous openings creating the interconnectedness. When dealing with ventilated crawlspaces and attics, you can also improve results at any of the levels of ZPD testing by temporarily covering over the vents to better evaluate the zone-to-outside barrier, although this is often impractical. With Level II and III tests, another way of getting more reliable results is to select the opening in the less leaky barrier (greatest pressure difference) when choosing which door or hatch to open (the house/garage barrier is typically a better air barrier than the garage/outside barrier).

It is important in all cases to use the proper blower door protocol in setting up the blower door, and in setting up a separate baseline hose for the gauge used for zonal testing, preferably where the reference hose to outside goes for the blower door, otherwise place it out through a nearby window, etc. Once the leakage rate for a zone is established, you may conclude that no sealing work be done on a house-to-zone barrier if the leakage rate is low. Some suggest the cutoff for recommending against ceiling work be 200 CFM50 or 20 square inches effective air leakage. Another question might be, do you establish a different standard for garages where the issue is not just leakage but also exposure to pollutants from the garage?

The American Lung Association Healthy House Standards require the garage-to-house basic ZPD test to not go below 40 Pascals when the house is depressurized to -50 Pascals. Another standard that some go by in cold weather areas is to limit the leakage from the house into the attic to 10% of the overall whole house leakage. For instance, if the whole leakage were 2000 CFM50, you would not bother to seal an attic that had 200 CFM50 or less going from the house into the attic. Another issue relates the previously mentioned problems with the misleading results that can occur when the house-to-zone pressure is greater than 45 Pascals or the zone-to-outside pressure is less than 5. If this is the case, try to temporarily seal intentional ventilation (such as attic or crawlspace vents), or obvious holes or leaks, and then retest the zone pressure until the readings are outside of those ranges. Otherwise, you may end up with misleading results.

**Table 4-17. Basic ZPD (Level I) and ZPD Open-a-Door and Open-a-Hatch (Level II) Step-By-Step Procedure Using the Gauge Inside (Conditioned Space) Using the Retrotec DM-2 or the EC DG-700 Gauges**

1. Make sure your gauge batteries are fully charged or that you have adequate outlets and extension cords to connect the pressure gauge to its recharger/transformer.

   Close the door or hatch in the house to the zone wall/barrier (for example, between the house and the attic, basement, enclosed unheated porch, kneewall, crawlspace, etc.) and doors, etc. in the zone-to-outside wall/barrier (for example, between the attic, basement, enclosed porch, kneewall, a cantilever, bay window, etc. zones that is covered under the "Gauge Outside" Procedures Table below.

2. **Set Up 1—House-to-Outside Pressure**

   Place a hose to outside (typically red for Retrotec and green for EC). Preferably use a longer hose that is placed along the outside wall through the blower door screen so you can use the gauge throughout the house to take zone pressure measurements. If you are unable to do this, place the hose out a window or door on the same level near where you are checking the zone pressures.

   *Retrotec*—Turn on the gauge by pressing and holding down the "On/Off" button. Once the gauge is on, press "Auto Zero" button until "On" shows in the upper right corner of the display. If both "PrA" and "PrB" do not show up on the left edge of the display, repeatedly press the

## EC Zone Pressure Diagnostic (ZPD) Testing
## Gauge Inside: Set-Up 1- House to Outside (H/O)

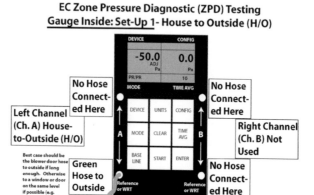

## Retrotec Zone Pressure Diagnostic (ZPD) Testing
## Gauge Inside: Set-Up 1- House to Outside (H/O)

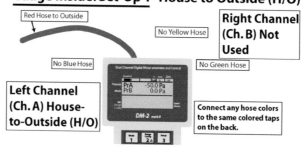

"Mode" button until they do. Connect the other end of the hose going to the outside to the red tap on the back of the gauge. Disconnect all other hoses to the gauge. Do a baseline as follows:

Press the "Baseline" button once.

Once the average shown near the bottom of the gauge seems to level off without as much change, press the "Enter" button once.

*EC*—Turn on the gauge—Press "On/Off." If "PR/PR" is not showing in the lower left corner of the display, keep pressing the "Mode" button until it appears. Connect the other (inside) end of the hose to outside onto the gauge's Channel A reference tap (lower left tap). Disconnect all other hoses to the gauge. Do a baseline as follows:

Press the "Baseline" button once.

Press "Start" once.

Once the numbers on the gauge seem to level off without as much change, press the "Enter" button once.

3. *Two Gauges Available*: Start the blower door and establish the -50 Pa pressure using a different gauge for the blower door. Follow the instructions for blower door testing, and use the "Set Pressure" (Retrotec) or "cruise control" (EC) settings to keep the pressure at -50 Pascals.

*One Gauge Available*: If you have only one gauge to conduct the ZPD tests, use the instructions for blower door testing utilizing the manual speed controls to set the fan rpm's so the gauge reads -50 Pascals. Once you have established the -50 Pascals using your one gauge, record the flow and pressure in the next step and disconnect the tubes going to the blower door. Keep the hose to outside connected to your gauge if it is long enough to reach the zones you will be testing.

4. Record the blower door flow below. Measure the difference in pressure between the outside and the inside near the zone by reading the number on the display next to "PrA" (it should be -50).

_____CFM50 (with any doors/hatches to zones, closed)

_____Pa H/O

Proceed to the zone you are testing with a hose to outside connected to the gauge as previously described.

5. **Set Up 2—House-to-Zone Pressure**

*Door or Hatch Available to the Zone*: Place one end of a hose in the zone by placing it through the door or hatchway and then closing the door or hatch on the hose. If you think you may be pinching the hose in the process, use a metal hose that tightly fits this end of the hose so that the door or hatch is pinched against it rather than the flexible portion of the hose. If you have to leave the hatch or door open to a significant degree that compromises the reading, consider using a smaller non-conducting rigid tube, adapted to fit the hose. Slip the non-conducting hose around the outside edge of an electrical box (after removing the cover) or other penetration through the barrier and into the zone to get a pressure in the zone. Since you have an opening to the zone, you can conduct an open-a-door or open-a-hatch test later.

## EC Zone Pressure Diagnostic (ZPD) Testing
### <u>Gauge Inside: Set-Up 2</u> - House to Zone (H/Z)

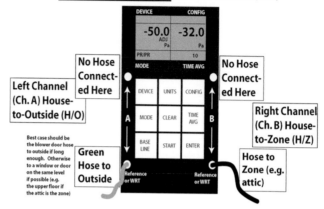

No Hose Connected Here

No Hose Connected Here

Left Channel (Ch. A) House-to-Outside (H/O)

Right Channel (Ch. B) House-to-Zone (H/Z)

Best case should be the blower door hose to outside if long enough. Otherwise to a window or door on the same level if possible (e.g. the upper floor if the attic is the zone)

Green Hose to Outside

Hose to Zone (e.g. attic)

## Retrotec Zone Pressure Diagnostic (ZPD) Testing
### <u>Gauge Inside: Set-Up 2 - House to Zone (H/Z)</u>

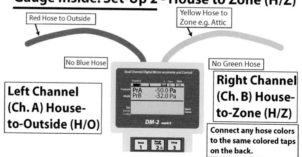

Red Hose to Outside

Yellow Hose to Zone e.g. Attic

No Blue Hose

No Green Hose

Left Channel (Ch. A) House-to-Outside (H/O)

Right Channel (Ch. B) House-to-Zone (H/Z)

Connect any hose colors to the same colored taps on the back.

*No Door or Hatch Available to Zone*: If you do not have a door or hatch through the barrier into the zone, consider using a smaller non-conducting rigid tube, adapted to fit the hose. Slip the non-conducting hose around the outside edge of an electrical box (after removing the cover) or other penetration through the barrier and into the zone to get a pressure in the zone. If there are no penetrations through the barrier, with the homeowner's approval, create a small hole with as little damage as possible in an area of the barrier where it will be less noticeable (see ZPD section text on ways to reduce the damage when accessing zones). Good locations for this include a closet ceiling, floor, or wall adjacent to an attic, crawl space or chimney zone respectively. Consider using the smaller non-conducting rigid tube. Slip the non-conducting hose through the small hole in the barrier to read a pressure in the zone.

Note: If you do not have an opening between the house and the zone, and you also do not have an opening between the zone-to-outside, you will not be able to do any advanced or Level I or II ZPD tests.

*Retrotec*—Connect the other end of the zone hose to the yellow tap on the back of the gauge. After making sure the pressure reading for PrA is stable at about -50 Pascals, record this House-to-Zone (H/Z) pressure in the zone you are testing below after reading the number on the display next to "PrB."

*EC*—Connect the other end of the zone hose to the lower right reference tap on the gauge (lower tap of Channel B). After making sure the pressure reading on the left side of the display (Channel A) is stable at about -50 Pascals, record this House-to-Zone (H/Z) pressure in the zone you are testing below after reading the number on the right side of the display (Channel B).

All Gauges: Record H/Z pressure here

_____Pa H/Z

If this pressure is greater than 25 Pascals, the barrier between the house and zone is tighter than the barrier from the zone-to-outside. The opposite is true if this pressure is less than 25 Pascals.

6. **Low Tech Air Leakage EfLA/Flow Test (Alternative to Further ZPD Testing)**

   *One Gauge Only*—Blower door used without a gauge to allow for "Set Pressure" (Retrotec) or "cruise control" (EC) setting: **THIS TEST NOT AVAILABLE**

   *Two Gauges Used*—Blower door used with a gauge to allow for "Set Pressure" (Retrotec) or "cruise control" (EC) setting while a separate gauge is used for H/Z tests, etc.: Proceed as follows:

   Slowly lodge open a door or hatch to the zone from the house (one way is to open a door or lift one edge of a hatch to one side of the opening while leaving the other edge of the hatch against the opening) until the house-to-zone (H/Z) pressure goes down to half of what it started at with the opening closed. Measure the opening you created to get half the pressure to find the effective leakage area of the H/Z barrier:

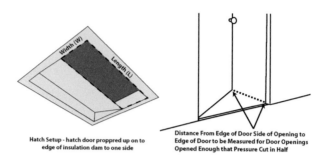

Hatch Setup - hatch door proppred up on to edge of insulation dam to one side

Distance From Edge of Door Side of Opening to Edge of Door to be Measured for Door Openings Opened Enough that Pressure Cut in Half

*Door Opening Over a Floor or Landing*: Measure the distance on the ground from the lower corner of the door opening to the door edge. Multiply this measurement by the multiplier based on how wide the door opening is: Door size = Multiplier: 36"=98; 32"=96; 30"=95; 28"=94; 24"=92. Record the measurement and the multiplier here and complete the calculation:

Distance x Multiplier = Area of opening (A)

_____ in. X _____ = _____ square inches

*Hatches with Insulation Dams Around Opening*: Measure the opening width and length and multiply them by each other (W x L = Area):

_____ in. X _____ in. = _____ square inches

Multiply this number by 10 to get the approximate CFM50 for the H/Z barrier.

_____ sq. in. X 10 = _____CFM

If you have chosen this alternative method, you have completed the ZPD testing on this zone and need complete only the next step of this process.

7. **Set Up 3—Zone-to-Outside Pressure**

To confirm the house-to-zone pressure you also check the zone-to-outside pressure. Check to see if the house-to-zone and zone-to-outside pressures add up to about 50 Pascals. For this set up keep the hose in the zone where you had it in the last set up but change hose connections on the gauge:

*Retrotec*—First remove the zone hose from the yellow tap and reconnect it to the green tap on the back of the gauge. Also, remove the red hose

## EC Zone Pressure Diagnostic (ZPD) Testing
## Gauge Inside: Set-Up 3 - Zone to Outside (Z/O)

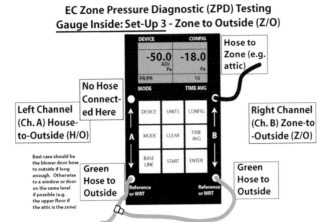

## Retrotec Zone Pressure Diagnostic (ZPD) Testing
## Gauge Inside: Set-Up 3 - Zone to Outside (Z/O)

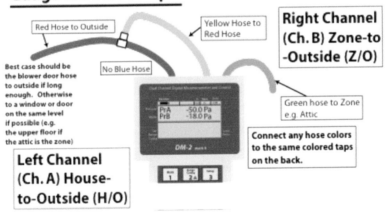

from the red tap on the back of the gauge. Then place a short stub of red hose onto the red tap instead. Add a "T" to the other end of the short red hose stub and then connect the long red hose to one of the legs of the "T" connection. Then connect a short stub of hose from the yellow tap on the back of the gauge to the last leg left on this "T" connection. After making sure the pressure reading for PrA is stable at about -50 Pascals, record this Zone-to-Outside (Z/O) pressure in the zone you are testing below after reading the number on the display next to "PrA."

*EC*—First remove the zone hose from the lower right reference tap on the gauge and then reconnect it to the upper right "input" tap on the gauge (upper tap of Channel B). Also, add a short stub of green hose to the lower left reference tap on the gauge (lower tap of Channel A). Add a "T" to the other end of the short green hose stub and then connect the long green hose to one of the legs of the "T" connection. Then connect a short stub of green hose from the lower right reference tap (lower right tap of Channel B) to the last leg left on this "T" connection. After making sure the pressure reading on the left side of the display (Channel A) is stable at about -50 Pascals, record this Zone-to-Outside (Z/O) pressure below after reading the number on the right side of the display (Channel B).

_____Pa Z/O barrier/wall

8. Add the house-to-zone (H/Z) pressure to the zone-to-outside (Z/O) to see if they add to close to -50 Pascals:

_____ Pa Z/O _____ Pa H/Z =

_____Pa (should be close to -50)

9. **Continue with these procedures from here only if you are planning to conduct an Open-a-Door or Open-a-Hatch ZPD test.** You must have a door or hatch in the barrier between the house and the zone or between the zone and outside to conduct these tests:

Open a door or hatch in the barrier that is the tightest. In the case of garages, the tightest barrier is typically, but not always, the wall between the house and the garage. If the house-to-zone (H/Z) pressure reading is between 25 and 50, it is the tightest barrier and it is preferred to open a door or hatch from the house to the zone if one is available there. If the house-to-zone (H/Z) pressure reading is between 0 and 25, it is the leakier of the zone barriers and it is preferred to open a door or hatch from the zone to the outside if a door or hatch is available there. These situations result in the best accuracy in evaluating zonal air barriers. In any case, you must have a door or hatch in either the house-to-zone or zone-to-outside air barriers to complete these tests, regardless of which provides more accuracy. You should get a change in pressure of at least 6 Pa for the results to be acceptable.

*Open-a-Door*: Once you have opened the appropriate door, momentarily disconnect the hose to the outside from the gauge for the channel connected on the right (on Retrotec gauges disconnect the yellow hose for the yellow tap and look for the PrB reading, on EC gauges, disconnect the hose to the lower right tap and look at the reading on the right of the display or Channel B). Confirm that the pressure difference between the house and the zone is now at 0 Pascals, (if you opened a house to zone barrier door) or 50 Pascals (if you opened a zone-to-outside door). If the reading is not zero, you must consider the test an Open-a-Hatch test. Then reconnect the tube.

10. *If you are only using one gauge*: Confirm that the blower door is still operating at -50 Pascals (on Retrotec gauges look for the PrB reading, while with EC gauges, look at the reading on the right of the display or Channel B). If the blower door is not providing -50 Pascals house-to-zone, set up your gauge for blower door readings again, use the manual speed control to bring the house-to-zone pressure back to -50 Pascals.

*All Tests*: Record below the new blower door flow reading, the barrier where the door was located, the type of zone tested, and, if you are conducting an open-a-hatch test, the zone-to-outside (Z/O) pressure (on Retrotec: the PrB reading; on EC: the right channel or channel B reading):

_____ CFM50 with door or hatch open

Barrier/Door/Hatch opened: H/Z Z/O

Zone Location: Attic Crawl Porch Other_____

Open-a-Hatch Test: Zone-to-outside pressure (Z/O):

_____ Pa Z/O barrier/wall

11. Subtract the first blower door flow from the last:

_____CFM50(open) –

_____ CFM50(closed) =

_____CFM50 Difference

12. *Open-a-Hatch Test*: Repeat Setup 2 shown above one more time and record the pressure readings here:

_____Pa H/O (PrA or channel A)

_____Pa H/Z (PrB or channel B)

Add the H/Z to the Z/0 and you should get about -50 Pa

_____ Pa H/Z _____ Pa Z/O =

_____Pa (should be close to -50)

13. **Open-a-Door:**

Opening made in H/Z barrier—If you have opened a door from the house-to-zone (such as the house-to-garage door), use Column A of the Open-a-Door Chart below to find the house-to-zone pressure before opening the door (circle this pressure in Column A).

Opening made in Z/O barrier—If you have opened a zone-to-outside door (such as the garage-to-outside door), use Column B of the Open-a-Door Chart below to find the house-to-zone pressure before opening the door (circle this pressure in Column B).

All Open-a-Door Tests: Record the multipliers in the appropriate chart across to the right in the same row as the house-to-zone pressure you circled for the "Internal" (house-to-zone leakage), "External" (zone-to-outside leakage), and "Combined" (leakage when both barriers considered) found:

Internal:_____ External: _____

Combined: _____

Now multiply each of the multipliers by the CFM difference you recorded above to find how much reduction, at a maximum, can be achieved by completely sealing both barriers (combined):

Internal:_____ (multiplier) x _____ CFM Diff =

_____ CFM50 H/Z barrier/wall

External:_____(multiplier) x _____ CFM Diff =

_____ CFM50 Z/O barrier/wall

Combined: ____(multiplier) x _____ CFM Diff =

_____ CFM50 max of both barriers/walls combined (maximum reduction possible)

**Open-a-Hatch:**

*Opening made in H/Z barrier*—If you have opened a hatch or hole from the house-to-zone (such as the house to attic hatch, house to crawlspace hatch, etc.), use *Column A* of the Open-a-Hatch Chart below to find the *house-to-zone pressure before opening the hatch* (circle this pressure in Column A). Then, use Row A to find the house-to-zone ending pressure after opening the hatch (circle this pressure in Row A). Find the multiplier at the intersection of these two circled items by going right across from the starting pressure and down from the ending pressure. *Opening made in Z/O barrier*—If you have opened a hatch or hole from the zone-to-outside barrier (for example, the roof to attic hatch, outside to crawlspace hatch), use *Column B* of the Open-a-Hatch Chart below to find the *house-to-zone pressure before opening the hatch* (circle this pressure in Column B). Then, use Row B to find the house-to-zone ending pressure after opening the hatch (circle this pressure in Row B). Find the intersection of these two circled items by going right across from the starting pressure and down from the ending pressure and record them in the equation below where the multiplier is expected to go.

*All Open-a-Hatch Tests*: Multiply the CFM difference calculated above (the change in flow from before opening the hatch and after opening the hatch) times the multiplier from the chart. The result gives the expected maximum reduction in leakage if you were to completely seal the house-to-zone or the zone-to-outside barriers:

_____ (multiplier) x _____ CFM50 Diff =

_____ CFM maximum reduction if both barriers perfectly sealed so no leakage

14. If you prefer results in effective leakage area, divide the total leakage numbers by 10 to get effective leakage area in square inches (numbers cannot be calculated for combined barriers, including the combined numbers that result from the Open-a-Hatch test):

*Open-a-Door*:

_____ CFM50 H/Z x 10 = _____ sq. in. ELA H/Z barrier/wall

_____ CFM50 Z/O x 10 = _____ sq. in. ELA Z/O barrier/wall

15. If you would like to calculate the percent error you might expect from these calculations, use the Excel files/sheets that may be available to do so.

## Table 4-18. Open-A-Door Chart

For <u>Door Opening in Barrier Between House to Zone:</u>
Use Column A to Match H/Z and Column B to Match Z/O
For <u>Door Opening in Barrier Between Zone to Outside:</u>
Use Column B to Match H/Z and Column A to Match Z/O

| Door closed pressure | | multiply CFM50 change by: | | |
|---|---|---|---|---|
| A | B | Internal | External | Combined |
| 48 | 2 | 0.14 | 1.14 | 0.14 |
| 47 | 3 | 0.20 | 1.19 | 0.19 |
| 46 | 4 | 0.25 | 1.24 | 0.24 |
| 45 | 5 | 0.31 | 1.29 | 0.29 |
| 44 | 6 | 0.37 | 1.34 | 0.34 |
| 43 | 7 | 0.43 | 1.39 | 0.39 |
| 42 | 8 | 0.49 | 1.44 | 0.44 |
| 41 | 9 | 0.56 | 1.49 | 0.49 |
| 40 | 10 | 0.63 | 1.54 | 0.54 |
| 39 | 11 | 0.70 | 1.60 | 0.60 |
| 38 | 12 | 0.78 | 1.65 | 0.65 |
| 37 | 13 | 0.87 | 1.71 | 0.71 |
| 36 | 14 | 0.96 | 1.78 | 0.78 |
| 35 | 15 | 1.06 | 1.84 | 0.84 |
| 34 | 16 | 1.17 | 1.91 | 0.91 |
| 33 | 17 | 1.29 | 1.98 | 0.98 |
| 32 | 18 | 1.42 | 2.06 | 1.06 |
| 31 | 19 | 1.56 | 2.14 | 1.14 |
| 30 | 20 | 1.71 | 2.23 | 1.23 |
| 29 | 21 | 1.88 | 2.32 | 1.32 |
| 28 | 22 | 2.07 | 2.42 | 1.42 |
| 27 | 23 | 2.27 | 2.52 | 1.52 |
| 26 | 24 | 2.50 | 2.64 | 1.64 |
| 25 | 25 | 2.76 | 2.76 | 1.76 |

Following Collin Olson and Anthony Cox, 2006, following Michael Blasnik
Internal = multiplier for zone to house leakage
External = multiplier for zone to outside leakage
Combined = multiplier for total/combined path leakage

## Table 4-19. Open-A- Hatch Chart

### Hatch/Hole Opened in House to Zone (H/Z) or Zone to Outside (Z/O) Barriers

For Hatch/Hole Opening in House to Zone Barrier: Use A Column and Row to Match H/Z and Column and Row B to Match Z/O

For Hatch/Hole Opening in Zone to Outside Barrier: Use B Column and Row to Match H/Z and Column and Row A to Match Z/O

Start Press — Ending Pressure After Opening Hatch or Adding Hole in Barrier

| A | B | 44 | 42 | 40 | 38 | 36 | 34 | 32 | 30 | 28 | 26 | 24 | 22 | 20 | 18 | 16 | 14 | 12 | 10 | 8 | 6 | 4 | 2 | 0 |
|---|---|----|----|----|----|----|----|----|----|----|----|----|----|----|----|----|----|----|----|---|---|---|---|---|
| 50 | 0 | 0.00 | 0.00 | 0.00 | 0.00 | 0.00 | 0.00 | 0.00 | 0.00 | 0.00 | 0.00 | 0.00 | 0.00 | 0.00 | 0.00 | 0.00 | 0.00 | 0.00 | 0.00 | 0.00 | 0.00 | 0.00 | 0.00 | 0.00 |
| 49 | 1 | | 0.35 | 0.29 | 0.25 | 0.22 | 0.20 | 0.18 | 0.17 | 0.15 | 0.15 | 0.14 | 0.13 | 0.12 | 0.12 | 0.11 | 0.11 | 0.10 | 0.10 | 0.10 | 0.09 | 0.09 | 0.09 | 0.09 |
| 48 | 2 | | 0.68 | 0.54 | 0.45 | 0.39 | 0.35 | 0.32 | 0.29 | 0.27 | 0.25 | 0.23 | 0.22 | 0.21 | 0.20 | 0.19 | 0.18 | 0.17 | 0.17 | 0.16 | 0.15 | 0.15 | 0.15 | 0.14 |
| 47 | 3 | | | 0.84 | 0.68 | 0.58 | 0.51 | 0.45 | 0.41 | 0.38 | 0.35 | 0.33 | 0.31 | 0.29 | 0.27 | 0.26 | 0.25 | 0.24 | 0.23 | 0.22 | 0.21 | 0.20 | 0.20 | 0.19 |
| 46 | 4 | | | 1.23 | 0.96 | 0.80 | 0.68 | 0.60 | 0.54 | 0.49 | 0.45 | 0.42 | 0.39 | 0.37 | 0.35 | 0.33 | 0.32 | 0.30 | 0.29 | 0.28 | 0.27 | 0.26 | 0.25 | 0.24 |
| 45 | 5 | | | | 1.30 | 1.05 | 0.89 | 0.77 | 0.68 | 0.62 | 0.56 | 0.52 | 0.48 | 0.45 | 0.43 | 0.40 | 0.38 | 0.37 | 0.35 | 0.33 | 0.32 | 0.31 | 0.30 | 0.29 |
| 44 | 6 | | | | 1.76 | 1.36 | 1.12 | 0.96 | 0.84 | 0.75 | 0.68 | 0.63 | 0.58 | 0.54 | 0.51 | 0.48 | 0.45 | 0.43 | 0.41 | 0.39 | 0.38 | 0.36 | 0.35 | 0.34 |
| 43 | 7 | | | | | 1.76 | 1.41 | 1.18 | 1.02 | 0.90 | 0.81 | 0.74 | 0.68 | 0.63 | 0.59 | 0.56 | 0.53 | 0.50 | 0.48 | 0.45 | 0.43 | 0.42 | 0.40 | 0.39 |
| 42 | 8 | | | | | 2.28 | 1.76 | 1.44 | 1.23 | 1.08 | 0.96 | 0.87 | 0.80 | 0.73 | 0.68 | 0.64 | 0.60 | 0.57 | 0.54 | 0.52 | 0.49 | 0.47 | 0.45 | 0.44 |
| 41 | 9 | | | | | | 2.20 | 1.76 | 1.47 | 1.27 | 1.12 | 1.01 | 0.92 | 0.84 | 0.78 | 0.73 | 0.68 | 0.65 | 0.61 | 0.58 | 0.55 | 0.53 | 0.51 | 0.49 |
| 40 | 10 | | | | | | 2.80 | 2.15 | 1.76 | 1.49 | 1.30 | 1.16 | 1.05 | 0.96 | 0.89 | 0.82 | 0.77 | 0.72 | 0.68 | 0.65 | 0.62 | 0.59 | 0.56 | 0.54 |
| 39 | 11 | | | | | | | 2.65 | 2.11 | 1.76 | 1.51 | 1.33 | 1.20 | 1.09 | 1.00 | 0.92 | 0.86 | 0.81 | 0.76 | 0.72 | 0.68 | 0.65 | 0.62 | 0.60 |
| 38 | 12 | | | | | | | 3.32 | 2.54 | 2.07 | 1.76 | 1.53 | 1.36 | 1.23 | 1.12 | 1.03 | 0.96 | 0.90 | 0.84 | 0.80 | 0.75 | 0.72 | 0.68 | 0.65 |
| 37 | 13 | | | | | | | | 3.09 | 2.45 | 2.04 | 1.76 | 1.55 | 1.38 | 1.26 | 1.15 | 1.07 | 0.99 | 0.93 | 0.87 | 0.83 | 0.79 | 0.75 | 0.71 |
| 36 | 14 | | | | | | | | 3.83 | 2.93 | 2.38 | 2.02 | 1.76 | 1.56 | 1.41 | 1.28 | 1.18 | 1.09 | 1.02 | 0.96 | 0.90 | 0.86 | 0.81 | 0.78 |
| 35 | 15 | | | | | | | | | 3.54 | 2.80 | 2.33 | 2.00 | 1.76 | 1.57 | 1.42 | 1.30 | 1.21 | 1.12 | 1.05 | 0.99 | 0.93 | 0.89 | 0.84 |
| 34 | 16 | | | | | | | | | 4.35 | 3.32 | 2.70 | 2.28 | 1.98 | 1.76 | 1.58 | 1.44 | 1.33 | 1.23 | 1.15 | 1.08 | 1.01 | 0.96 | 0.91 |
| 33 | 17 | | | | | | | | | | 3.98 | 3.14 | 2.61 | 2.24 | 1.97 | 1.76 | 1.59 | 1.46 | 1.34 | 1.25 | 1.17 | 1.10 | 1.04 | 0.98 |
| 32 | 18 | | | | | | | | | | 4.86 | 3.70 | 3.01 | 2.54 | 2.20 | 1.95 | 1.76 | 1.60 | 1.47 | 1.36 | 1.27 | 1.19 | 1.12 | 1.06 |
| 31 | 19 | | | | | | | | | | | 4.42 | 3.49 | 2.89 | 2.48 | 2.18 | 1.94 | 1.76 | 1.61 | 1.48 | 1.38 | 1.29 | 1.21 | 1.14 |
| 30 | 20 | | | | | | | | | | | 5.38 | 4.09 | 3.32 | 2.80 | 2.43 | 2.15 | 1.93 | 1.76 | 1.61 | 1.49 | 1.39 | 1.30 | 1.23 |
| 29 | 21 | | | | | | | | | | | | 4.86 | 3.83 | 3.18 | 2.72 | 2.38 | 2.13 | 1.92 | 1.76 | 1.62 | 1.50 | 1.41 | 1.32 |
| 28 | 22 | | | | | | | | | | | | 5.89 | 4.48 | 3.63 | 3.05 | 2.65 | 2.34 | 2.11 | 1.91 | 1.76 | 1.63 | 1.51 | 1.42 |
| 27 | 23 | | | | | | | | | | | | | 5.30 | 4.18 | 3.46 | 2.96 | 2.59 | 2.31 | 2.09 | 1.91 | 1.76 | 1.63 | 1.52 |
| 26 | 24 | | | | | | | | | | | | | 6.41 | 4.86 | 3.94 | 3.32 | 2.87 | 2.54 | 2.28 | 2.07 | 1.90 | 1.76 | 1.64 |
| 25 | 25 | | | | | | | | | | | | | | 5.75 | 4.52 | 3.74 | 3.20 | 2.80 | 2.49 | 2.25 | 2.06 | 1.89 | 1.76 |
| 24 | 26 | | | | | | | | | | | | | | 6.92 | 5.25 | 4.25 | 3.57 | 3.09 | 2.73 | 2.45 | 2.23 | 2.04 | 1.89 |
| 23 | 27 | | | | | | | | | | | | | | | 6.19 | 4.86 | 4.02 | 3.44 | 3.01 | 2.68 | 2.42 | 2.20 | 2.03 |
| 22 | 28 | | | | | | | | | | | | | | | 7.43 | 5.64 | 4.55 | 3.83 | 3.32 | 2.93 | 2.63 | 2.38 | 2.18 |
| 21 | 29 | | | | | | | | | | | | | | | | 6.63 | 5.21 | 4.30 | 3.67 | 3.21 | 2.86 | 2.58 | 2.35 |
| 20 | 30 | | | | | | | | | | | | | | | | 7.95 | 6.02 | 4.86 | 4.09 | 3.54 | 3.12 | 2.80 | 2.54 |
| 19 | 31 | | | | | | | | | | | | | | | | | 7.07 | 5.55 | 4.58 | 3.91 | 3.42 | 3.04 | 2.74 |
| 18 | 32 | | | | | | | | | | | | | | | | | 8.46 | 6.41 | 5.17 | 4.35 | 3.76 | 3.32 | 2.97 |
| 17 | 33 | | | | | | | | | | | | | | | | | | 7.51 | 5.89 | 4.86 | 4.15 | 3.63 | 3.23 |
| 16 | 34 | | | | | | | | | | | | | | | | | | 8.98 | 6.79 | 5.48 | 4.61 | 3.98 | 3.51 |
| 15 | 35 | | | | | | | | | | | | | | | | | | | 7.95 | 6.24 | 5.14 | 4.39 | 3.83 |
| 14 | 36 | | | | | | | | | | | | | | | | | | | 9.49 | 7.18 | 5.79 | 4.86 | 4.20 |
| 13 | 37 | | | | | | | | | | | | | | | | | | | | 8.39 | 6.58 | 5.42 | 4.63 |
| 12 | 38 | | | | | | | | | | | | | | | | | | | | 10.00 | 7.56 | 6.10 | 5.12 |
| 11 | 39 | | | | | | | | | | | | | | | | | | | | | 8.83 | 6.92 | 5.71 |
| 10 | 40 | | | | | | | | | | | | | | | | | | | | | 10.52 | 7.95 | 6.41 |
| 9 | 41 | | | | | | | | | | | | | | | | | | | | | | 9.27 | 7.26 |
| 8 | 42 | | | | | | | | | | | | | | | | | | | | | | 11.03 | 8.33 |
| 7 | 43 | | | | | | | | | | | | | | | | | | | | | | | 9.71 |
| 6 | 44 | | | | | | | | | | | | | | | | | | | | | | | 11.54 |

Following Anthony Cox and Collin Olson, 2006 following Michael Blasnik

**Table 4-20. Basic ZPD Step-By-Step Procedure Using the Gauge Outside the Building and Using the Retrotec DM-2 or the EC DG-700 Gauges**

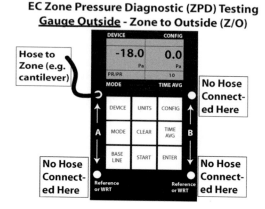

**EC Zone Pressure Diagnostic (ZPD) Testing**
**Gauge Outside - Zone to Outside (Z/O)**

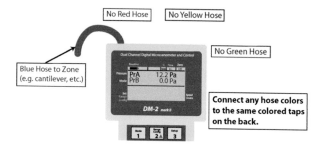

**Retrotec Zone Pressure Diagnostic (ZPD) Testing**
**Gauge Outside - Zone to Outside (Z/O)**

1. Close the door or hatch in the house to the zone wall/barrier (attic, basement, enclosed unheated porch, kneewall, crawlspace, etc.) and doors, etc. in the zone-to-outside wall/barrier before starting the test.

2. **Set Up A—House-to-Outside Pressure**

   Place a hose to outside (typically red for Retrotec and green for EC). Preferably use a longer hose that is placed along the outside wall through the blower door screen so you can use your gauge throughout the house to take zone pressure measurements. If you are unable to do this, place the hose out a window or door on the same level near where you are checking the zone pressures.

*Retrotec*—Turn on the gauge by pressing and holding down the "On/Off" button. Once the gauge is on, press "Auto Zero" button until "On" shows in the upper right corner of the display. If both "PrA" and "PrB" do not show up on the left edge of the display, repeatedly press the "Mode" button until they do. Connect the other end of the hose going to the outside to the red tap on the back of the gauge. Disconnect all other hoses to the gauge. Do a baseline as follows:

Press the "Baseline" button once.

Once the average shown near the bottom of the gauge seems to level off without as much change, press the "Enter" button once.

*EC*—Turn on the gauge—Press "On/Off." If "PR/PR" is not showing in the lower left corner of the display, keep pressing the "Mode" button until it appears. Connect the other (inside) end of the hose to outside onto the gauge's Channel A reference tap (lower left tap). Disconnect all other hoses to the gauge. Do a baseline as follows:

Press the "Baseline" button once.

Press "Start" once.

Once the numbers on the gauge seem to level off without as much change, press the "Enter" button once.

3. *Two Gauges Available*: Start the blower door and establish the -50 Pa pressure using a different gauge for the blower door. Follow the instructions for blower door testing using the "Set Pressure" (Retrotec) or "cruise control" (EC) settings to keep the pressure at -50 Pascals.

   *One Gauge Available*: If you have only one gauge to conduct the ZPD tests, use the instructions for blower door testing utilizing the manual speed controls to set the fan rpm's so that the gauge reads -50 Pascals. Once you have established the -50 Pascals using the one gauge, record the flow and pressure in the next step and disconnect the tubes going to the blower door and to the outside.

4. Record the blower door flow below. Measure the difference in pressure between the outside and the inside near by reading the number on the display next to "PrA" (Retrotec) or the right side of the display (EC). It should be -50.

   _____CFM50

_____Pa H/O

Proceed to the zone outside you are testing with the gauge.

### Set Up B—Zone-to-Outside Pressure

Connect to a hose a smaller non-conducting rigid hose that has been adapted to fit a hose. Slip the non-conducting hose around the outside edge of an electrical box (after removing the cover) or other penetration through the barrier and into the zone to get a pressure in the zone. If there are no penetrations through the barrier, with the homeowner's approval, create a small hole with as little damage as possible in an area of the barrier where it will be less noticeable (see ZPD section text on ways to reduce the damage when accessing zones). Consider using the smaller non-conducting rigid tube. Slip the non-conducting hose through the small hole in the barrier to get a pressure in the zone.

*Retrotec*—Connect the other end of the zone hose to the green tap on the back of the gauge. Record the pressure below for the zone you are testing after reading the number on the display next to "PrB."

*EC*—Connect the other end of the zone hose to the upper right input tap on the gauge (upper tap of Channel B). Record the pressure below for the zone you are testing after reading the number on the right side of the display (Channel B).

*Both Retrotec and EC gauges:*

_____ Pa Z/O

If this pressure is greater than 25 Pascals, the barrier between the zone to the outside is tighter than the barrier from the house to the zone. The opposite is true if this pressure is less than 25 Pascals.


## INFRARED THERMOMETERS AND THERMOGRAPHY

Infrared (IR) thermometers and IR thermography (cameras) (Figure 4-61) can also be useful tools in diagnosing energy issues in homes. They work in much the same way with a few differences. Infrared cameras of today, for the most part, allow you to "see" the differences in temperature on a wall, etc. all at once and very quickly, while the infrared thermometer (also know as the spot radiometer) gives an average temperature over a smaller area, thus providing less detail and taking more time to cover a given area. It is important to note that the IR spot radiometer actually averages out the temperature over the entire "spot" covered, which is a much wider area than the time laser point that shows on the wall when you pull the trigger. Thus, spot radiometer/thermometer measurement can be extremely misleading if it is not understood that it is averaging a much larger area than the time laser spot, and that you cannot see how large the "spot" covers unless the thermometer is a type that shows the coverage area using multiple laser points. It is very common for the novice to make egregious errors in this regard using a spot "laser pointer" thermometer and this has sometimes led to potentially disastrous consequences. However, you can use an IR camera effectively to scan the interior of the home both before and while a blower door creates a negative pressure in the home. In fact, IR cameras can represent the most effective tool, when combined with a blower door, in finding insulation voids and air leaks in a home.

Taking exterior images with an infrared camera can be done. However, unless you are scanning at night, in the early morning before the sun comes up, on a surface not yet exposed to the sun, or on an extremely cloudy day without rain, there will be too many interferences from the sun and moisture, making it difficult to collect much valuable information from the images. This does not mean you can never take exterior images, but rather, it means you are probably better off primarily relying on interior infrared images. If you take exterior images, they are probably best taken in the morning before the sun comes up so the effects of any remnant solar exposure from the day before are gone.

For interior images, if you use infrared before turning on the blower door, you will more likely be looking at insulation gaps where there are differences in temperatures. If you use infrared while the blower door is on and then compare the thermal image results with the originals taken without the blower door on, the differences will more likely be where leakage sites exist. The thermal differences between the images taken before and after the blower door is on, typically appear as "streaking" variations on the interior wall side of the leakage areas.

**Figure 4-61.**
**An Infrared Camera**

**Figure 4-62. Infrared Images of the Solar Effect** Upper: The sun beating down on the outside of an aluminum sided house to the left of the tree that shades the house, results in the same inside wall of the house being heated where the sun is hitting it on the outside (the lighter shades to the right on the inside wall in the lower image). Clearly, the shading of the tree causes the temperature of the wall on both the outside and inside to be cooler.

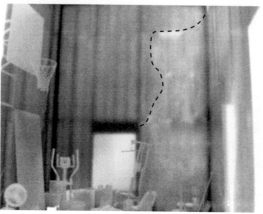

For instance, if you are looking at a wall during the cold weather-heating season, the insulation voids will appear colder in the image than the well-insulated portions. Likewise, the areas where there are leaks will show up as colder in the second set of images (during depressurization with the blower door on). These tests can also be done before and after retrofitting a home to see where the major air sealing and insulation faults are and to see how well the retrofit has been done.

Having a larger temperature differential between the inside and outside of the home, such as when heating the house in the winter or cooling the house in the summer, can help bring out insulation and leakage issues more readily if done during the IR imaging. Many modern infrared cameras are capable of detecting adequate thermal differences on an interior wall with less than a 10° F differential between inside and outside ("delta T" or "ΔT") when doing the insulation check prior to starting the blower door. However, most standards suggest no less than an 18°F differential or to be able to see the studs to do the insulation inspection. These same cameras have been reported to detect adequate thermal differences on interior walls with 3°F or less differential when a blower door protocol was later used. It may be easy to find such a low differential working in a laboratory with the wall in steady state and no solar influence using a good camera (50mk or better). However, it is often difficult to assure such a small differential will reveal itself in the real world where temperatures may vary significantly from one wall, floor, or space to another. Lower temperature differences subject the evaluation to greater risk of the regional influence of capacitance of materials in the wall. This is especially true when the interior temperature is artificially raised or lowered by involving manipulation of the building's HVAC system.

The typical interior infrared protocol begins with the conduction or insulation inspection/evaluation and involves turning the HVAC system off for at least 5 to 15 minutes, doing a quick overview, taking thermal images of the inside face of exterior walls, ceilings, floors etc. Once this insulation inspection has been done, do the air leakage inspection/evaluation by turning on the blower door to roughly 20 Pascals and, within about 30 minutes, taking thermal images of those same interior components. This way you can see if there is any differential or "streaking" when comparing the pre-blower door with the post-blower door thermal images. This protocol would require being able to see the studs in the wall or having a 12°F temperature difference between inside and outside for the insulation inspection. Any air leakage evaluation, whether conducted in conjunction with an insulation evaluation or not, only requires a 3°F temperature differential.

If you do not use a blower door to do the insulation evaluation, you should be able to see the studs in the walls or have an 18°F temperature differential between inside and outside. Using the blower door to conduct an air leakage evaluation after having done an insulation evaluation helps identify the insulation defects. That is the reason for a lower, 12°F differential when adding the blower door/air leakage evaluation to the conduction or insulation evaluation.

If there is not an adequate temperature differential between indoors and outdoors, you can manipulate the HVAC system to increase the temperature differential. Under these circumstances, to do the insu-

 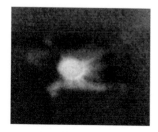

**Figure 4-63. Infra-red Images**

Left—A light fixture before turning on the blower door reveals insulation void as a square shape.

Right—The separate similar fixture in the image on the right does not show as significant an insulation void as it does "streaking" radially out from the fixture. This indicates an air leak around/through the fixture after the blower door has been turned on.

lation evaluation, whether you can achieve the appropriate temperature differential or not, turn on the heat or the air conditioning for about 15 minutes, and then let that heating or cooling "soak in" for about 15 more minutes before you start taking infrared images without the blower door on. Then take a second set of images using the blower door protocol at 20 Pascals to do the air leakage evaluation. By manipulating the HVAC system, infrared testing can be used virtually year-round. A generally accepted test for whether you have adequate temperature differential between interior and exterior on a typically built wood frame wall is whether you can distinguish the framing behind the drywall in the infrared images. If you are not able to identify temperature differences between the framing in the drywall and the stud bays, you probably do not have an adequate temperature differential to do IR imaging.

Some common aberrations you need to be aware of are the temperature differential between 1) the top plate and the middle of the wall, and 2) the mudsill and the middle of the wall. These temperature variations are due to the convective loop that occurs in many walls. Because of the limited opportunity for circulation of air where the ceiling meets the wall and the floor meets the wall, the top plates and mudsills appear to be at significantly different temperatures than the middle of the wall. Top plates and mud sills tend to be higher risk leakage areas. However, the convective loop effect makes it more difficult to identify leakage in these areas unless you can see differential temperature "streaking" or "feathering" emanating from those areas once you turn the blower door fan on. Infrared scanning can also expose faults in the air barrier or insulation in vaulted ceilings, flat ceiling joist cavities, outside walls, basement ceilings, etc. It is not uncommon to identify where insulation has settled in wall stud bays, missing or lower density insulation in stud bays, poorly blown insulation in a sloped ceiling, wind driven infiltration in the ceiling near exterior walls, blower door driven infiltration in dropped soffits, tongue and groove coverings, wall ceiling junctions located just inside of soffit venting, and poorly sealed fan units or canned lights. Many other direct or indirect leakage sites can be found using an infrared camera.

Be aware that infrared can fool you when it comes to low emissivity surfaces such as unpainted metals and highly reflective surfaces. Low emissivity surfaces fool the camera because, by definition, they do not emit IR radiation at the same rate as the typical high emissivity material you are usually viewing in a building. Since you must set the IR camera to a specific higher emissivity, you will be "seeing" surface temperatures on unpainted metals that are lower than their real surface temperatures because not as much IR radiation is being emitted from its surface. Interpretation is made even more difficult if the low emissivity surface is also reflective. Reflective surfaces present a different problem because you cannot "view" the surface temperature under the reflection. Taking an infrared image of a windowpane is more likely to show temperatures that are reflections off the glass rather than the temperature of the glass. One way to find out if you are looking at a reflection or a real temperature on any surface is to move a little and see if a reflection on the surface moves as well. If the temperatures on the surface move as you move, then you are most likely looking at a reflection and not temperature differences on the surface. Unpainted metal surfaces can also be interpreted to be hotter than they really are by reflecting IR radiation from hotter sources. Thus, temperatures indicated on surfaces you know have low emissivity or high reflectance cannot be trusted. One method of overcoming low emissivity and reflectivity is to place a high-grade electrical tape (3M Scotch 33 or 88) on, for instance, the metal or glass surface, to obtain a reliable temperature of the material. Be sure that the area of the infrared camera spot you are using to obtain the temperature in this case does not extend off of the edge of the electrical tape or you may be reading the wrong temperature.

Conditions that can affect an accurate infrared camera evaluation

**Figure 4-64.**
**Example of Reflection**
Upper: If you look closely at the ceiling in the upper picture, you can see a "hot spot" dot on the ceiling.
Lower: When the IR camera was moved up higher in the room, the "hot spot" moved on the ceiling to a location closer to where a duct served the upper floor in this home (the duct runs diagonally from the center top to the right center edge of this picture). Because the temperature changed when the camera was moved, this is a reflection.

include solar loading on the outside that "soaks" through to the inside, wind, moisture, curtain walls and insulation breaks (such as brick fascia, insulated sheathing or vinyl siding).

Many experienced infrared camera operators prefer taking images on a standard gray scale image index or palette, such as these images, using the colder temperatures as the darker gray and the warmer temperatures as the lighter gray even though images can be produced using false color palettes on almost all infrared cameras. With the gray scale palette, in the winter the studs in the wall look darker than the rest of the wall while in the summer the studs look lighter than the wall. Likewise, insulation voids in walls will look darker than the rest of the wall in the winter and lighter than the rest of the wall in the summer. Using the gray scale seems to help most people not only distinguish temperatures on the image but also seems to help with focusing the image. Carefully focusing the image not only makes it easier to see temperature differences, but it also provides greater accuracy in temperature readings. Using grayscale can also help to reduce the risk that colorblindness will affect interpretation of the IR images.

Some advertisements suggest that IR imaging can be used to automatically determine the R-value of a wall. This is generally not accepted by scientists in the field as accurate, since a steady-state R-value on a wall can only be done under unchanging, or steady state, circumstances over many hours of time. Often this test can only be easily reproduced in a lab setting.

At one time, certification in thermography of homes was limited to a protocol established by the American Society for Nondestructive Testing (ASNT). However, this is a more general thermography certification in which buildings are only one of many applications for infrared thermography covered by the established curriculum. RESNET has developed a designation for residential thermography if the applicant is already certified as a RESNET HERS Rater, a type of energy auditor classification. The National Association of Home Inspectors (NAHI), on the other hand, has developed a designation for residential infrared thermography that is available even to those that are not already NAHI members. The NAHI program also allows those that are ASNT Level I Thermographers to be "grandfathered in" to avoid duplicative training requirements.

# Chapter 5

# Weatherization Requirements And Similarities in the Private Arena

## DEFERRAL OF SERVICES

Deferral of services means that conditions exist where you may have to either deny or postpone weatherization services to an otherwise qualified client/homeowner. While this is a weatherization program definition, any of the same principles could be applied in the private arena. Some deferral of service decisions are left to the discretion of the local weatherizer. Under some circumstances, spending DOE weatherization money on certain homes is forbidden. As a rule, even in the private arena, be sure to have a good reason for not following the restrictions of the weatherization program.

The local weatherizer is prohibited from weatherizing a dwelling that:

1)   has already been weatherized;
2)   is vacant;
3)   is scheduled for demolition within 12 months;
4)   has been condemned;
5)   has serious structural problems/costs that outweigh the energy savings opportunities; or
6)   is a mobile home with poor structural supports.
7)   has an owner who refuses weatherization services;

Some of the circumstances that require at least postponement of weatherization services include:

1)    unsanitary conditions or other health and safety concerns;

2)    threatening animals or clients;

3)    the home is being remodeled and the weatherization work is not coordinated with the remodeling effort;

4)    enough weatherization measures are refused by the client that the weatherization work would not be cost-effective or violates state policy;

5)    illegal activities, including but not limited to the cultivation or distribution of illegal drugs, are ongoing at the house; or

6)    other unusual situations like a client who has a respiratory ailment alleviated by the use of the humidifier and the client refuses to discontinue the humidifier use with the result that the humidifier may cause moisture problems in the home.

For weatherization work, be sure to follow proper procedure in deferring or denying weatherization to an otherwise qualified client. Inform them in writing of the reason for the deferral of services, what corrective action must be taken to allow the weatherization work to proceed, and some deadline within which that corrective action must take place. The homeowner has a right to appeal to the local weatherization director's office. The client must be told about this right to appeal in a written letter. It should also be mentioned that most areas have some alternate assistance programs that may be available locally. You can refer a homeowner to that service for assistance to make repairs, etc. on the home. This can be especially productive if you are aware that the cost of incidental repairs would be too large to allow the weatherization to proceed.


ACCEPTABLE MEASURES

In the weatherization program, it is important to know which retrofit measures are acceptable. The list of acceptable measures could be found in any of the following:

1)    the audit software approved for their state,

2)    the state "priorities list" for the area,

3)    the State weatherization guide,

4)    Appendix A,

5)    the Code of Federal Regulations or CFR,

6)    any other related regulations or guidelines,

7)    applicable state and local laws and ordinances that apply to modifying a building, including obtaining a permit and following all the codes relating to the retrofit work on the home.

Appendix A is found in the DOE WAP Rule 10 of the CFR Part 440. The appendix lists the standards that specific products must satisfy to be used as part of a weatherization retrofit measure. For example, if a specific type of caulk satisfies an ASTM standard that is the standard in Appendix A, then Appendix A is satisfied. As another example, in Appendix A, loose fill cellulose insulation to dense pack an outside wall is listed under "Insulation-Organic Fiber" in the insulation category. To use a specific product under this category, it must satisfy ASTM 739-00. If the insulation does not show on the packaging that it satisfies this standard, then it cannot be used in the program.

Some measures that might be considered for energy savings may not even be listed as an option in Appendix A. For instance, Appendix A lists approved measures and standards for replacing refrigerators but not chest freezers. Thus, the cost of replacing chest freezers is not even an option in the WAP program.


SAVINGS TO INVESTMENT RATIO (SIR)
IN THE WEATHERIZATION PROGRAM

Appendix A is not the only limiting factor. To install retrofit measures to help the home become more energy efficient, the retrofit measure must pass a financial test. It must have a savings to investment ratio, or SIR of greater than or equal to one, also known as the benefit/cost ratio or BCR. This ratio is the amount of energy savings over the lifetime of the retrofit measure divided by the cost to install the retrofit measure. Each individual retrofit measure and the package of all the retrofit measures as a whole, must have an SIR greater than or equal to one. The SIR for a measure is typically calculated by entering audit data into a software program that is DOE approved, such as NEAT for regular homes or MHEA for manufactured housing.

As part of the SIR calculation, both present value of money and fuel escalation rate are taken into consideration. For instance, $10 spent today is probably worth more than $10 saved 20 years from now. Integrating present value, inflation adjustments, value of current cash over future savings, and energy cost escalation into this equation can be quite complex. Therefore, SIR calculations are usually performed using a software program.

With regard to the SIR, here is an example of calculations for a refrigerator replacement. First, find out the energy consumption of the existing refrigerator. Either get that by actually measuring it with a wattmeter or find its expected electrical demand from a database of refrigerators. Next, find the energy consumption of the replacement refrigerator. Typically, this is on the EnergyGuide label of the new appliance. To calculate the total cost of replacement, include the price and installation costs of the new unit, including delivery costs. Also include the cost of removal and decommissioning of the old unit, including any fees that the local solid waste/decommissioning facility charges. You also need to know what the local utility rates are so you can calculate the projected energy savings from the difference in electrical use between the old and the new refrigerator. You must also take into account the fuel escalation rate and the present value of money or discount rate by referring to the "Annual Supplement to NIST Handbook 135 and NBS Special Publication 709." If you are doing an approved software-based individual audit on a home or using priority lists instead to determine appropriate SIR, the details of the fuel escalation rate and the present value of money will already be taken into account. Otherwise, without the software it would be very difficult to calculate SIR manually.

If you do not have the software, you can use a DOE approved priority list that has the SIR built-in for a given area with a given climate. Often these priority lists are developed at the state level for use by local installers. They can vary greatly from state to state.

Under the old weatherization system used prior to July 1989, such a big emphasis was placed on doors and windows which have such a poor payback, that the overall energy savings for that period was only 5 to 10% per home with anywhere from a 20 to 50 year payback. Since July 1989 with less emphasis being placed on windows and doors and more being placed on advanced air sealing, high density wall insulation, and heating improvements, the energy savings are often between 17 and 24% and the payback period between 10 and 17 years.

## PAYBACK OF RETROFITS IN THE PRIVATE ARENA

Determining cost-effectiveness in the private arena involves using one of several possible ratios. Divide the total initial cost of a retrofit by the annual savings the retrofit will produce to obtain the "payback" in years. Thus, if it costs $2000 for a retrofit measure and it saves $100 per year in utility bills, it will take 20 years to get back the initial investment. Take the annual savings and divide by the initial costs to obtain the annual return as a percent. Thus, divide $100 per year by the $2000 and retrofit costs to get a 5% annual return.

Lifecycle costing involves comparing the lifecycle cost of conducting the retrofit against the lifecycle cost of not conducting the retrofit. The lifecycle of equipment is the length of time that the equipment is expected to continue to be usable. If the lifecycle costs of the retrofit are less than the lifecycle costs of not conducting the retrofit, then you should conduct the retrofit. As an example, say an older style boiler system is expected to last 20 years and use $1000 per year in fuel for a total of $20,000. Through research, it is found that despite costing $3000 to install, a new type boiler system would only cost $500 per year in utility costs over the next 20 years or a total of $10,000. Thus, if both boiler systems are expected to have a lifetime of at least 20 years, you can save $7000 over the next 20 years by installing a new boiler system whose total costs over the next 20 years will only be $13,000 rather than the $20,000 in utility bills using the old boiler.

Simple return on investment (ROI) analysis involves dividing the difference between projected savings and installed cost, by the installed cost, and then multiplying that number by 100. Thus: Simple ROI= ([projected savings - installed cost]/installed cost) x 100. For example, if a furnace costs $3000 to install, and saves $400 per year over the 15 year life of the furnace for a total of $6000 in savings, the equation would look like this: ([$6000 - $3000]/$3000) x 100 = ($3000/$3000) x 100 = 100%. "Discounted ROI" takes into account the interest rate or time value of money and is most easily taken into account using a software model.

The concept of "interactive savings" recognizes that savings is a moving target: if you tighten up and insulate a home as part of the retrofit process, you make the installation of a new furnace less cost-effective. Why? Because savings from future bills by adding a new furnace will be lower because of the sealing of leaks and the added insulation that must be done at the same time, and vice versa. Thus, adding the new furnace reduces the savings from the added insulation.

## AUTHORIZATION FOR WEATHERIZATION WORK

One last hurdle affecting whether or not a retrofit measure is installed in a home is that the homeowner must authorize the installation. If the client does not want to give up his old refrigerator, then the auditor must decide whether the weatherization should continue or if weatherization services should be deferred to a later date. Any given home can only be weatherized once by the weatherization program, so it is important to make sure to do all possible approved measures in a home to best complete the job. In the private arena, there is more flexibility in letting the client choose retrofit measures. However, you must still follow all the codes related to the retrofit work and typically obtain a permit as well.

One caution when retrofitting homes, either for a weatherization program or for a private individual: most local codes do not require an upgrading to current code for the entire home when the retrofit only involves a small portion of the home. However, if, in the judgment of the inspector and building department, a large enough portion of the home is being changed, the inspector and building department can potentially require the entire home to be brought up to current code, including the parts that are not being retrofitted. Perhaps it is more likely that this will happen with a private client than it will in the weatherization program, but it still represents a cause for concern.

Many of the retrofitting methods described in this book are typical; however, several factors can affect how a retrofit is installed on a particular home. These factors include: the methods used by similar contractors in the area, the local code, the interpretation of the local building inspector, the existing condition of the home and site, and common sense. As a result, the ideas in this book or modifications thereof are only suggestions as to how the work might be completed. At the same time, a good-faith attempt will be made to give the reader an opportunity to distinguish the different levels of installation.

## REPAIRS

Repairs that are necessary for the effective performance or preservation of weatherization materials, often referred to as "incidental repairs," are not exempted from being included in the overall SIR calculation for the entire project. A good example of incidental repairs are making roof repairs or adding framing so that the attic insulation is kept from getting wet. Complete roof replacement is never covered. Fixing a cold-water leak could be considered an incidental repair in a mobile home since it will help keep the insulation dry below the leak. On the other hand, repairing a toilet drain would fall under health and safety. When calculating the SIR for the entire home retrofit package including all the repairs, the cost of incidental repairs goes into the denominator or bottom half of the ratio. Incidental repairs are limited because this makes it more difficult to get an SIR of greater than or equal to 1 if the combined cost of all of the incidental repairs becomes large enough.

## HEALTH AND SAFETY

Two very important principles direct the installation of retrofit measures: cost-effectiveness, and health and safety. Cost-effectiveness was discussed earlier. Health and safety work costs related to energy savings is not included in calculating the SIR. Energy-related health and safety work has no mandated upper limit of cost. Historically however, states set their upper limit for health and safety funds at about 6 to 7% of the total cost of all retrofit measures on the home. Lead safe weatherization or LSW, often an energy-related health and safety effort, has over time driven this percent up to higher values in many states.

Another area of common concern for all retrofitters is employee safety as determined by the Occupational Safety and Health Administration (OSHA). OSHA determines what is safe or unsafe for employees whether specifically written into an OSHA rule or not. Thus, if you have any doubt about whether a situation is safe or not, contact OSHA to be sure.

## PRIORITIES IN THE PRIVATE ARENA

The priorities of retrofit workers in the private arena are up to the homeowner. To determine priorities in the weatherization program, use a DOE approved audit software program or a priority list. While priority lists may vary from state to state, they are all based on the cost effectiveness of a measure with the most cost-effective measures at the top of the list. Unless serious, unsolvable problems prevent the consideration of a measure, skipping measures and going down to the next one is discouraged. Examples of serious, unsolvable problems include things such as the need for a completely new roof to protect the attic insulation or there is a flooded basement caused by poor drainage that would preclude air sealing. It should also be mentioned that even though window and door replacement might be last on the list, if the windows are in such great disrepair that their replacement can be justified as an infiltration measure, then replacement of some windows might be justified.

In the private arena, you can evaluate how far to go in air sealing a home using "air sealing economic limits." First, set a spending limit, say $50 per 100 CFM. If the home has a blower door reading of 3000 CFM50 and a calculated MVR of 1000 CFM50, you can only seal up to 2000 CFM50, the difference between these two numbers. If you divide 2000 by 100 CFM increments, you get 20 100 CFM increments. Since economic limits in this case were set at $50 per 100 CFM, you should only spend $50 times 20 or $1000 on the air sealing work, both materials and labor. The leakier a home is, say above 5000 CFM50, the more cost-effective the initial air sealing work. For just a few dollars, you may be able to seal off large leaks in the house representing hundreds of CFM50s. On tighter homes like those below 1500 CFM, you may decide to not do any air sealing work at all because it is not cost effective.

## ACCEPTED HOME ENERGY DESIGNATIONS

Once a home meets certain minimum qualifications, it may be eligible for one of several energy-related designations:

HERS index-the efficiency of a home is related to a specific standard by modeling software. A score on an index identifies that efficiency level.

Energy Star Program for Homes-this EPA program requires the home to satisfy the International Energy Conservation Code to be eligible for this designation. This designation requires an evaluation by a third-party "rater" or auditor of these homes.

LEED for homes-this US Green Building Council program is a flexible sliding scale approach to home evaluation. It allows for a lower energy savings/upgrade if more environmentally related issues are satisfied with the construction or retrofit, and vice versa.

## NAHB Green

This National Association of Homebuilders program is not just based on energy efficiency, but also on water usage, homeowner education, and indoor environmental quality.

# Chapter 6

# Sealants, Insulation and Barriers and How to Install Them

## RETROFIT MATERIALS

Materials used in retrofitting not only need to be safe to place in a home but also safe to work with. They need to have a useful life that will last as long as the building materials already in place. Various retrofit materials are discussed below.

### Tapes and Weatherstrip

Tapes and weatherstrip are used to seal areas. Duct tape is never used to seal the joints or leaks on ducts. As a general rule, tapes, including UL 181 tapes, are at risk of failing due to the force of gravity and the possible failure of their adhesives. Weatherstrip often provides the best seal in many sites, such as attic hatches, especially if the weatherstrip is permanently attached by an occasional staple, nail, etc.

### Sealants: Caulk and Mastic

Sealants such as caulk or mastic are also used for sealing. Typically, caulk is used for cracks and openings 1/4 inch or less. The three types of caulks most often used include siliconized acrylic latex caulk, pure or 100% silicone caulk, and polyurethane caulk. In the ¼" to ½" range of cracks, apply a backing, such as tape or a foam rod, to span the width before applying caulk.

Pure silicone caulk is typically used to seal concrete, metal, or wood together. Because it cannot be painted, its use is often avoided for

**Table 6-1. Caulks and Applications**

| Caulk | Application |
|---|---|
| Pure Silicone | Seal concrete, metal, or wood together |
| Siliconized Acrylic Latex | Visible areas inside and outside the home |
| Polyurethane | Visible areas inside and outside the home |

exterior applications that are visible. Siliconized acrylic latex caulk can be painted and so it is more frequently used on visible or conspicuous areas both inside and outside the home. However, with acrylic caulk, backing must be used if the gap is greater than 3/16 inch because it shrinks more than pure silicone caulk. Polyurethane caulk can also be painted. These caulks also have properties that allow them to be used as adhesives in a number of applications.

Mastic is critical for appropriately sealing duct leaks and joints as well as other cracks in air barrier materials. Apply one layer of mastic to the duct joint, then the mesh tape, and then a final coat of the mud-like mastic. You can also use pure silicone caulk for sealing duct joints and leaks. Some suggest that polyurethane caulk can substitute for silicone. Be sure the caulk is compatible with the materials, etc.

Be aware that many sealing materials, including caulks, cannot be exposed to low

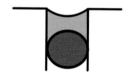

**Figure 6-1. Backing up a wide crack with foam backing helps provide a foundation for caulk as well as way to keep it from "running" before it dries**

or freezing temperatures. In addition, before applying caulk or other adhesives, the surfaces must be appropriately cleaned. Cleaning solvents are used for cleaning metals, removing old paint and wiping down brick and wood. Be sure to check with the manufacturer's instructions on how to prepare surfaces for caulk, including whether or not a primer coat is necessary.

**Rigid Materials and Films**

Rigid materials such as plywood, foam board, aluminum, steel, and rigid plastic sheeting can help span openings greater than 3 inches in width. Rigid materials can also be used as blocking to serve as a top plate on a wall, in between floor joists, and in other large areas.

**Table 6-2. Rigid Materials**

| |
|---|
| Plywood |
| Foam Board |
| Aluminum |
| Steel |
| Rigid Plastic Sheeting |

If the rigid material is too air permeable, air barrier papers or films can be used. Staple the air barrier to the rigid material or the wood around it to seal the air barrier at the edges and seams. Cross-linked woven polyethylene films are examples of this type of air barrier. Polyethylene films and regular cross-linked polyethylene serve as both air and vapor barriers. Be sure to seal the overlaps and edges of the air barrier with the appropriate tape or sealant as specified by the manufacturer. If possible, sealants are preferred because tapes typically lose their adhesiveness over time.


INSULATION

Insulation is categorized as either mineral or organic based. The mineral-based insulations are perlite, vermiculite, fiberglass and rock wool. Of these, perlite and vermiculite insulations are typically no longer installed in homes. They typically look like popcorn because they form little pellets when they expand by being heated. If you are in an older home, you could see other older style insulations such as redwood bark, asbestos and even UFFI (urea formaldehyde foam insulation—a white foam insulation that crumbles to dust when it is touched). UFFI was discontinued in the late 1970s because it contains hazardous formaldehyde. Because it would have been installed so long ago, any home it is in now should not be giving off formaldehyde anymore.

Organic based insulations include cellulose and the foams. Cellulose and fiberglass insulations are the most commonly used for retrofitting homes because they exist as loose fill material that can be easily installed in attics and walls.

### Fiberglass

White, yellow, or pink insulation is usually made of fiberglass. It is not fire resistant because it has an adhesive binder to hold the fibers together. Fiberglass comes in batts, compressed rigid boards, and blankets. Fiberglass batts can be purchased at most home improvement stores. Because it is easy to compress, fiberglass batt insulation is usually considered inferior to blown-in insulation.

Fiberglass batt insulation can usually only be used on walls without drywall or other cover already over them. If fiberglass batts are not correctly installed, there can be significant voids and other weaknesses in the insulation. This is a common problem. Fiberglass batts are typically unfaced—they do not have a Kraft or foil backing to them, although they can be surrounded with an air barrier material from the factory. The facing can serve as an air moisture barrier and allows for fastening at walls and ceilings at the studs, joists, or rafters. When using faced batt, make sure the faced side of the batt faces the warm side of the wall. Thus, in cold weather areas, the facing should face inside while in hot humid areas the facing should face outside. At the same time, there are some areas where facing should not be used. Contractors in cold weather areas often make a mistake with floor joists and install the batts with the facing on the crawlspace or basement side rather than against the underside of the floor. The map in Figure 6-2 shows the locations in the U.S. where the vapor barrier should be placed on the inside of the

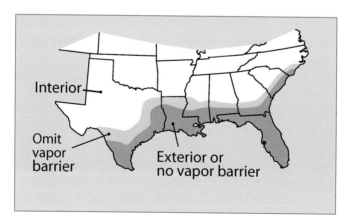

**Figure 6-2. This map shows the dividing line for whether the vapor barrier goes on the inside or outside face of exterior walls, or is left out entirely. All other states not shown have their vapor barriers on the inside of the wall.**

wall, outside of the wall, or where no vapor barrier should be installed.

Often fiberglass batts are compressed into rolls or shrink-wrapped piles so they take up less space in transportation, etc. Before installing batts, unwrap or unroll them long enough to allow them to fully expand. Choosing medium or high-density batts rather than low-density batts can help get more insulation value per inch on a wall or floor space. This typically means using an R-13 or R-15 batt in a 2 x 4 wall rather than an R-11, or using an R-21 instead of an R-19 for 2 x 6 walls. For floors, this might mean using an R-38 rather than an R-30 in cold-weather areas depending upon the width of the floor joists.

Regardless of where insulation is being installed, try to follow some important guidelines. Make sure the air sealing for the wall, floor, or ceiling area has been completed. After allowing the packaged compressed batt insulation to adequately "fluff," first friction fit the batt insulation into place if possible. Be sure that the insulation fits snugly, but uncompressed, against all the surfaces in the stud, floor joists, or rafter space. Avoid compressing, leaving voids, rounding, or bunching batt insulation as this greatly reduces its insulation capacity.

Cut the batts as closely as possible to the exact length of the stud space, otherwise there will be a void if it is too short, or a compression it if it is too long. Cut around large obstacles such as electrical boxes and drainage plumbing pipes. Put thinner scrap pieces behind these obstacles. For smaller obstacles, such as electrical wires and smaller supply plumbing, it is best to attempt to cut through the batt to allow the batt to surround the smaller obstacles rather than allowing the obstacles to compress the insulation. For floor installations, once the insulation is properly in place, hold it in place using wood lath, adequately sized metal push rods/wires, or other pinch-in-place systems.

Strong polyester or polypropylene twine can also be stapled in using a zigzag pattern with no more than a 12- to 18-in spread between twine staples along the length of the floor joist space, depending whether the joists are 24 or 16 inches on center, respectively. Typically, use a minimum 5/8-inch staple made of copper, brass or stainless steel to avoid corrosion that could result from exposure to moisture. You can also use a perforated foam board to support the insulation. In this case, use wood strips over the foam board that are no more than 24 inches apart running perpendicular to the underside of the floor joists, although local codes may require that the foam be completely covered by drywall instead.

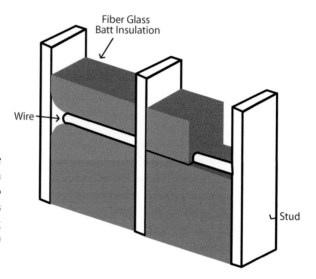

**Figure 6-3. Electric Wire with Batt**
Cut through the side of the insulation (right) in order to keep wires or pipes from compressing the insulation (left) when it is installed.

Another type of batt insulation that was originally designed for metal buildings is manufactured in 2- to 6-ft-wide rolls or batts. A common application in the residential arena is to attach it at the sill plate above the foundation wall in basements or crawlspaces, and drape it down to the floor, as in Figure 6-4. It is not recommended to use fiberglass insulation on the walls of basements or crawl spaces due to the possible buildup of condensation inside the insulation. The condensation is made even worse if the fiberglass has any kind of Kraft or foil fac-

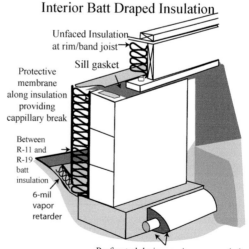

**Figure 6-4. Batt insulation can be hung from the top of the foundation to the floor below in crawlspaces or basements. A protective membrane between the insulation and the foundation wall is critical in keeping moisture out the insulation.**

ing. Foam board or spray foam is the insulation of choice for these areas.

Rock wool or mineral wool batt insulation is used and installed much like fiberglass batt insulation, except that since it has a fire resistance rating, it can make contact with high temperature exposures such as flues and chimneys. Another important advantage of some rock or mineral wool batt insulation is that some types appear to have a "memory." In other words, even if it is compressed when installed, it will decompress soon after being installed. Fiberglass batts are not known for this.

### Blown-in/Dense Pack Insulation

Some reports suggest that blowing in insulation and dense packing closed-in cavities and walls can reduce the overall leakage in the home by 30 to 50%. Simply insulating the walls to a very dense level not only provides a good, appropriate, cost-effective R-value of insulation, it also helps seal leaks in walls that otherwise would be very difficult and cumbersome to seal.

Loose fill insulation is often made of cellulose that has been fire treated. However, this fire treatment does not allow it to be placed against high temperature items, such as flues or chimneys. It is often made from recycled paper and sometimes from wood waste. Blown-in insulation can also be made of fiberglass or rock wool. Loose-fill cellulose insulation is the most commonly used insulation for walls and attics. However, cellulose blown-in insulation tends to be a sponge for moisture in humid conditions. This is the reason fiberglass insulations are often preferred for mobile home installation. In addition, older fiberglass and rock wool fibers irritate the respiratory system and skin more than cellulose. Some of the newer fiberglass insulations are not as irritating.

Cellulose blown-in insulation has enough small fibers that can plug cracks and small holes in the building shell when it is densely packed. Fiberglass or rock wool blown-in insulation requires more inches of insulation than cellulose to get a desired R-value. The rock wool variety will likely be denser and less porous to air than the fiberglass type. In addition, it will take even more inches if it is not "virgin" blown-in glass insulation.

Virgin is blown-in fiberglass insulation that is not recycled from batt material. With recycled blown-in insulation, the fibers are longer and a binder not only thickens them but also keeps them closer together.

**Figure 6-5. This shows an example of how a fill tube being used to fill a wall with cellulose insulation is wrapped with a cloth around it where it goes into the wall to keep the cellulose from blowing back out of the stud cavity.**

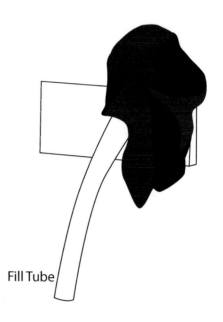

Fill Tube

They also may be somewhat compressed due to the chopping process to convert batts to blown-in insulation. The glass fiber blown-in insulations are more suitable for manufactured homes because of the moisture issues and the fact that manufactured home walls and exteriors are thin enough that they will not allow a dense pack insulation to be installed in them without distorting the wall finish materials. Blown-in fiberglass insulation is more difficult to properly fluff to a higher density than cellulose. Cellulose and fiberglass blown-in insulations have a binder that helps the insulation stick together and hold it up inside open wall cavities (a wall without covering over both sides that allows full access to the stud bays).

Whether it is cellulose, fiberglass or rock wool, the insulation is commonly blown-in with a gas or electric blower machine that uses compressed air to drive the small pieces of insulation through a tube and into the wall. With the blower machines, a bale of insulation is thrown into an agitator that breaks the insulation into smaller pieces so it can blow through the tube lines, see Figure 6-6.

Some of the tools that dense pack installers use include: 1) safety items such as gloves, dust masks, eye protection, and a mister or sprayer to conduct lead safe weatherization; 2) a special

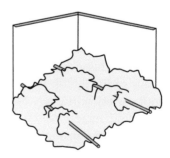

**Figure 6-6. Insulation Slowly Feeding into an Agitator**

siding remover for removing vinyl siding, 3) a special vibrating grind-er/combination saw for cutting and removing aluminum siding, 4) a hammer and pry bar for removing wood siding, 5) a minimum 1/2 inch drive or bit drill gun on a cord along with different drill bits for differ-ent wall drilling applications, 6) a smaller special flexible clear plastic or tygon plastic type fill tube attached to the end of the larger blower machine flex hoses, 7) enough bags of insulation to adequately fill the walls to the proper density, and 8) plugs to seal off the holes (created in the wall sheathing) before replacing the siding.

Proper installation of dense pack insulation includes providing the proper R-value, packing the cellulose insulation to a density of 3.5 to 4.5 pounds per cubic foot, and providing this density uniformly through-out the wall cavity. Steps for installing dense pack insulation include:

1)   make sure you have the proper equipment and make sure the equipment is working correctly

2)   inspect the walls to determine the best strategy for installation,

3)   remove the siding if you are filling from the outside of the wall,

4)   drill the access holes in between each stud bay,

5)   fill the wall cavities with insulation,

6)   put plugs in the access holes and re-cover with siding,

7)   repair or replace damaged materials.

The more the equipment is checked and maintained, the more smoothly the work goes on the jobsite. Make sure the moving parts of the equipment and any belts on the equipment are operating properly. Check the filters on the equipment before every use. Check that the static pressure on the blower machine is at least 2.9 psi at both the take-off and the end of the hose.

Check any air seals in the system to make sure that they are sealing properly. Finally, make sure there is a long enough length of tube for the particular job and that the tube is not leaking.

### Rigid Insulation

Rigid fiberglass or rock wool insulation is typically used in commercial applications. Because it is more durable and less likely

**Figure 6-7.**
**Blower Machine Gauge**

to rot than less dense glass insulations, it can be used as a slab edge or foundation insulation without having to worry if it drains properly. Rigid foam board insulation is made from petrochemicals and typically comes in large sheets. It can have either air or some special gases entrained in it to act as insulators. Unfortunately, over time these special insulator gases typically diffuse out of the foam board, reducing its insulation value. The greatest energy losses in slabs in cold weather areas are on the edge of the slab within four feet of its edge. In non-cold weather areas, the heat loss is primarily just from the edge of the slab.

Before using foam board, check with the local municipality to find out if it needs a cover such as drywall. Other cover materials that can satisfy local requirements include galvanized steel at a minimum of 26-gauge thickness, hardboard, particleboard, spray-on mineral fiber insulation (rock wool) at a minimum of 1 1/2 inches thick, and plywood.

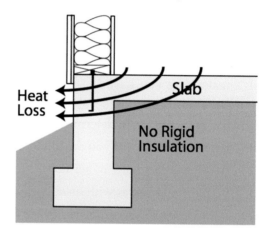

**Figure 6-8. Energy Loses Through the Edge of Slabs**

In the last few years, plastic foam has come pre-applied with sheathing sandwiched on both sides. This can strengthen the foam and the wall. These "SIPS" (structural insulated panel system) (shown in Figure 6-9) can actually serve as the wall where the foam is sandwiched between two sheathing boards. Because the framing studs in SIPS walls often only occur every 4 to 8 feet, thermal bridging caused by framing is significantly reduced. SIPS walls can come in thicknesses of up to 12 inches, making a highly insulated wall.

Foam board is one of the best choices for insulation in crawl spaces and basements because it not only insulates, but also resists moisture and air infiltration. The foam board typically has to be covered with a fire resistant material. Moisture resistant drywall is probably the cover of choice in basements, while plywood or OSB is preferred in crawl-spaces.

Foam board insulation can also be used 1) as a thermal break for thermal bridging caused by framing in outside walls, 2) as support for holding in blown-in insulation in knee walls or floors above basements and crawl spaces, 3) for insulating the inside or outside of foundation walls 4) for insulating slab foundations on the outside face, and 5) on top of cold slab floors before installing a wooden floor "sleeper" system, Figure 6-10.

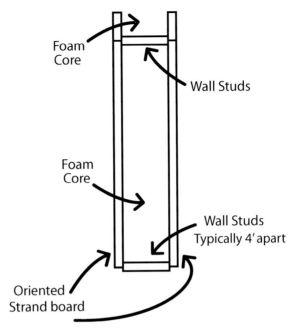

Foam Core

Wall Studs

Foam Core

Wall Studs Typically 4' apart

Oriented Strand board

**Figure 6-9.**
**Structurally Insulated Panel ("SIPS") (Top View)**

To install foam board behind drywall on a concrete wall, metal or wood furring strips typically 1" x 3" are needed. The strips must be spaced to match up with the joists for the drywall, etc. To help avoid corrosion or rot, it is best to place some waterproof material such as "ice shield" between the furring strip and the concrete or block before nailing it. After installing the foam to fit, screw the cover material over the foam with the joints lined up with the furring strips. Some experts suggest using cement board instead of drywall on the lower foot or two of the wall to help mitigate the effects of moisture issues instead. Electrical can even be installed in these walls if the local building official approves a shallower electrical box. Another option is to hold the foam board against the concrete or block by building a frame wall on the interior side of the foam board and using the studs in the wall to press the foam board against the foundation.

Regardless of which technique is used to install foam board, be sure to seal any gaps in the foam board with one-part foam to seal out air and moisture. You should also use an adhesive to preliminarily attach the foam board to the foundation wall and to attach the drywall to the foam board. If you use metal furring strips, a special self-tapping metal screw is used rather than the typical drywall screw to attach the frame to the metal furring strips. The metal furring strips themselves

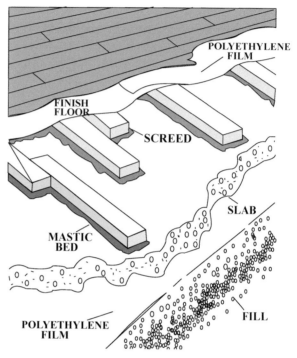

**Figure 6-10. Sleeper Boards for Flooring Over a Slab**
Sometimes the poly film is placed on top of the concrete. If foam board is used, it is typically placed just above the concrete.

can be attached to the wall with concrete screws or by "shooting" concrete nails through the metal furring strips into the foundation with a powder-activated nail gun. If you use the gun, be sure the foundation material is suitable for use with the nail gun because some block, for instance, will break when a nail gun is used.

When you try to seal air leakage gaps larger than 3 inches, a rigid material must be used as backing and the joints are sealed around the rigid material with foam. Unfaced fiberglass board, rigid foam board, cardboard, etc. can serve this purpose very well. You can roughly cut pieces of the rigid material around pipes and support or attach them with drywall screws or other temporary attachment as long as the gaps or joints are filled well enough that a one or two-part foam at the joints will adequately seal the bypass.

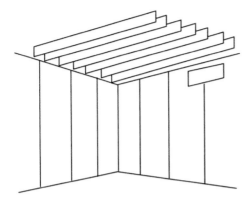

**Figure 6-11. Foam Board Panels Attached to a Basement Concrete Wall**

In some cases, to a limit, large sized chunks of fiberglass can be stuffed into openings of more than 3 inches. A good example would be sealing a wall without a top plate, as in balloon frame construction. Once you have packed the fiberglass, cover it with a thin layer of foam to seal it off. This can be more cost-effective and faster than using a rigid board and foaming around the edges, or spraying foam into the opening around the edges first and then in the middle until it fills without using any backer. Because foam is expensive, it is important to remember to not use any more foam than necessary for air sealing, and no more than needed to match the proper R-value.

**Spray Foam**

Before spray foam became readily available, retrofitters would use cardboard, plastic bags stuffed with fiberglass batts, or plywood to seal bypasses. Cardboard is still used today but typically, any rigid board application involves sealing the bypass rigid board joints and edges with spray foam. Spray

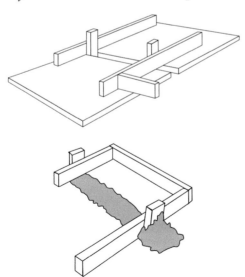

**Figure 6-12. Stuffing Fiberglass and Foaming Over It for a Balloon Frame Wall Opening to the Attic**

foam is commonly used to insulate both horizontal and vertical surfaces such as attic hatches, underneath floors, and knee walls.

Several spray foams can be used in retrofitting. Tripolymer and Icynene are both organic, water-based, injectable foams for closed wall or ceiling cavities. The specialized equipment needed to use these foams is only available from the supplier.

Polyurethane spray foam also requires the use of proprietary equipment along with special training. It is useful for filling open wall or ceiling cavities or open cellar or basement walls. It is also known as liquid plastic foam and comes in different types of kits or canisters. It can be purchased as expanding or non-expanding foam. It is not appropriate to use expanding foams in constricted spaces like door or window jambs. In these cases, expanding foam can distort the jambs and thereby affect the proper operation of the doors and windows.

Spray foam is especially appropriate in areas where there are significant temperature fluctuations, such as unconditioned spaces like basements and attics. Spray foam is available in either one-part or two-part formula.

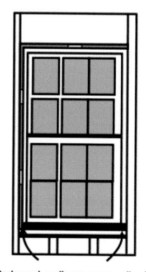

Backer rod, caulk, or nonexpanding foam

**Figure 6-13. Using Non-Expanding Foam Between Windows and Trim**

*One-part Spray Foam (see Figure 6-14)*

One-part foam can be applied directly from a spray can, such as "Great Stuff," or by using spray guns that allow for screw-on cans of foam. Typically, one-part foam is best utilized for filling holes and gaps that are less than 3/4 of an inch. One-part foam can be purchased in many hardware stores; however, the spray patterns are not adjustable. Gun type spray systems, typically only available through commercial suppliers, allow for different thicknesses and spray patterns, and can be used repeatedly if properly maintained.

Advantages of one-part foam include that it is portable and lightweight, that it can provide a very good seal for smaller holes and gaps, and that it is easy and quick to use for the smaller gaps and holes.

Some of the disadvantages of one-part foam include that it is not as cost-effective for larger jobs, it typically does not attach well to surfaces overhead, and it has a longer curing time of typically over two hours. This last problem can be especially important in cold-weather areas in winter if the foam is exposed to temperatures below the manufacturer's stated limits over the two hours it takes for the foam to set. This is probably most likely to be a concern when the foam is applied near the end of the day when the temperature begins to drop.

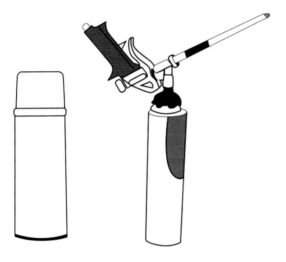

**Figure 6-14. Examples of a One-Part Foam System**
Some spray directly from the can without a gun (although some use a plastic tube from the spray gun tip) and some can be equipped with a gun to provide for better placement of the foam.

*Two-part Spray Foam*

Two-part foam is used to seal gaps that are three quarters of an inch to 3 inches and for insulating large surfaces. Some of the advantages of two-part foam include that it has better adhesion, it cures quickly, it can be applied quickly so labor costs are reduced, and it seals more thoroughly than most other types of insulation. Spending less time in hot or cold cavities of the building by using two-part foam can be a big benefit to the morale of the crew. Using two-part foam can take 1/3 to ¼ the time as compared with one-part foam.

Some of the disadvantages of two-part foam include the fact that it is expensive (it can cost upwards of $500 per kit), it is relatively easy for material to be wasted, it cannot be applied across all temperatures, and more intensive and expensive personal protection equipment is required when it is used. In addition, more significant training is necessary to operate a two-part foam kit than to install foam using a one-part foam system.

Two-part foam is available in small, medium, or large two-tank kits that are portable and can be truck mounted. Two-part foam is more appropriate for areas with numerous air leaks and for larger areas such as crawlspaces, foundations, walls, under floors, attics, or knee walls in attics. In addition, two-part foam is a triple expanding foam. The typical two-part foam process involves mixing two different chemicals from different pressurized tanks at the tip of the spray gun or a frothing chamber. A tube runs from each of the two tanks to a spray gun that can have various sizes of spray tips. The mixture is sprayed into the open air and creates a triple expanding, closed cell, polyurethane foam when it makes contact with a surface. Two-part spray foam equipment (Figure 6-15) can put out large volumes of foam in a short period of time with a hard shell cure and a curing time of under one minute.

If you are learning to apply 2-part form, it can help to practice in the outdoors, where not as much cumbersome safety equipment is necessary. This way you become proficient before you start installing it in homes.

Typically a portable type two-part foam system is intended to be "one-time use only." Guns and hoses are all disposed of once the tank is empty. This is why they are referred to as kits. Typically, the yield on a kit is designated in terms of board feet. There are about 16.5 ft³ of coverage in a 200 board foot kit. The larger 600

**Figure 6-15.
2-Part Foam Tanks**

board foot kits have two tanks each that look like the portable propane tank for a barbecue grill. Depending upon the size of the application, the larger kits may be a better value based on the cost per board foot of spray foam. However, these larger tanks are more difficult to carry and manipulate on the jobsite. Longer hoses may be needed to overcome this issue.

### Disadvantages of Foam

There are some disadvantages to foam. Foams are easily physically damaged and they can be damaged over time by sunlight as well. If they are applied on the exterior, other material must protect them. If they are heated, they can lose their thermal properties and even begin to melt if made hot enough. They also produce toxic gases when they catch fire.

In many regions, foams need to be covered with a fire-resistant material, such as drywall, if they are used inside the house. Likewise, Kraft, paper, tarpaper, flexible plastic, or other backed batt insulations are also flammable and need to be covered much like the foams are. Only foil facings alone are non-flammable.

Furthermore, instead of a fire barrier, a less intensive cover called an ignition barrier must be installed to help avoid ignition of the foam and to cover it on the inside of the home that is not living space, such as crawlspaces or attics. Both the fire barrier and the ignition barrier will typically help avoid ignition from a spark, but only the fire barrier will protect the foam from a fire for an extended period. Some foams do not require an ignition barrier because of their independently tested properties. In addition, some spray coatings and paints claim they can serve as an ignition barrier. Always check with the municipality for more information on whether foam needs to be covered in all areas and what kind of cover is acceptable.

**Safety Issues with Foam**

Whenever chemicals are used on site, safety procedures must be followed including MSDS, proper ventilation, and proper safety equipment. The MSDS is a material safety data sheet that must be at the jobsite. The MSDS will explain precautions to take for the safe use and handling of chemicals such as foam. It also lists first aid and emergency procedures. A well-ventilated area that exhausts fumes and provides cross flow of fresh air must be provided using ventilation fans. Enough ventilation should be provided to keep the chemical levels in the air below the allowed exposure level.

Appropriate personal protection equipment (PPE) (Figure 6-16) must be used including eye protection, rubber or latex gloves, kneepads, a headlamp, a bump cap or hardhat, and coveralls, such as disposable Tyvek suits. Each employee working with two-part foam should be wearing an appropriate personal monitoring device.

Even if chemical levels do not exceed allowed exposure levels, use an appropriate NIOSH-approved air-purifying respirator with a particle filter and an organic vapor sorbent. If the allowed exposure levels are exceeded, all employees should wear a supplied air, positive pressure respirator as recommended by the MSDS.

These positive pressure supplied-air respirators use a hood or mask and have their own air compressor. The air compressor should be located in a fresh air area to assure good air is being provided to each employee. The hoods and masks have removable protective screens on them so if they are scratched or sprayed with foam to the point where visibility is significantly reduced, the protective screen can be replaced. The hoods cost about $50 each while the protective screens are only about 50 cents. Total cost for all equipment per employee can run up to $1000 for two-part foam installation. (Here is a helpful hint--by tucking the hood into your coveralls, it will keep you cooler in hot attics by allowing the cooler

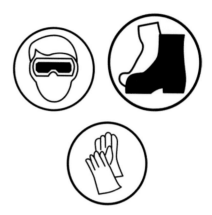

**Figure 6-16. Safety glasses, steel toed shoes, and gloves are some of the basic safety tools that should be used almost all the time**

outside fresh air going into the hood to circulate through the suit).

## FACTORS AFFECTING THERMAL PERFORMANCE OF INSULATION

The true R-value of installed insulation is a function of more than its laboratory tested R-value.

**Table 6-3. Material R-Values**

| Material | R-Value |
|---|---|
| Fiberglass | 3.5 per inch |
| Cellulose | 3.0 per inch |
| Bead Board or expanded polystyrene | 3.6 per inch |
| Styrofoam or extruded polystyrene | 5 per inch |
| Polyisocyanurate Board | 5.6-7.6 per inch |
| Glass | 1 per layer |
| Wood | 1 per inch |
| Concrete | 1 per 8 inches |

For instance, how densely insulation is packed into a wall will affect its R-value. Figure 6-17 shows that most types of insulation have an optimal density. In other words, if the insulation is packed to a density that is lower or higher than its optimal density, you are not getting the best R-value out of the insulation.

In fact, a low density of insulation, as well as the extreme case of low-density caused by voids or openings in the insulation, can be a cause for convection within the wall. "Wind washing" of air through the insulation is also more likely to occur if there are leaks in the air barrier, or if there are other exposures of the insulation to the wind. A void of only 7% in the surface area of the insulation can result in the R-value of the insulation being reduced to half of what it would be if there were no voids, even when there are no leaks in the air barrier. Air leakage results in even greater losses of R-value (see wall R-values in Appendix G).

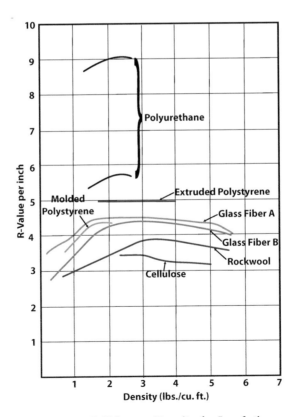

**Figure 6-17. R-Value vs. Density for Insulation**

Because water conducts heat much better than air, if the insulation has absorbed moisture, it will also lose the benefit of some of its R-value. High humidity air leaking into the wall cavity, whether from inside or outside, can result in condensation forming within the insulation, thus reducing its effectiveness. Even if there are no air leaks into a wall cavity, moisture can still pass through permeable materials, such as brick or gypsum board, and enter the insulation. This is referred to as vapor diffusion. (See vapor barrier section)

Thermal bridging, as seen in Figure 6-18, also reduces a wall's R-value. The "clear wall" R-value is the average R-value of insulation between the wood framing. The "whole wall" R-value takes into account the insulating characteristics of the framing as well. This is also known as the area weighted R-value or weighted average wall R-value. Both of these R-values assume "steady state" or constant temperature conditions and do not take into account changes in temperature on the inside or outside.

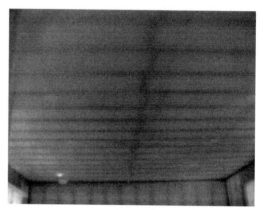

Figure 6-18. An example of thermal bridging through ceiling joists on the top floor of a home as seen using infrared imaging.

Table 6-4 is a chart of whole wall R-values (See the thermal bridging and calculating envelope energy loss sections).

Because of the lag time of heat transmission through particularly dense or massive walls such as concrete, brick, earth, solid wood, logs, etc., the traditional steady state R-value will not adequately

**Table 6-4. Whole Wall R-Values**

| Wall Type | Whole Wall R-Value |
|---|---|
| 2 x 6 perfect installation | 12.8 |
| 2 x 6 poor installation | 11.0 |
| Insulating Concrete Foam | 26-44 |
| Lightweight Concrete Block | 10-30 |
| Standard 2 x 4 | 9.7 |
| Steel frame wall C-Stud | 5.6 |
| Steel stud wall with EPS sheathing | 10.5 |
| Structural 6 inch EPS insulated panel | 21.6 |
| Stucco covered straw bale | 16-28 |

describe the true R-value of the wall over a day or a season. The so-called "thermal mass effect" can determine the true dynamic energy performance of a wall in a dynamic or changing temperature situation. This is called the wall's dynamic energy performance. In cooling-dominated climates, the dense materials in these massive walls absorb heat slowly throughout the day, and then give that heat off again during the night. Thus, the home is not as hot during the day as it is outside because of the "mass factor." For some thermal mass-insulation configurations, the mass of the wall essentially slows the heat transmission through the wall during the day. Ventilating this home with outside air at night keeps the interior of the home from becoming too hot as heat dissipates from the high mass outside walls. In other situations, the heat stored in the building envelope components might be released during the night to the outside. This mass factor lag effect improves the dynamic performance of the wall over and above what it might be otherwise. With optimum circumstances, the increased dynamic thermal performance can be over 200% of what it might otherwise be for a building.

Determining the dynamic energy performance of a building is too complicated to describe here. However, some of the factors that affect dynamic energy performance include:

1.   Climate—the hotter climates benefit more from the mass effect.

2.   Steady state R-value of the wall—a low steady state R-value mass wall in a cold weather area can cause an increase in energy usage.

3.   Building size—Size of the building also has an effect. For instance, the thermal barrier surface area to building volume ratio determines how much of an effect the mass can have on the building.

4.   Orientation of the building—how much solar exposure the building gets will affect how much heat and how much benefit the mass wall will have on the dynamic energy performance.

5.   The configuration of building components in the wall—typically placing the insulation towards the outside of the building relative to the massive portions of the wall offers more energy savings than placing insulation on the inside of the masonry, etc.

Another example of using the benefit of mass in increasing dynamic energy performance of the thermal barrier of buildings that allows for even greater ability to control the temperature involves the use of phase change materials (PCMs). These not only have great mass but also allow for absorbing more heat through the material changing phase from the solid to the liquid phase. While these could also be easily integrated into walls, PCMs have also more recently been tested as attic thermal barriers where heat is an even more significant issue than in walls. An example of such a material that could be used as a PCM heat sink would be a wax or paraffin that has been developed to melt at a specific temperature but is contained in a sealed box-like container.

Needless to say, due to its complexity, accurate analysis of dynamic energy performance of a given home or building requires the use of a computer program that is not readily used. Thus, this type of analysis is not typical for auditors to take into account until the software becomes commonly available to the public for auditors to use in evaluating a home for dynamic energy performance.

## RADON

Radon can be a potential problem in any home regardless of its location. If you seal off the air leaks in a home as part of a weatherization or retrofit effort, you can run the risk of increasing the radon level in the home after it is finished. By no longer allowing the air inside the home to be naturally diluted by the air leaks that have been sealed off, a higher concentration of radon can occur in the house than existed before the retrofit effort.

It is therefore recommended that a radon test be conducted at least after, if not also before, weatherizing a home. If it is found that the radon level exceeds the recommended 4 Pico Curies per liter limit, radon mitigation efforts should be completed to reduce the level below the limit. The potential cost of any radon mitigation effort, identified before air sealing has been completed, may keep a home from being a candidate for weatherization.

## VAPOR BARRIERS

Vapor barriers, unlike air barriers, serve to hold back water vapor from getting into the insulation of walls, etc. One of the most common vapor barriers currently used is polyethylene sheeting stapled to the

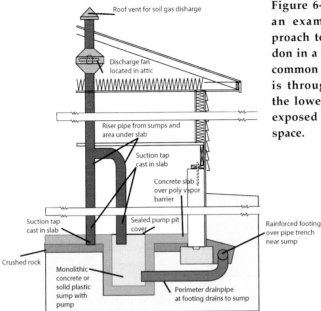

Roof vent for soil gas discharge

Discharge fan located in attic

Riser pipe from sumps and area under slab

Suction tap cast in slab

Concrete slab over poly vapor barrier

Suction tap cast in slab

Sealed pump pit cover

Crushed rock

Monolithic concrete or solid plastic sump with pump

Perimeter drainpipe at footing drains to sump

Rainforced footing over pipe trench near sump

Figure 6-19. This shows an example of an approach to mitigating radon in a home. The most common source of radon is through the slabs of the lowest floor, such as exposed dirt in a crawl-space.

studs inside the home after the insulation is installed, but before the drywall is installed. In the past, the facing on batt insulation typically served as the vapor barrier with the facing stapled over the edges of the studs on each side of a wall cavity. The facing could be Kraft backed, foil backed, soft plastic (vinyl), tarpaper, etc. Even an oil-based primer painted on the inside of the house or vinyl wallpaper on the drywall can serve as a vapor barrier.

Remember that cold air holds less moisture and so anytime air is cooled, there is the potential for condensation

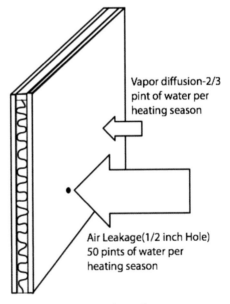

Vapor diffusion-2/3 pint of water per heating season

Air Leakage(1/2 inch Hole) 50 pints of water per heating season

Figure 6-20. Significantly more water vapor travels through a wall by air leakage than by diffusion.

water droplets forming in the cooled area. In a cold weather area, the vapor barrier is more inclined to be placed on the inside wall of the home. This prevents the relatively warmer, moister air inside the home from leaking into the wall where it would cool down and deposit water droplets inside the wall and insulation. On the other hand, in a warm weather area where air conditioning is used extensively in the summer, the vapor barrier is usually placed on the outside wall of the home. This prevents the relatively warmer, moister air outside the home from leaking into the wall where it would cool down and deposit water droplets inside the wall. The closer the air gets to the inside of the house, the colder it becomes if the inside of the house is being air-conditioned. In some areas, it is recommended that no vapor barriers be installed (see map in fiberglass section).

# Chapter 7

# Auditing, Planning, and Retrofitting

Even though auditing, planning and retrofitting appear to be separate topics, they are best discussed in combination with each other because they are closely related. Auditors cannot communicate to retrofitters what work is needed to complete a retrofit if they are not aware of how to properly retrofit. Likewise, retrofitters will not understand the reason for the auditor asking for specific retrofit work to be done if they do not understand auditing. Furthermore, retrofitters need to do much of the same testing as auditors when they finish their work at the end of each day and after all the retrofit work is done. Specific information on how to conduct tests mentioned here, such as blower door and duct leakage tests, can be found in Chapter 4.

## WORK FLOW SEQUENCE

This section reviews a typical workflow sequence in a weatherization project. Generally, the following sequence could be considered a good guideline for the stages of evaluating and weatherizing a home, but it is only one possible sequence. In no event should any sequence, including this one, take precedence over any indoor air quality, health and safety, or other related environmental or life safety issue (see the appropriate sections for more details on each of these steps in the sequence). It should also be mentioned that software-based energy models are commonly used in conducting audits. Information inputted into these programs typically includes construction materials and type, the square footage and/or volume of the home, the kind or type of mechanicals used in the home to heat and cool (including heating water), the leakiness or infiltration rate of a home as determined by a blower door test, and the climate of the area where the house is located. Even carefully designed models will have limitations due to the inability to predict occupant behavior, errors in inputs, and features of the home that go unaccounted for in the software. "Truing up" can help. This is the process of comparing the model results with the actual energy bills from the home being audited.

1) Conduct an indoor air quality and health and safety inspection of the home including conducting a worst-case draft test. Use your personal CO monitor as you walk around the home to evaluate for carbon monoxide levels. Look for fuel leaks. Evaluate the condition and clearances of combustion appliance chimneys and vents, and evaluate exhaust and ventilation in the home.

2) Interview the homeowner to learn more about how the house is used, what it costs to heat and cool the house, and where there might be issues with thermal comfort in the home.

3) Evaluate the house for its construction and framing type.

4) Evaluate the home to locate the primary zones and where thermal boundaries exist.

5) Conduct an infrared scan of the home to find insulation gaps.

6) Conduct an additional infrared scan of the home with a blower door set to -20 Pascals to find air leakage sites.

7) Conduct a blower door test at -50 Pascals.

8) Conduct basic zone pressure tests with all the primary zones in the house (attic, basement/crawl, garage, attached porch).

9) Use the pressure ratio charts available to evaluate for relative leakiness of different surfaces in the home next to intermediate zones.

10) Basic ZPD tests on secondary zones such as chases, soffits, floor/wall cavities, cantilevers, outside porch ceilings, etc.

11) Evaluate zones for interconnectedness with other zones – how pressure in other zones is affected by opening a door or hatch in a primary zone with blower door on.

12) Do a pressure pan test for duct leakage. Be especially aware of which ducts are in unconditioned zones and therefore potentially outside of the pressure or thermal boundaries.

13) Conduct advanced zone pressure diagnostics on garages that are attached to the house, and on crawlspaces that have mold or moisture issues.

14) Conduct advanced zone pressure diagnostics on attics in colder climates particularly if the house is a relatively "tight" home with the potential for high moisture.

15) Identify leaks in the pressure boundary using smoke or visual cues.

16) Confirm which zones are "outside" of the house and which zones are "inside" of the house.

17) Evaluate whether the thermal boundary is adjacent to or lines up with the pressure boundary.

18) Evaluate the best strategy for air sealing and insulating knowing where the thermal and pressure boundaries are located.

19) Explain to the client what weatherization procedures you are recommending in terms of what areas will remain or become comfortable because they will be treated as inside of the house and which areas will remain or become uncomfortable because they will be treated as outside of the house.

20) Install retrofit measures for all indoor air quality and health and safety issues.

21) Complete the duct sealing and insulation.

22) Air seal and insulate the boundaries of the home.

23) Conduct a second blower door test to evaluate effectiveness, or to test out at the end of the workday.

24) Recheck the zone pressure diagnostic readings to see if they change in the expected direction (outside zones are closer to 50 Pascals WRT the house, inside zones are closer to zero Pascals WRT the house). Confirm that pressure pan testing of ductwork located outside the house resulted in a reading of 1.0 Pascals or less.

25) Conduct additional advanced zone pressure diagnostics to evaluate what leakage remains in the primary zones that were previously measured.

26) Evaluate whether pressure relief is necessary for rooms that are significantly pressurized or depressurized during furnace or fan operation.

27) Conduct a final infrared scan without the blower door on (insulation check), and then immediately after turning on the fan to -50 Pascals (air leakage check).

28) Conduct the final blower door test and retest for pressure pan readings.

29) Use a checklist to confirm that all health and safety issues have been addressed and resolved, if necessary.

30) Educate the client on issues relating to operating the retrofitted home, including explaining which behaviors can increase or decrease energy usage in the home.


## AUDITING AND RETROFITTING

### Operating a Business

If you are providing services outside of the government-related weatherization arena (for example, in the private arena), then you must become familiar with how to adequately operate a business. Some concepts of business principles are discussed in the New Construction Chapter in this book. Furthermore, more extensive information is available on running an energy-related business at www.bpi.org/quality-management.html, the BPI website.

### Communicating with Clients

When you audit or retrofit a home, you interact with clients on a very personal level. You could be working with them for days. You need to be a detective when it comes to how best to weatherize the home, but you also need to be a diplomat in working with the client. You may have questions for the client, or you may have things you want to teach the client, like how to replace a furnace filter. Here are some tips for communicating with clients.

Remember that actions can speak louder than words. Some studies suggest that as much as 55% of communication is through body language used while speaking. Even the tone of a voice could represent as much as 30% of the communication with the client. This is without changing any of the words that are spoken. Thus, the words themselves may represent only 7% of communication (Table 7-1).

**Table 7-1. Communication**

| Form of Communication | Percent of Communication |
|---|---|
| Body Language | 55% |
| Tone of Voice | 30% |
| Spoken Words | 7% |

Be very respectful of your clients and their property. Their homes may not be much, but it may be all that they have. Be sure to preserve the privacy of your client as well by making sure that you knock before entering a room. Also, never handle any of your clients' personal belongings. Be sure not to discuss anything relating to politics or religion and never use profanity.

Be aware of the comfort zone of your client. If your clients seem uncomfortable or continuously back away when you approach them, give them a little more space. There are wide variations between cultures with regard to what a typical comfort zone involves. Understanding is an important component of communication. Avoid the use of technical jargon. Educate your clients about maintenance or other items, etc. They are far more likely to remember and appreciate the lesson if you tell them just how it benefits them.

Culture, language, and disability can be barriers to effective communication. Culture may dictate who it is you should speak with in the home, so be flexible. If you do not have a common language with your client, consider trying to have a relative or neighbor help translate. Be patient with those with disabilities, such as poor sight or impaired hearing, and be mindful of their needs.

## Interviewing

The first step in assessing the building is to meet with and interview the client. In some cases, you may need

a translator to help communicate with the client. Show understanding and respect and be honest in your dealings. Use the client interview process to learn important information to help evaluate the home for health and safety, energy efficiency and comfort. Find out how many people live in the home to better understand what the minimum ventilation requirements will be. Find out where the hot or cold rooms are in the home to help focus on potential problem spots. Also find out the typical thermostat settings, whether the fireplace or any unvented space heaters are used, and what the fuel bills are. If you notice that the client uses window shades to help control solar heat gain, then you know you may have an easier time educating the client about possible retrofit measures.

Ask the client questions about health and safety such as whether any of the occupants have asthma and whether headaches seem to be common in the heating season. If headaches are common in the winter, test for carbon monoxide right away.

| **Winter Headaches = Test for CO Immediately** |

If an occupant has asthma, there should be a note on the work order for special precautions if dense pack sidewall insulation is specified.

Spend some time explaining to your client what is involved in the audit process and, once the audit is complete, discuss the options for retrofitting the home. Explain about assessing the home for heating and cooling, base load analysis, air leakage testing, and some of the conspicuous existing conditions you notice in the home that may affect health and safety or integrity of the building. Perhaps above all, stop to explain how the client's life and the home will be improved through this retrofit process.

EXTERIOR

Once the interview is completed, begin the physical assessment of the home by conducting an exterior walk-around inspection. Some start this process at the entry door and consistently move in a clockwise direction. As you go along, take pictures of everything that is worth noting and then ultimately return to the starting point at the entry door.

Determine the framing type of the home—if it is a framed home— such as whether it is a platform, balloon, or plank style of construction. Learn to distinguish platform framing from balloon framing. Since about the 1940s, platform framing has involved the use of top plates on every level and building the floor before the walls. Balloon framing allows open access between levels down the outside walls of the home. If, in the attic, the gable end walls or other outside walls are open to the framing of the outside walls in the floors below (without a top plate), then it is probably balloon frame construction.

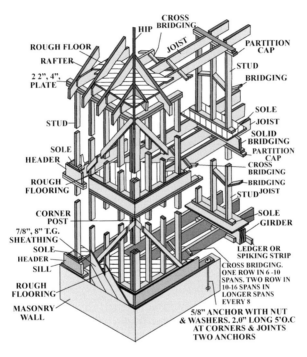

**Figure 7-1. Western Framing**

Specifically, while you are outside, note the following:

1) the types, numbers, orientation, and condition of all windows and doors,
2) the siding condition and type,
3) the roof covering type and condition,
4) adequate height on chimneys,
5) any apparent porches, additions, cantilevers, or tuck under or attached garages, etc. (see ZPD section),

6) exhaust fans, chim-
neys, and possible
safety problems,

7) water management
issues such as the
slope to grade around
the home (including
the driveway), flash-
ing, and the pres-
ence and condition
of downspouts. Re-
member that mois-
ture problems in
basements often are
traced to poor drain-
age on the outside of
the house.

8) the degree of peeling
paint on the outside
walls—this could in-
dicate high moisture
on the inside,

9) the presence of gut-
ters,

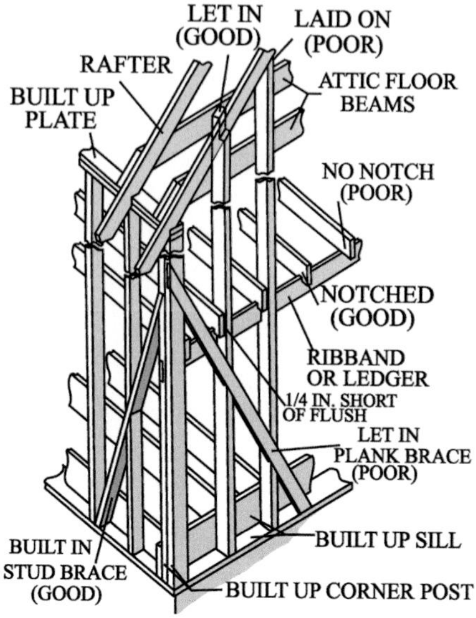

**Figure 7-2. Balloon Framing**
The important difference between balloon
faming and western or platform framing is
in the fact that often some of the exterior
and interior walls on balloon frame homes
can communicate with the crawlspace, basement, or attic.

10) whether there are storm doors
11) where passive attic ventilation or bath fans could be installed and
12) whether there are auto-closing hinges on the house-to-garage door.

Draw footprint and side view or elevation sketches of the building along with dimensions. Either sketch these elevations showing the
window and door locations, or take a picture of each wall of the home.
Document where magnetic and/or true north is relative to the home's
position, determine the percent above grade for the foundation and the
foundation type, and document the framing and other major characteristics of the home. On the footprint sketch, provide the exterior dimensions, differentiate the unheated from the heated sections of the house,
and make a note of anything else that may seem of interest.

It is important to take accurate notes during the visual inspection. The elevation sketches such as Figure 7-3, should include the di-

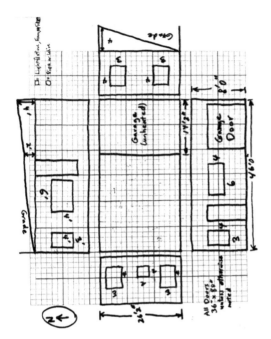

**Figure 7-3. Sample Sketch**

An example of a footprint and elevation sketches on the sheet. Keeping proper proportions is not as important as getting the dimensions right.

mensions of doors and windows and record the height of the foundation that is exposed so that basement wall insulation can be specified.

Record the width and height of each wall of the heated areas in the home so you can later estimate the wall insulation needed above the foundation level. Also, note the siding type so the retrofitters will know what tools to bring with them to adequately insulate the outside walls. This will also tell them, for instance, whether there is asbestos siding and whether the insulation must be fed from the inside of the home. Specify any replacements or repairs, and estimate the amount of insulation materials required.

EVALUATING INTERIORS

When evaluating the inside conditioned space of the home there are a number of things to check for including:

1) hazards such as electrical problems,

2) possible presence of paint that is lead based,

3) unvented space heaters whether fixed or portable,

4) whether any mechanical ventilation is operational,

5) sources and evidence of moisture,

6) possible indoor air quality problems,

7) the building components comprising the thermal and pressure boundaries based on the client's use of the interior of the home,

8) both the volume and area of the conditioned space,

9) significant air leaks such as those due to missing covers over access ports such as attic accesses, large penetrations, damaged wall or ceiling coverings, and broken glass.

10) the presence of wall insulation,

11) air leakage around fixtures and outlets,

12) the operation and condition of doors and windows,

13) whether the fireplace has a glass door and if the damper is operable,

14) the thermostat setting and type,

15) the location and number of furnace registers or radiators,

16) recessed light fixtures,

17) leaks around window air conditioner units and whether they are covered,

18) evidence of excessive moisture as indicated by peeling wallpaper,

19) mold on walls and ceilings or condensation on any building surfaces,

20) air leakage problems above open dropped or "T" ceilings—open a few to evaluate,

21) air leakage bypasses that might exist around plumbing and behind access doors, and

22) unvented space heaters.

SPACE HEATERS

**Figure 7-4.**
**Portable Space Heater**

In the weatherization program, there is a special policy on space heaters, such as in Figure 7-4, referred to as WPN 08-4. The weatherization program prohibits any space heaters from having an input rating in excess of 40,000 Btu per hour and prohibits them from being located in sleeping rooms, boiler rooms, bathrooms, or storage closets.

Unvented liquid or gas-filled space heaters are prohibited as a primary heat source. Weatherization funds can be used to replace a primary unvented space heater with a permanent, vented, code compliant system. Funds cannot be used to replace secondary systems. Thus, no weatherization work can proceed unless the client has agreed to allow his unvented space heater to be replaced by a permanent vented heating system. Minimum ventilation requirements must be calculated and satisfied with the assumption that a space heater will be used if it is in fact being appropriately used.

A special rule that prohibits all unvented space heaters applies to mobile homes. Only vented, fuel burning, heat-producing appliances can exist in mobile homes. BPI standards do not allow the use of any unvented space heaters in any home.

## REDUCING MOISTURE IN THE HOME

If your initial audit of the home reveals that moisture in the home is a significant problem, you can recommend any of the following:

1) Educate the homeowner. If you notice a clothesline by the washer or wood drying in the home, let your client know that these activities increase moisture problems. Tell your client to close windows and doors during warm, humid periods. If possible, remove materials that are susceptible to moisture issues.

2) Make repairs. Repairing and clearing downspouts and gutters typically solves some of the minor drainage issues. Repairing an existing sump pump can be more useful in removing condensation than adding mechanical dehumidification.

3) Close the vents of the crawlspace to the exterior and convert the crawlspace to a conditioned area to prevent condensation.

4) Place a vapor retarder on the ground in the crawlspace, such as in Figure 7-5. This will bring the surface temperatures in the crawlspace above the dew point.

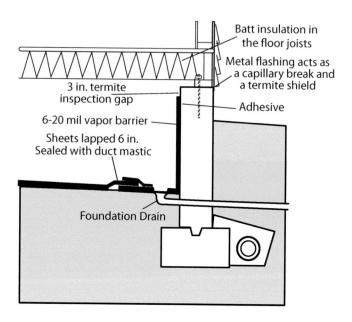

Batt insulation in
the floor joists

Metal flashing acts as
a capillary break and
a termite shield

3 in. termite
inspection gap

Adhesive

6-20 mil vapor barrier

Sheets lapped 6 in.
Sealed with duct mastic

Foundation Drain

**Figure 7-5. This demonstrates the placement of a vapor barrier on the ground of a crawlspace with overlapping and sealing of the barrier at joints and edges. The foundation drain is often considered optional and often drains to the French drain next to the footing.**

5) If the problem is severe, you can excavate around the outside perimeter of the basement and install a drain tile to collect water. This solution is extremely expensive, usually only done as a last resort in the private arena and is beyond the scope of the weatherization program.

6) Install a dehumidifier to remove moisture. Dehumidifiers pull the humidity from the air using a refrigeration process to cool the air (cold air holds less moisture than warm air). Then the dehumidifier heats the air, which by definition reduces its relative humidity. Since air conditioning cools air and dehumidifiers heat air, these should not be operated simultaneously for efficiency reasons.

7) Install fans to circulate air throughout the house. Unfortunately, using fans to circulate air, and mechanical dehumidification will only help with light to moderate condensation.

If your state has no specific requirement for fans, then you can use either ASHRAE 62.2 of 62.1. If ASHRAE 62.2 is properly followed, moisture problems can be practically eliminated while keeping air transported heat loss to a minimum. Relatively speaking, the air change rate for ASHRAE 62.2 is roughly 50% of that required by ASHRAE 62.1, which will reduce the client's heat cost significantly. Find a way to relate fan operation to high interior humidity but only when exterior humidity is low. The best option is probably client education. One of the best indicators of excessive moisture in a home is frost build up or fogging on windows. It is best to instruct the client to run the fans until the moisture on the windows goes away.

## AIR SEALING

Air sealing activities should always be the number one priority. Building envelopes or air barriers in buildings serve to keep the conditioned air (heated in the winter, cooled in the summer), from escaping outside and thereby losing the benefit of the energy paid to heat or cool a building. Air leakage through gaps or openings in the envelope is typically a much bigger culprit for utility costs and energy usage than poorly installed insulation. For instance, a 7% insulation gap cuts the R-value of an attic to 50% of its design. However, an increase of only one air change per hour of air leakage through the envelope on a tight building can increase utility costs and energy use up to 300% (when there is a 40 degree differential in temperature between outside and inside). The size of the hole in terms of Effective Leakage Area (ELA) is only 0.01% of the envelope, as compared to the 7% gap in insulation, even though the costs of heating are three times greater under these conditions. Thus, a gap 700 times smaller causes a loss of energy three times greater. This amounts to an overall effect that is 2100 times greater when comparing the effects of insulation void sizes to leakage site sizes.

---

**Insulation Void x 2100 = Air Leakage Opening**
Comparison of Insulation Gap Area with Air Leakage Area

---

Air leakage is clearly far more important than insulation and this is yet another reason for making air sealing a priority. (See Appendix D for calculation of ELA required to get 1 ACH per hour in a sample home). Some typical leakage sites include attic accesses, kitchen and bathroom soffits, windows and doors, flue and chimney chases, outlets and switches when there are bypasses in the top or bottom plates of the wall, light fixtures, top story sub floor leaks when knee walls are present, poor foundation to framing seals, etc.

Studies have determined where the primary air infiltration sites are in homes, see Table 7-2. The suggestion is that floors, walls, and ceilings represent 36%, while fireplaces represent 16%, plumbing penetrations 15%, doors 13%, windows 12%, fans and vents 5%, and electrical outlets 2%. While this study did not separate out attics, it was noted that attics are the largest piece of the "floors, walls, and ceiling" category, probably making it the largest single category of all.

**Table 7-2. Air Infiltration Sources**

| Category | Percent |
|----------|---------|
| Floors/Walls/Ceilings | 36 |
| Fireplaces | 16 |
| Plumbing Penetrations | 15 |
| Doors | 13 |
| Windows | 12 |
| Fans/Vents | 5 |
| Electrical Outlets | 3 |

In retrofitting a home, technology stresses blocking the "high holes" to the attic that exist in the envelope, but economics stress to seal "low holes" that exist through the basement walls to the exterior because it is so inexpensive and effective. The point is to stop air-driven heat loss since that is typically where air-sealing measures are concentrated.

Direct leakage occurs at direct openings from inside to outside while indirect leakage describes air entering at one location, moving through building cavities, and then exiting at a different location. Some examples of direct leakage include doors and windows, dryer vents, and any other penetrations in the building envelope. Examples of indirect leakage include attic hatches, plumbing chase ways, electrical outlets, interior walls in a balloon frame home with no top plates, where a porch roof joins the side of the house, etc.

There are several general approaches to air sealing a home. One is simply blower door directed air sealing. In this case, auditors or install-

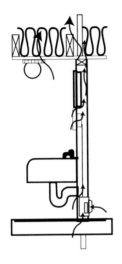

**Figure 7-6. This shows how leaky many homes are and how air can travel all the way from a basement or crawlspace up into the attic space.**

ers use a smoke stick or other smoke source to locate and seal air leaks with the blower door running. They can also narrow down which rooms are exposed to the leakiest parts of the shell by turning on the blower door and 1) opening each room door one at a time about an inch and trying to estimate which rooms have more airflow going through them, 2) opening each room door at a time (including attic, basement or crawlspace accesses, etc.) and seeing how much the blower door CFM increases, or 3) have all the room, etc. doors open at once, close them one at a time, and use a gauge to read the differences in pressure between the room and the hall, etc. The greater the increase in flow or the bigger the difference in pressure, the more significant the leaks are in the shell for that room.

Another approach to air sealing uses a checklist for where typical leakage sites are and then air sealing the leakage sites. Some typical hotspots for air leakage include flues and plumbing vents, wire pathways, recessed fixtures (including both canned lights and fans), and chimney penetrations. Yet another approach is to use an infrared camera in combination with a blower door at -20 Pascals to find leakage spots (see infrared section).

Once the air sealing work is completed, a blower door test is conducted to make sure there is a significant airflow reduction compared with pre-retrofit work and to make sure you do not need to add ventilation to satisfy a minimum ventilation rate (MVR). Another method is to combine the use of an IR camera and a blower door to find the leakage.

WALLS

**Inspect**

Exterior walls tend to be much less leaky than attics or basements. In most homes, the most significant leaks in the exterior wall tend to be around penetrations through the wall such as where gas or water pipes enter, or where electrical wiring, telephone, or cable utilities enter into the building. The other areas that tend to be leaky are where there are joints in the wall, such as corners of the building, cantilevers, porch roofs, bay windows, additions, and other offsets. Be sure to check all around these areas.

Knowing whether any insulation exists in the walls is helpful in determining whether insulation should be added. Probe the walls in the house to evaluate for the presence, type, and depth of wall insulation and to determine the depth of the wall cavity. To do this, remove a receptacle cover plate on an exterior wall and stick a non-conducting plastic crochet hook or another type of non-conducting hook behind the drywall and pull out some insulation. You can also push the probe in the gap all the way to the backside of the interior of the wall to determine the depth of the wall. Be sure to subtract out for the thickness of the drywall, lath and plaster, etc. Use this method to check several locations along different exterior walls in the home. If additions have been made to the home, be sure to take samples in at least two or three locations in each of the additions.

The presence of insulation in the wall can also be determined by using infrared thermography (see IR section) or by removing some of the siding.

If you plan to blow in insulation from the exterior, check to make sure there are no openings, damage, or weaknesses in the interior walls such as holes in the walls, inside faces only covered with weak paneling, cardboard, etc. Look for areas where insulation might spill when it is installed. These areas might include suspended ceilings, closet spaces, cabinets, and pocket doors, under sinks, etc. Check for openings around drainage and supply plumbing underneath sinks on exterior walls. Check for interior or exterior wall materials or repairs that

Figure 7-7. This example of voids in insulation in a wall approximates a Class II wall under the RESNET Standards.

were made with substandard or weaker materials than a typical wall, such as cardboard, thin wood paneling, and loose plaster. Check for drywall that has not been nailed into the framing adequately. Check for good sealing around windows and trim on the outside before insulating the walls. If some exterior stud cavities are used as return supply ducts, you will need either to reroute these ducts or avoid insulating these wall spaces, at great energy loss.

**Wall Strategy**

After inspecting the walls, determine the best strategy for insulating. For instance, decide whether dense pack should be blown in from the interior or exterior of the wall. With exterior walls of vinyl siding, insulation can be blown in from the exterior. This will involve removing siding, drilling access holes, etc. With exterior walls of asbestos siding, stucco, veneer brick, or other masonry type siding, the only option is to blow in dense pack insulation from the interior of the home. This will be messier and require more cleanup than an exterior filling.

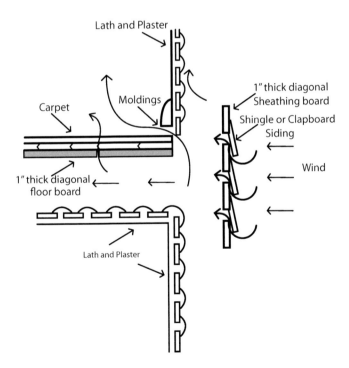

**Figure 7-8. Balloon frame homes allow heavy leakage due to their communication of the inside with the leaky outside walls of the home.**

The type of framing will also influence retrofit strategies. Balloon frame homes are notoriously leaky, see Figure 7-8. If the house is a balloon frame home (no top plates in the exterior or interior walls), it may be easier to blow in insulation from the top attic space openings into the stud spaces, rather than removing siding and drilling holes. If you take this approach, make sure there are sill plates in the exterior walls. If there are no bottom sill plates, plan to tightly stuff batt fiberglass or flexible foam insulation at the bottom of the stud spaces to keep the dense pack insulation from simply flowing out of the wall next to the bottom of the stud space into the basement or crawlspace when it is filled.

You may not have to drill holes at the wall tops or bottoms to install wall insulation in some balloon framed homes. However, you may have to drill holes in the wall for each floor joist bay in the second floor to blow insulation into the bag in between the floor joist bays to plug them off once you reach the middle floor level with wall insulation. If you do not, you are wasting your time blowing insulation into a middle floor.

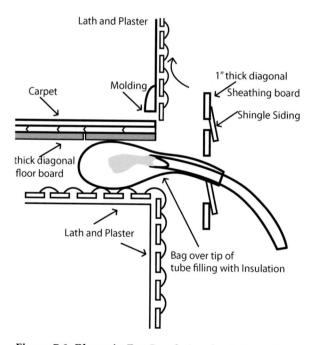

**Figure 7-9. Blown in Bag Insulation for Balloon Frame**

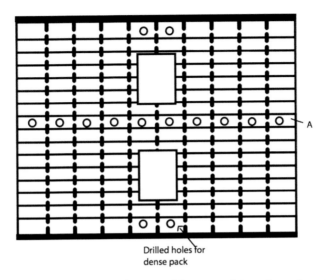

Drilled holes for
dense pack

**Figure 7-10. Balloon frame with top and bottom plate using a long fill tube. These holes (A) must be drilled to be able to plug the space between floor joists with bags filled with insulation.**

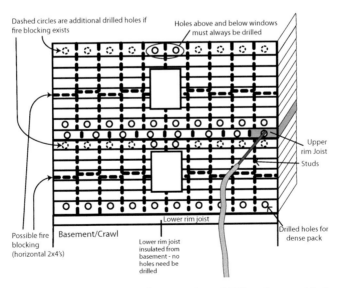

Dashed circles are additional drilled holes if fire blocking exists

Holes above and below windows must always be drilled

Upper rim Joist

Studs

Possible fire blocking (horizontal 2x4's)

Basement/Crawl

Lower rim joist

Lower rim joist insulated from basement - no holes need be drilled

Drilled holes for dense pack

**Figure 7-11. Platform frame often requires drilling the most holes.**

If there are penetrations through the top and bottom (sill) plates (such as pipes or electrical wiring), seal them off with fire-resistive spray foam, as in Figure 7-12.

Also, apply caulk at the joints between the top plates and the exposed drywall (from the attic side). This is a common place for air leakage into the attic. Air can enter into the attic from the basement, crawlspace, around electrical boxes, through penetrations at the base of the wall, etc. This space opens up due to natural gaps that occur from unsealed drywall at installation as well as from shrinkage in the wood after installation.

Concrete block walls are very difficult to insulate. An extreme measure would be to install a framed wall on the inside. A less costly option to reduce air leakage through concrete block would be to coat the inside wall of the concrete block with a primer and two layers of latex paint. Dense packing the inside cores of concrete block is another possibility to reduce air leakage, but often the opening is blocked by a top plate that probably should not be compromised.

Older homes that have "back plastered" walls can also be a problem, Figure 7-14. These walls have an additional layer of lath and plaster on the inside face of exterior walls and therefore, are not usually easily adequately insulated.

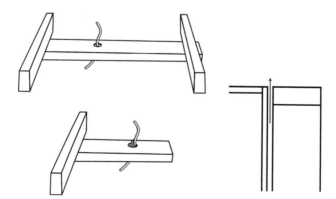

**Figure 7-12. Foam Added Around Electric Wire Through Top Plate.** (Lower Diagram). Be sure it is fire resistant foam.

**Figure 7-13. Drywall/Top Plate Joint Leakage Can Be Significant**

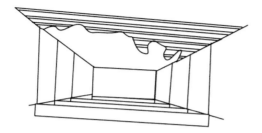

**Figure 7-14. An example of a back plastered wall looking down into the wall cavity between studs from above. Note that the interior of the wall has had plaster applied to it**

## Calculate

Calculate how many bags of insulation will be used if the insulation is properly packed to 3.5 pounds per cubic foot density for cellulose. Not only does this help ensure there is enough insulation to finish the job, but it also allows for a bag count at the end of the job to make sure enough bags were used to provide the proper density. Calculations are shown in Appendix I.

## Blowing into the Wall from Exterior
### Removing Siding

Before installing dense pack insulation in walls on the outside, first place plastic on the ground around the house to make cleanup easier. If you are installing insulation from the exterior, remove the siding. The vinyl siding courses can be detached from each other using a wire hook made from clothes hanger wire. Bend the other end around back on itself and tape it to make handle. Once a final course edge is unhooked from another, prop up the vinyl siding away from the house. If you are concerned about damaging the vinyl siding, pull the nails loose with a nail pry bar on the course below to remove a complete course of siding. You can also use special vinyl "zip" tools to separate different levels of siding without cutting it. Aluminum siding can be removed with a special zip tool, but often the aluminum is dented in the process. As a result, some installers will make a cut along the bottom outside corner of an entire course of aluminum siding with an oscillating saw (Figure 7-15) to remove the siding without inflicting as much damage.

Be careful not to damage siding, it can be difficult to match new siding with old if it needs to be replaced. If you are re-nailing aluminum or vinyl siding, "float" the nails so they do not press the siding against the wall. Because of the high expansion rates of siding, if you tighten the siding against the wall, it can ripple or bend.

Use a pry bar to remove wood siding by carefully lifting the siding off the wall and then pulling it free—Figure 7-16.

Otherwise remove the wood siding just as you would wood shingles by cutting along the bottom edge of a siding course or shingle above, carefully prying the siding loose from the wall, and carefully sealing this joint with an appropriate caulk when reinstalling.

*Drilling Holes*

Once the siding has been removed, drill access holes for the insertion of the blower tube in the underlying sheathing in the space between the studs on every stud bay in the home. Typically, you want to drill these access holes from either the top or bottom of the stud bay. Most

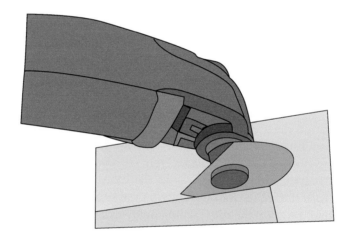

**Figure 7-15. Oscillating Saw**

**Figure 7-16. An example of using a pry bar to pull a segment of wood siding far enough away from the wall to snip the nails with a nipper.**

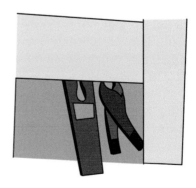

retrofitters prefer to drill the hole at the bottom of the stud bay so lower floor stud bays can be accessed from the ground level without having to use a ladder. If you are working in a two-story home, you have to drill separate holes for each story because 1) the floor of each level interrupts the stud bays from being continuously open on platform framing and 2) you need to place insulation filled bags in between floor joists between floors. Make the access holes slightly larger than the vinyl fill tube used to feed in the insulation so the tube can be moved around to adequately insulate inside the wall. Making the hole too large will allow too much cellulose insulation to spill out of the hole. Whether drilling on the interior or exterior of a pre-1978 home, appropriate lead safe practices for the area must be followed if you find lead in the paint.

Set the "stop" on the drill so that the bit does not penetrate into the stud bay space any farther than necessary as this may damage pipes or wires inside the wall. The stop is simply an adjustable rod that is anchored to the side of the drill that hits the wall and stops the drill from going in any farther than it has been set. Some prefer to drill the hole at an angle. If you are working from the bottom of the wall, you might drill the hole upwards into the sheathing to allow for easier access for the fill tube. On two or more story balloon frame construction, you may also have to drill a hole right at the floor level between stories. This is to create a "plug" just behind the wall at the floor cavity to keep the floor from becoming a bottomless pit for blown-in insulation.

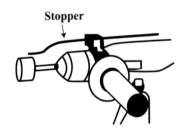

**Stopper**

**Figure 7-17. Drill Gun Stop**

Once the hole is drilled, probe to the left and right with a hanger wire to identify where the neighboring studs are in the wall cavity. Probe upwards and downwards with the end of the vinyl fill tube to determine if there are any fire-rated blocking boards (blockers) or diagonal wall supports. The fill tube should not be any shorter than to allow it to reach at least within 1 to 2 feet of the longest length of the stud spaces, or the space cannot be dense packed.

If the fill tube is more than 2 feet shy of reaching the end of any of the cavities, drill another hole in that space to adequately reach the distant areas of the cavity, or get a longer vinyl fill tube. If you find blockers or diagonal wall supports that would prevent you from completely filling a stud space, drill another hole for the area on the other side of the blocker or wall support to make sure there are no gaps in the wall insulation. If you decide to drill in the middle of the wall, be sure to fill the top portion first with insulation, and then the bottom.

*Blowing Insulation into Walls*

Before starting to insulate, you may want to select a few fully open, unobstructed wall stud cavities to adjust the air/material mixture on the blower machine, to make sure the vinyl fill tube is long enough, and to make sure the insulation has the proper density. Most workers use a towel to hold their hand around the fill tube where the fill tube enters the hole. This keeps the insulation from blowing back out of the hole during the blow-in process. Now everything is prepared and you

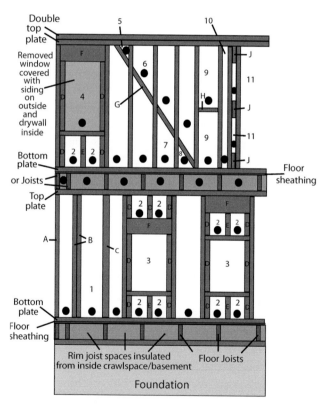

**Figure 7-18a. This diagram shows possible hole drilling locations based on where the framing is, including in some unusual situations.**

can start insulating. As you are filling an area with cellulose, when the blower machine sounds like it is becoming sluggish or slowing down, pull the tube out one foot at a time. With a faster blower, you may be able to pull it out two feet at a time. The blower machine starts to slow down when it has packed insulation about as tightly as it can. If the tube is not pulled back soon enough, the tube will plug and you will have to stop the machine and clear the tube before work can begin again. Clogs typically occur where the tube changes diameter, such as where a smaller vinyl fill tube is attached to the end of a larger flex hose. If you are clearing more clogs than blowing insulation, adjust the blower machine to reduce the cellulose feed or increase the amount of air by opening up the air feed.

In addition, some workers will mark numbered lines in one foot increments on the fill tube, as shown in Figure 7-18b, so they can see how far they are into the wall and when they are getting close to finishing filling each stud bay. When a stud bay is filled, shut off the blower machine with the remote switch but leave the tip of the tube in the hole until there is no longer any pressure in the hose. It is packed tightly enough if you are unable to push your finger through the dense pack cellulose insulation at the hole, as shown in Figure 7-19.

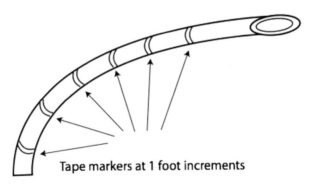

Tape markers at 1 foot increments

**Figure 7-18b. Fill Tube**

**Figure 7-19. Checking For Adequate Density with Dense Pack Insulation.** If you cannot push your hand into the insulation, it is packed tightly enough.

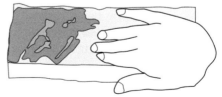

Another technique to check for density is to bend a heavy wire or utility flag into a Z shape, place it in the insulation so the middle section is perpendicular to the wall, and then see if the wire can be turned. If the wire can turn around the inside of the dense pack so that it disturbs the insulation, then the cellulose is not dense packed well enough.

Typically, an 8-ft stud space cavity can be filled in 2 to 4 minutes. If it takes longer than that, it is time to stop and reevaluate. Look for places inside the house where the cellulose may have been going or overflowing. You many want to have an assistant watching from the room side of the wall to make sure the insulation is not overflowing or damaging the wall. If you are pumping insulation, for example into a middle floor space, and too much insulation is being used, try the following. Cover the end of the fill tube with a plastic bag, hold onto the bag opening around the tube, stick the fill tube into the floor cavity through the hole, and after the bulk of the bag is inside the wall or floor cavity, blow insulation into the bag. If this works, you can keep from pumping excessive amounts of insulation into these floor areas.

After blowing in the insulation, plug the holes with wooden pre-cut tapered plugs (Figure 7-20) available from the cellulose supplier, repair any damage, and then reinstall the siding before doing any touchup painting, etc. Be sure the house is put back together so that it is almost as if nobody had been there.

### Blowing into Wall from the Interior

If you are blowing in insulation from the interior of the home, the process is much the same except that there will not be any siding to remove from the wall before drilling the holes through the drywall or lath and plaster. If you suspect possible asbestos on the drywall or drywall mud, you may not be able to blow in insulation from the interior. When you are finished, special foam plugs are available to finish over the drywall holes. If the holes have been drilled in the lower part of the wall that is more exposed to being hit by feet, etc., you can use special sheet metal drywall

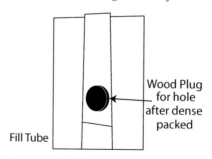

Figure 7-20. Wood plugs are typically used to fill drilled wall holes after blow in insulation is complete.

patches covered with drywall tape netting to help the drywall mud adhere. This can allow you to patch these holes quickly and more sturdily than with a foam plug. You can also try to patch the holes with an actual drywall plug, but this is likely to require the most work and not necessarily provide any additional benefit.

### Thermal Bridging in Walls

When a material has a significantly lower R-value than the insulation, it is referred to as a thermal bridge. An example is studs in a wood frame wall as shown in Figure 7-21. Metal studs create an even greater thermal bridge because they conduct heat and cold far more easily than wood. In addition, these studs can corrode inside the wall as a result of condensation that can form on the studs.

A thermal break is a lower conductivity material that helps reduce the effect of thermal bridging in a wall. For instance, you can create a thermal break over an entire wall by adding an intact layer of foam insulation on the entire outside wall of the house. Another way to reduce thermal bridging from framing in a house being newly built is to reduce the amount of framing. One example is using "ladder" framing instead of double studs as backing where interior walls intersect exterior walls. Other examples include placing studs 24 inches on center instead of the usual 16 inches, using two-stud corners instead of three-stud corners, and packing foam or other insulation into headers where there is a gap between the built-up joists. Different methods of making thermal breaks are shown in Figures 7-22, 7-23, and 7-24. Be sure that the reductions in framing are allowed by the local municipality as this typically reduces the structural strength of the wall. A "double wall" construction can also be framed in

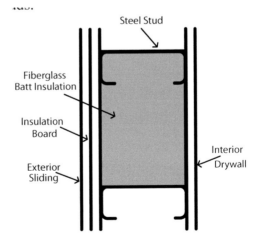

**Figure 7-21. Steel Studs in a Wall (top view)**
Steel studs represent a special thermal bridge concern because of their high thermal conductivity

the wall. This makes it about 50% or more wider than a regular wall by alternating studs or putting in two completely separate walls to serve the interior and exterior faces of the wall (without the sides of any stud touching both faces). Double walls are very expensive. Of course, changing the framing in an existing home is not practical.

**Pocket Doors**

Pocket doors are doors that slide into wall cavities. The older styles that are very leaky can be found in some older homes with balloon frame construction. If the pocket doors are on the story just below the attic, you need to do good air sealing above the pocket door in the attic if the interior wall containing the pocket door does not have top plates. In addition, any pocket door slot that abuts an exterior wall needs to have good air sealing between it and the exterior wall even if the pocket door is on the lower level.

Pocket doors that slide into a wall that is perpendicular to an exterior wall can wreak havoc when it comes to installing the exterior sidewall insulation.

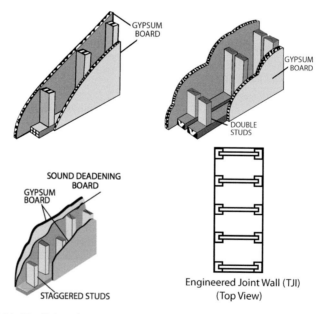

**Figure 7-22. Traditional construction (top left) has more thermal bridge effect than various configurations that are used to reduce the thermal bridge effect of studs in a wall.**

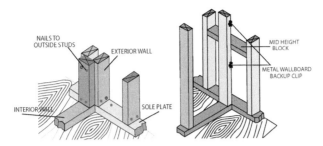

**Figure 7-23. Blocking (right) can break up the more traditional thermal bridging (left) caused by closely connected (or touching) studs where an interior wall ties into an exterior wall.**

**Figure 7-24. The special 2-stud corner (far right diagram) allows for a lessened thermal bridge effect because there are only 2 studs in a corner instead of the usual 3 (left). The middle diagram shows an approach that reduces some thermal bridging over the traditional 3-stud corner on the far left.**

In the older balloon framed homes, as shown in Figure 7-25, a perpendicular interior wall can openly communicate with an exterior wall and insulation could simply end up filling in the interior wall where the pocket door is stored, blocking the pocket door from sliding into the wall.

To insulate around pocket doors that abut an exterior wall, first remove the door completely from the doorway and set it aside. Pinch or pack fiberglass batt or flexible foam insulation in between the walls on each side of the door back behind where the door slides into the wall. Since this is often a long narrow opening inside the wall, using a broomstick or other long narrow tool can help in maneuvering this fiberglass backing material into place. Once the fiberglass backing material is properly placed in the wall, dense pack the exterior walls as usual. As you are dense packing, check to see how well the stuffed fiberglass backing holds back the dense pack insulation. If it looks like the back-

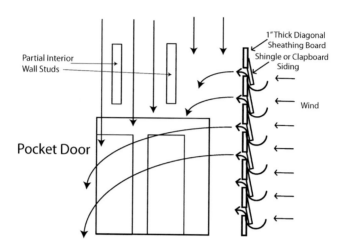

**Figure 7-25. Pocket Doors on Balloon Frame Construction**
Air easily leaks in from the outside when pockets doors meet the outside walls.

ing properly serves its purpose, you may only need to vacuum up a few errant pieces of insulation that may be lying on the floor inside the pocket door before reinstalling the pocket door. Make sure the pocket door operates properly after installing and that it can be pushed all the way back inside the wall as it did before it was insulated.

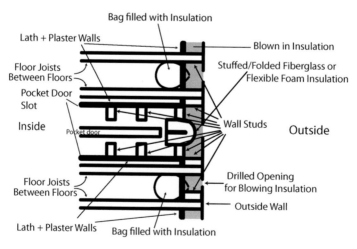

**Figure 7-26. This is a top view of the pocket door air leakage issue and its treatment.**

If the pocket door is located on the floor just below the attic, additional air leakage easily occurs from the attic. Before insulating the attic above the pocket door, stuff unbacked fiberglass insulation or nail in foam board that is sealed around the edges all along the top opening to the wall above the pocket door. This keeps the attic insulation from falling down into the pocket door from above. It is probably easier to stuff from the attic side, although you can stuff from the pocket door slot. In any balloon frame construction home without top plates on interior or exterior walls, stuff unbacked fiberglass insulation or flexible foam insulation into the top of all these walls to seal off the walls as a top plate normally would. In addition, if there are no bottom plates in the walls, similarly stuff insulation into the wall at the bottom from the crawlspace or basement.

## ATTICS

### Inspection

Because the attic seems to be the place where most of the conditioned air is lost, some of the most important retrofit work is needed there. Attic bypasses allow loss of heat and conditioned air from the house. Bypasses can also allow warm moist air from the house to get into the attic. This moisture can condense on the insulation and deteriorate its R-value. Therefore, be sure to thoroughly inspect the attic. If the attic is not accessible, try accessing it through an existing exterior vent or even consider asking the client's permission to create an interior access in a closet ceiling or a gable end of the home.

Some of the typical locations for air leakage bypasses into the attic include: around attic hatches, dropped soffits, the tops of walls with missing top plates or wiring and plumbing penetrations, combustion appliance flues, recessed lights, attic doors and windows, and attic sill plates. It is helpful for you to draw a simple sketch of the attic space noting specific spots or bypasses to seal and to include this sketch in the work order. Figure 7-27 shows such a sketch.

Because it is difficult to find air leaks by randomly walking around the attic, it is best to check for leaks in a systematic way. Start by inspecting the upper floor below the attic to identify where all interior walls and partitions are including closets, halls, and stairwells. Note any changes in ceiling height, such as dropped soffits that might ex-

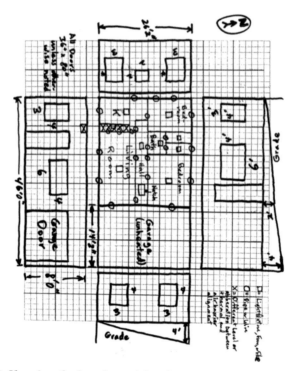

**Figure 7-27. Showing the location, with a diagram, of bypasses in the attic that need to be sealed can help installers properly follow through with the work auditor's request.**

ist above kitchen cabinets, built-ins, etc., or lower ceilings in kitchens, bathrooms, or closets. If there are lower ceilings or dropped soffits in the upper level of the home with an attic above it, this is probably a substantial source of air leakage. A change in the ceiling height typically represents a change in the support plane for the insulation above. Note the location of light fixtures mounted in the ceiling, especially if they are recessed lights that are not located in the center of the room.

You can also use a blower door to identify leaks by pressurizing the home to 50 Pascal pressure and then perusing the attic area using your hand for the feel of air coming into the attic. Be sure to take proper precautions like putting out any fire in the fireplace, etc. You can also use an infrared camera to identify leaks from inside the home when you are creating a negative pressure in the home. If it is warmer outside, the leaks coming into the house will look warmer than the walls and vice versa (see IR section). You can also check using smoke to see if it is drawn into bypasses when depressurizing the home if you check the attic side of the ceiling.

When you are in the attic, make a note of

1) the location and components of the thermal boundary or insulation,

2) whether the attic has a floor,

3) whether the ceiling seems to be in good condition,

4) whether occupant belongings are present, Virtually all storage in an attic space must be removed before retrofit work can begin. Be sure to let the client know this in advance of the retrofit.

5) the type of attic hatch and whether it is well sealed,

6) the current attic insulation levels,

7) the major air bypasses between the house and the attic. Such bypasses can be around chimney or combustion appliance flues or vents, changes in attic floor elevation, knee walls, wire penetrations through top plates, and recessed fixtures that protrude into the attic area such as fans and canned lights. One indicator of air leakage is discolored or dirty insulation because the insulation filters air passing through the leak in the air barrier,

8) the presence and degree of or absence of attic ventilation,

9) electrical problems including junction boxes that are uncovered, wire that is frayed, wire connections and splices that are not confined to a junction box, knob and tube wiring, if there is a voltage drop, and junction boxes without covers.

10) any duct system's condition and insulation level,

11) plumbing and mechanical terminations and whether they exit to the outside,

12) any evidence of roof leaks (such as in the flashing) or condensation-based moisture issues,

13) any other hazards such as vermin droppings, vermiculite insulation, possible asbestos insulation, etc.

14) whether the ductwork is properly connected and sealed with mastic at the joints (Figures 7-28 and 7-29),

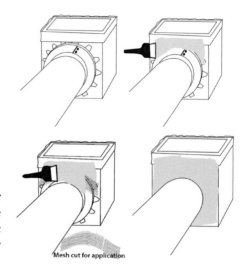

**Figure 7-28. Placing mastic over duct leaks may require the use of a mesh to help the mastic keep its sealing properties over time.**

Mesh cut for application

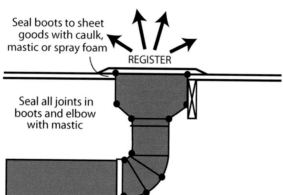

Seal boots to sheet goods with caulk, mastic or spray foam

REGISTER

Seal all joints in boots and elbow with mastic

**Figure 7-29. Many joints must be sealed to better assure that the ducts are not leaking significantly. Mastic is the best material to use for duct joint leaks.**

15) whether brick chimneys are in contact with the framing or are coated with tar, and

16) which attic ducts are exhaust fan ducts. Check for this by turning on all the exhaust fans in the home before surveying the attic. This also helps identify which fans may not be operating properly. Make a note to replace inoperable or poorly performing fans before air sealing the attic because you need to seal all

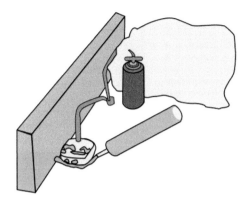

Figure 7-30. Sealing around fans, light fixture boxes, canned lights and other penetrations in ceilings with foam or caulk is critical for avoiding many problems that arise from air leakage to the attic space. However, caulk or foam should not be injected into the inside of the junction or fixture boxes, etc.

around the exhaust fans from the attic side before insulating. Sealing around a fan that is not working is wasted time if you have to replace it later.

### How Much Insulation?

Tables 7-3 and 7-4 show cost-effective insulation levels depending on the climate zone where the home is located and heating fuel and equipment. To determine the climate zone for an area, see the map in Appendix J that shows the eight different DOE climate zones. Notice that cost-effective insulation levels for attics range from R-30 to R-60. The savings from cooling equipment is not as great as from heating equipment, as evidenced by the lower insulation levels required in the warm zones 1, 2 and 3. This is because the $\Delta T$ is typically lower in cooling climates. Thus, the southern regions of the country have lower recommended R-values for insulation. If there is less insulation than recommended, then the cooling and heating equipment has to work harder, and, as a result, the energy bills may be higher. However, putting in more insulation than is recommended will extend the pay back time in the cost-effectiveness evaluation. It is even possible that with extensive insulation over and above the recommendations, that the energy saved from the additional insulation would never equal the cost of installing the insulation.

For an attic with walkable flooring, the depth of the ceiling joist cavities restricts the amount of insulation that can be placed in the attic unless the storage area is abandoned and the floorboards removed. Having floorboards means restricting the depth of insulation to only 6 to 10 inches, depending upon the width of the ceiling/floor joists. With

### Table 7-3. R-Values for Gas, Heat Pump and Fuel Oil

| Zone | 1 | 2 | 3 | 4 | 5 | 6 | 7 | 8 |
|---|---|---|---|---|---|---|---|---|
| Attic | R30 - R49 | R30 - R60 | R30 - R60 | R30-R60 | R30-R60 | R49 - R60 | R49 - R60 | R49 - R60 |
| Cathedral Ceilings | R22 - R38 | R22 - R38 | R22 - R38 | R30-R38 | R30-R38 | R30 - R60 | R30 - R60 | R30 - R60 |
| Wall Cavity | R13 - R15 | R13 - R15 | R13 - R15 | R13-R15 | R13-R15 | R13 - R21 | R13 - R21 | R13 - R21 |
| Wall Insulation / Sheathing | No | No | No | R2.5 -R6 | R2.5 -R6 | R5-R6 | R5-R6 | R5-R6 |
| Floor | R13 | R13 | R25 | R25-R30 | R25-R30 | R25 - R30 | R25 - R30 | R25 - R30 |

### Table 7-4. R-Values with Electric Furnace

| Zone | 1 | 2 | 3 | 4 | 5 | 6 | 7 | 8 |
|---|---|---|---|---|---|---|---|---|
| Attic | R30-R49 | R30-R60 | R30-R60 | R38-R60 | R30-R60 | R49-R60 | R49-R60 | R49-R60 |
| Cathedral Ceilings | R22-R38 | R22-R38 | R22-R38 | R30-R38 | R30-R60 | R30-R60 | R30-R60 | R30-R60 |
| Wall Cavity | R13-R15 | R13-R15 | R13-R15 | R13-R15 | R13-R21 | R13-R21 | R13-R21 | R13-R21 |
| Wall Insulation/ Sheathing | No | R2.5-R5 | R2.5-R5 | R5-R6 | R5-R6 | R5-R6 | R5-R6 | R5-R6 |
| Floor | R19-R25 | R25 | R25 | R25-R30 | R25-R30 | R25-R30 | R25-R30 | R25-R30 |

this limitation, specify dense pack insulation at the proper density. For instance, if cellulose is the chosen material, dense packing should occur at 3.5 pounds per cubic foot. You should also suggest that bypasses that exist under the attic floor, such as electrical and plumbing bypasses into interior walls should first be sealed before dense packing under the attic floorboards. (See "Attic Floors" section for more details).

## The Law of Diminishing Returns

The Law of Diminishing Returns should be used when evaluating how much insulation to place in a home. This law says that as more insulation is added to a home, the less the benefit received from each additional inch of insulation. A general rule of thumb states that each successive increase in R-value by one in any assembly saves approximately 50% of the energy saved by the immediately preceding R-value. The break-even point in a heating environment is in the R-38 to R-40 attic insulation level.

Adding insulation to walls that already have insulation is expensive. The high cost and low potential to further reduce heat loss in the walls of a home like this results in an SIR below 1. Walls with no insulation have much more potential. For example, consider a home with a semi-conditioned basement with no wall insulation. At best, the concrete would have an R-value of 1. Since it is relatively inexpensive to install basement wall insulation, bringing the wall from R-1 to R-10 creates a much higher return than adding the same insulation to any surface that already has an R-value greater than 1. This combination of low-cost and high return produces an excellent SIR for installing insulation in uninsulated, semi-conditioned basements.

In weatherization, anytime it is proposed to install insulation in a home, use the SIR to determine the optimum insulation to add. The weatherization auditing software can help determine this, or the predetermined priority list parameter for determining the best level of insulation to install can be used. Either way, SIR must be taken into account.

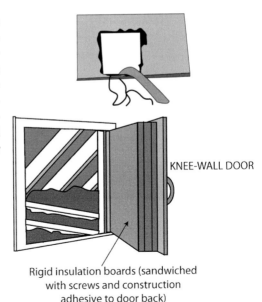

**Figure 7-31. Sometimes attic spaces such as a kneewall, have to be opened to adequately seal and insulate hidden spaces (upper diagram). Once the retrofit process is complete, it is necessary to adequately seal and insulate the knee wall door (lower diagram).**

KNEE-WALL DOOR

Rigid insulation boards (sandwiched with screws and construction adhesive to door back)

## Strategies for Homes With Different Attic Levels

If you find a knee wall, a porch roof, a one-story addition, or an attic over a garage, make a note that you might have to cut an access port to get into these areas for air sealing and insulating. If these spaces are too small or too tight, you may need to just fill the entire space with blown-in insulation.

Where two large attic areas exist at different levels, as shown in Figures 7-32 and 7-33, there is probably a wall at the change in ceiling height. This can also mean there might not be top plates in the wall at the ceiling height change. It is not as cost-effective to try to keep both of these large levels at the same height. Thus, installing the equivalent of top plates in the form of foam board inside the wall opening at the same level as the

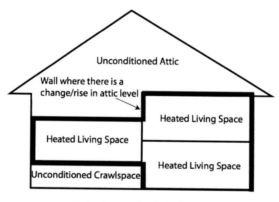

Darker lines are insulation lines

**Figure 7-32. Attic areas at different levels.**

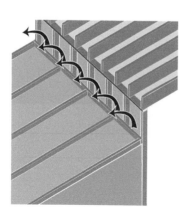

**Figure 7-33. Changes in level in the attic will promote bypasses to the wall below where the two levels meet if the stud bays in the wall are not blocked off where the lower level meets that wall.**

backside of the ceiling below is often the best method of sealing this type of ceiling height change. This prevents filling the lower attic ceiling area with insulation unnecessarily and allows for consistent blown-in attic insulation all across the attic. Be sure to order insulation all along the wall that exists where the ceiling height takes a step up in height.

If there are dropped soffits in the upper level of the home with an attic above it (Figure 7-34), there could be a substantial source for air leakage because the air barrier is not aligned with the thermal barrier.

Completely filling the dropped soffits and the areas above them so the insulation matches the height in the rest of the attic would be unnecessarily wasting insulation. With a smaller drop-down recess in the attic floor, such as a dropped soffit, you could insulate the small drop-down floor level and the walls surrounding this level to properly insulate the living space from the attic. However, this is typically a great deal more work than simply installing foam board or other similar support material between the rafters or trusses in the attic above the dropped ceiling to provide a continuous floor level for the attic and then insulating over this just like the rest of the attic.

Sometimes existing batt or other insulation can cover dropped soffits and make them difficult to identify from the attic side. Surveying from the story below will help you know where they exist, but you can also look for telltale signs of them on the attic side. For instance, if a true kitchen vent exists along the dropped soffits in the story below, you should see a vent duct coming up out of the insulation in the attic just above the dropped soffits.

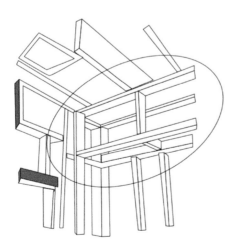

**Attic Preparation**

To properly prepare the attic for insulation, place appropriate shielding around the flues and fixtures to provide a proper clearance, install insulation baffles down at the eaves between the rafters or trusses, build a sturdy dam

**Figure 7-34. Example of a Dropped Soffit at the Framing Stage**

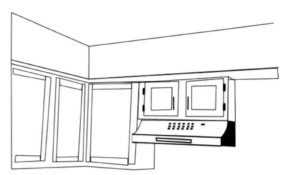

**Figure 7-35. A dropped soffit at the finish stage allows for the attachment of kitchen cabinets from above in this kitchen.**

from plywood or OSB around the attic access hatch, add ventilation to the attic if necessary, make sure there are not any roof leaks, make sure all the kitchen and bath fans are vented to the outdoors, make sure an appropriate attic hatch has been installed, and mark all the recessed lights fans and junction boxes with flags on the rafters or trusses above them. These location flags will help future electricians more easily find trouble spots. It will also help keep the electrician from having to disturb any more insulation than necessary.

Make sure all the wiring is safe. Make sure all the wiring splices and connections are contained in junction boxes and that all the electrical junction boxes have an appropriate cover. Look for frayed or otherwise compromised wiring. Insulation cannot be placed over any knob and tube wiring in the attic. Make sure that any splices of wiring, other than original soldered knob and tube wiring connections in some jurisdictions, are properly enclosed in an electrical junction box. If you find any non-original wire tied into knob and tube wiring, such as when someone has connected modern Romex wire into the older knob and tube wire, those connections must be enclosed in a junction box. In addition, the electrical system itself should be properly grounded. Unless they are not IC rated recessed lights, seal the electrical boxes for light fixtures to the backside of the ceiling material, making sure no foam is inside the electrical box itself. You can seal directly against an "IC" rated recessed light, but not against a recessed light fixture that is not "IC" rated. You can also seal around all the joints of exhaust fans and where they meet the drywall or plaster.

**Attic Insulation Ruler**

To prepare the attic for insulation, staple rulers that face the exit about every 15 feet or so from each other. The rulers help the insulator keep track of the appropriate depth required to provide the recommended level of insulation, as shown in Figure 7-36.

Be sure these rulers are facing the attic hatch or exit to the attic. The installer will be first blowing in insulation in the far corners of the attic and filling in the insulation in front of him as he backs his way over towards the attic hatch or exit. If the rulers are all faced so they can be seen from the attic hatch, then the insulator will always be able to see them as long as he is not "painting himself into a corner."

**Figure 7-36. An insulation ruler installed in an attic space assures adequate thickness of blown-in insulation. Rulers should face the attic access so they can be read without having to walk over to them and disturb the insulation in the process.**

## Sealing

Start at one end of the house, perhaps a gable end if present, and take care of any top plate penetrations or openings first. Seal around any wiring, plumbing, and vent openings with the appropriate materials. Seal the top plate to the plaster, drywall, or sheathing on both sides of the top plate. Also, remember if there are no top plates on the outside walls, place a plug at the bottom of each exterior wall stud bay in the crawlspace or basement, if it is open, so that when you blow insulation into the stud bay from the attic it will not simply flow down into the basement. If there is no top plate on an interior partition wall, the strategy is different. You can pinch fiberglass batts down into these walls tightly enough that the batts will not move when the regular attic insulation is blown in over them.

Another option is to pinch in flexible foam board or staple in cardboard in place and then inject foam around its edges to provide an adequate seal where the top plate normally would be in a balloon framed home. In a multilevel home with different ceiling heights, or where a

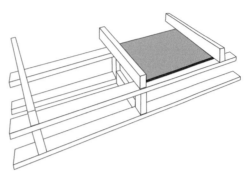

**Figure 3-37. Drywall over an open kitchen dropped ceiling (looking from the attic side) allows for a consistent support and thickness for the blown-in insulation.**

lower ceiling meets up with a taller wall and there is no blocking in the stud spaces, foam or cardboard should be placed to match the level of the lower ceiling inside the stud cavity and then foam around the edges to seal it. If you find built-up joists, where multiple 2-inch joists have been tied together to provide a major structural support, seal at the joints between each of the 2-inch joists.

The next step could be locating the center of the attic where a load-bearing wall might exist. Start at one end of the attic and try to tell where other partition walls start from the main load bearing wall. When sealing, if you see any such partition walls, temporarily leave the center load-bearing wall and follow the partition wall all the way over as far as you can get towards the eave. As you are doing this, if you notice another partition wall taking off from this partition wall, follow it until it reaches another intersecting wall. Once it reaches another intersecting wall, return to the previous partition wall and complete the air sealing on that wall.

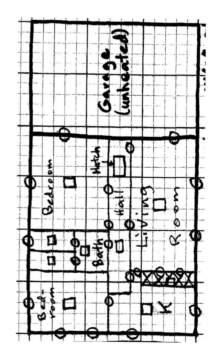

**Figure 7-38. Basic Sketch of Attic Top Plates Indicating Where Sealing is Needed**

Thus, as you seal above the walls that extend off the main center load-bearing wall to the eave, you ultimately return to the main load-bearing wall to continue the journey towards the other end of the house.

## Batt or Blown In

Trying to add batt insulation to an attic is not typically the method of choice. Cutting batts to fit around electrical fixture boxes, plumbing, etc. is cumbersome and difficult to do well. However, some types of rock wool or mineral fiber batt insulation have been found to "uncompress" after installation. If there is batt insulation already placed in between the attic joists, even if there are gaps, adding blown in insulation is a very effective way to add insulation and still receive the benefit of your existing batt insulation.

If you decide to install fiberglass batt insulation, avoid using faced batts such as Kraft or foil backed batts unless you place the facing on the heated side of the batt (on the house side of the batt in cold-weather areas, and vice versa in warm weather areas). Also, make sure you have a tight fit between batts.

Blown-in insulation is usually the first choice for attics. When you are ready to begin insulating, be sure to wear the appropriate safety equipment as mentioned previously. It is helpful to bring a 5-ft length of 1" x 12," or similarly sized plywood, to keep from falling through the drywall or plaster of the ceiling below and to have a more comfortable sitting or kneeling arrangement while working in the attic. If you are using two-part foam you need to bring up the two-part foam kit as well as a second similar support board for the tanks.

It is suggested to keep the tip of the blower tube pointed slightly down and close to the attic floor to avoid generating dust and fluffing. Fill to 2 inches above your proposed insulation levels to allow for settling and make a special effort to pack the insulation a little tighter right next to the baffles near the eaves. Run the tube through an opened attic, gable end hatch window or vent for easy access. To keep from having to walk over and compress insulation that was just installed, make sure you can always exit to the hatch without stepping on insulation.

Always use the remote switch (Figure 7-39) near the end of the blower tube to turn off the blower machine when you have finished blowing in insulation in one area and you need to move to a new area of the attic. The delivery rate for blown-in insulation in an attic is higher than the delivery rate used blowing in wall insulation. In addition, if

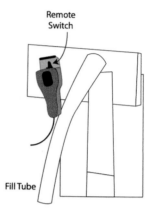

Figure 7-39. A remote switch allows workers directing the blower tube to control when they want the flow of insulation to stop or go.

there is any PVC pipe in the blow-in process, there may be a risk of static electricity shock. To avoid this, either add small amounts of water into the hopper from time to time, or make sure the blower tube is grounded if PVC pipe is used.

Avoid allowing the insulation to come too close to high temperature items such as chimneys, flues, or recessed canned light fixtures that are not "IC" rated. At the same time, you should blow insulation right up against the insulation baffles between the rafters or trusses near the eaves, making sure to blow the insulation to the proper density without "fluffing" it up at a lower density.

## SPECIAL SITUATIONS IN ATTICS AND BASEMENTS

### Recessed Lights

Recessed light fixtures, can lights, "pot lights," "top hats," or "down lights" have become more common in modern homes, yet they can represent a very big source of air leakage, as shown in Figure 7-40. They are especially leaky if they are placed in ceilings that have attic space above them.

If the can lights can be accessed from the attic and they are not IC rated, place a fabricated or commercially available code-compliant cover over them from the attic side. If a proper cover is not used for a canned light that is not IC rated, the life of bulb in that fixture is shortened and, in the worst case, it could cause a fire.

For the non-IC rated recessed light fixtures (Figure 7-41), seal a box

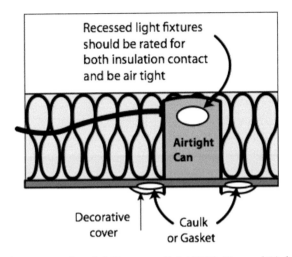

**Figure 7-40. Special Concerns Exist With Canned Lights**

around the fixture that has a minimum of 3 inches clearance away from the fixture on the sides and 4 inches away from the top. The sides of the box need to be at least as high as the blown-in insulation will be. For most applications, a 20-inch height should be adequate. To fit inside of 16 inch on center ceiling joists or trusses, it should not be any wider than about 14 inches and typically does not need to be any longer than about 14 inches. While the sides of the box can be made of foil-faced duct board, metal, or drywall, the top of the box should be one half-inch drywall or noncombustible material with a greater insulation value than one half-inch drywall. As a result, many simply place ½ inch drywall on the top portion of the box. If you decide to use metal to build this box, consider using round galvanized duct with a galvanized duct cap for round recessed light fixtures. If the light fixture is square, consider bending a larger diameter (for example, 18 inch or larger) galvanized round duct to create four corners to establish a box shape and then add metal or drywall as a cap and seal the joint with mastic. Regardless of what materials are used, any joints or seams in the box should be covered with duct mastic embedded in mesh. No blown-in insulation should be placed on top of these boxes that cover canned lights that are not IC rated.

In the private arena, if it is not too expensive for the client, the preferred approach is to replace the non-IC rated recessed light fixtures with an airtight IC rated recessed light fixture. Not all IC rated fixtures

are airtight and so the label should
reference either ASTM™ E283 – 91,
the Washington State Energy Code, or
the MEC (Model Energy Code) if it is
an airtight IC fixture.

If these recessed light fixtures are
IC rated, then they can be completely
covered with insulation on the attic
side without having to place a special
cover over them. If they are IC rated,
you will see the bold letters "IC" from
the living space side on the inside of
the cylinder that surrounds the bulb.
Seal all the joints on IC rated recessed

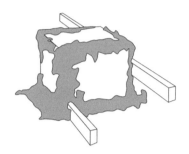

**Figure 7-41. Non IC-Rated
Canned Light Covered With Drywall and Tied Together with Fire-
resistant Foam**

lights with fire resistant foam or caulk, including where it meets the
drywall or plaster of the ceiling below.

Even ordinary light fixtures can allow for significant leakage from
the upper floor into the attic. Good sealing with a fire resistant foam
from the attic side around the box, but not inside the box, is one method
of handling this situation. Fire resistant foam may be available in your
area either as a regular foam application or one that is Portland cement
based.

### Attic Floors

Attic floors that can be walked on, especially in older homes
where the joists may be only 4 inches wide/high, can limit the ability to insulate the space below the attic
floorboards. If there is a built-in floor to
walk on, the most common way to insulate under the floor is to dense pack
as you would a wall. If there are floor
planks, perhaps the easiest way is to lift
a plank or two about halfway between
the center-bearing wall and the outer
edge of the floor nearest the eaves. This
way insulation can be blown in both directions about the same length until you
either run into blocking that might exist
over the central load bearing wall in one

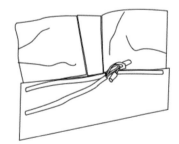

**Figure 7-42. Electrical con-
nections should not be made
outside of approved junction
boxes.**

direction, or the tube gets to the edge of the flooring, usually near the lowest head clearances in the attic towards the eaves. You will probably have to cap off the joist space with foam board or other rigid materials under the eave edge of the flooring if you are going to have something to dense pack against. This board can serve a purpose similar to a top or bottom plate in a wall—it allows you to push the insulation in more densely. Otherwise trying to dense pack may only blow insulation out freely from the open joists spaces at the edge of the flooring.

If the flooring is plywood or OSB, blow in dense pack insulation by either drilling a hole through the flooring for each joist space under the floor, or pull-up entire sheets of plywood or OSB just over the central load-bearing wall to blow in the insulation and pack it toward each edge of the house. One way to put more insulation than would normally fit under the floor in an attic is to insulate between the old joists and then build up a new floor frame that is perpendicular to the old joists, as shown in Figure 7-43. You can then install flooring to this upper deck of new framing and fill the new perpendicular floor joists with insulation. Insulating is limited only by the height of the two layers of joist. Be sure to have a structural evaluation done before attempting this because the ceiling joists in many homes, especially older ones, may already be inadequate structurally.

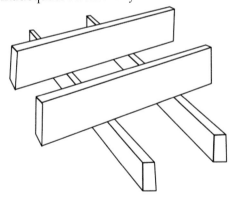

**Figure 7-43. Adding perpendicular joists to existing ceiling joists on the attic side is one way to increase the space for installing insulation in a walk-in attic. Precautions should be taken by having experts evaluate adding the extra weight to ceiling joists that may already be compromised.**

The alternative already mentioned would be to first air seal at the attic side of the upper floor ceiling level. Sealing at this level provides for easier alignment of thermal and pressure boundaries without having to fill the entire cavity below the floor level of the attic with insulation. If the air barrier is not placed at the attic side of the ceiling using rigid board at changes in level, an exorbitant amount of insulation may have to be used to fill the gaps below the normal ceiling level in the dropped soffit, dropped ceilings, etc. Unfortunately, if the space has floorboards for storage, etc., it may be more costly to temporarily pull up the walkable flooring to seal the attic side of the ceiling than to leave the floor boards in place and just fill the dropped soffits, etc., with insulation and air seal at the floor boards or dense pack below the floor boards. If you seal the ceiling in areas where there are no floorboards, remember to provide a continuous air seal by providing a rigid board connection between the ceiling air seal and the floorboard seal (place some foam board across each joist space much like you do below the knee wall on a Cape Cod attic.)

### Knee Walls

Knee walls require special attention because they can have hidden defects and can create "cold walls" and even "cold ceilings" and other air leaks. Special problems include the following:

1)  Spaces between the attic are not sealed off from the unconditioned knee wall area. This is often taken care of by placing foam board horizontally between the attic floor joists right below the knee wall. Placing the board underneath the conditioned attic floor space can allow for more insulation in this sometimes otherwise susceptible area.

2)  There is no top plate in the knee wall. Thus, you need to provide an air barrier using foam board or similar material to seal off the top of the wall from the space between the angled ceiling rafters (that support the finished attic ceiling) above the knee wall and the part of the angled roof deck above it. It is common to find that there is no top plate in this section of the angled rafters. A top plate must also be installed before dense packing these angled rafter sections surrounding conditioned attic space. Figure 7-44 shows above the knee wall in a Cathedral ceiling section.

3)  The knee wall itself is often not insulated or it is under insulated. Thus, installing wall insulation, such as fiberglass batt into the knee wall, and covering it with house wrap or other air barrier material is often the best way to remedy this deficiency. All access doors through knee walls should be weatherstripped and insulated to as high an R-value as the knee wall, probably a minimum of

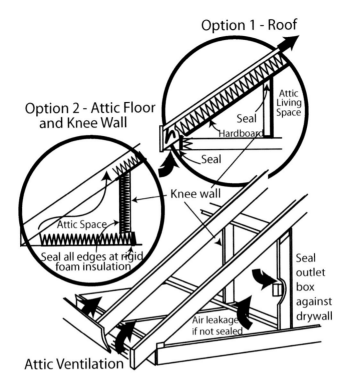

**Figure 7-44. Above the Knee Wall "Cathedral Ceiling Section"**

R-15. This is most easily accomplished using foam board insulation or fiberglass batts wrapped in house wrap around all six sides.

You want to insulate the knee wall to as high an R-value as you do the unfinished attic floor. Spray-on foam has the advantage of better R-values per inch and of being easier to install to a thickness greater than the wall studs. When using foam spray to get the proper R-value, it is best to spray the foam in successive layers and allow each layer to cure fully before applying another layer. Expect that extensive applications of two-part foam will create an increase in temperature of the foam. This is a reason to avoid applying two-part foam in just one pass.

### Walk-up Attics/Walk-down Basements

All attic hatches must be weatherstripped and insulated to the same level as the attic. Thus, an ordinary attic hatch should be carefully weatherstripped around its edges where it seals to the opening. Figure 7-45 shows a method of sealing an attic accessible with a foldout ladder.

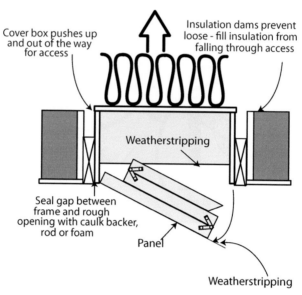

**Figure 7-45. A Method of Sealing an Attic Accessible with a Foldout Ladder**

Walk-up attics and walk-down basements represent a special challenge for weatherization. These accesses have permanent, built-in stairs with a stairwell to access the attic or basement. The real question is where to align the air and thermal boundary in such a case. The most important consideration is how often do the occupants use the attic or the basement. If they use these areas often for storage, make it easier for them to get into the attic. On the other hand, if they do not use these areas very often, then do not worry too much about making it easy to access to the attic or basement. In this case, less involved retrofit measures could be used.

If the occupant seldom goes into the attic or basement, simply place an airtight cover at the opening to the attic or basement from a permanent stairwell to seal the attic from the rest of the house, as shown in Figure 7-46.

You can either build a custom cover or use a prefabricated foam cover available commercially. Make sure it is airtight and will provide an adequate R-value. Optimally, it should have an R-value roughly

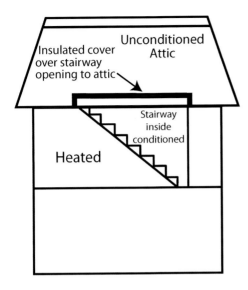

**Figure 7-46. Example of Insulated Cover Over Stairway**

equal to the R-value of the rest of the attic or basement insulation. While this may be more cumbersome for the owner, since they do not access the attic very often, it is a very cost-effective way to deal with attic stairwells. In this case, the attic staircase is "inside" because the stairwell will be heated, which makes it conditioned space.

On the other hand, if the attic or basement is used often, a more complicated approach is typically recommended for insulating the stairwell walls, the underside of the stairs, and the door from the living space into the stairwell. Insulate and air seal the walls inside the attic stairwell with dense pack insulation, at least down to just below the stairs themselves. Also insulate the under side of the stairs and place foam board insulation on the stairwell side of the door at the bottom of the stairs. Weatherstrip the stairwell door as well. This is more work than a simple cover at the top of the stairwell, but it keeps it a traditional door access. You are effectively putting the attic staircase "outside" because you are not conditioning that space. See the example in Figure 7-47.

If the walls in the staircase go up above the attic floor, you can also bring the staircase "inside" with a similar approach but insulating everything above the attic floor level (see Figure 7-48).

If it is determined to establish the thermal and pressure boundaries at the stairwell walls, stairs, and attic or basement stair doors, a less efficient barrier will be created. The additional surface area allows for more heat loss and the narrower stairwell walls limit the R-values as compared with the thicker blown-in insulation that can be installed in the attic or basement ceiling. However, if this is the choice, then weatherstrip and insulate the stairwell side of the door, blow in dense pack insulation in the stairwell walls to the appropriate density, and provide an adequate R-value insulation under the stairs. To the extent practical, attempt to match the R-values in these locations to that of the blown-in insulation in the attic floor or basement ceiling. In addition, in older homes with lath and plaster walls, some installers have found it useful to not drill the access holes along the stairwell wall at the same horizontal level. Instead, drill the holes for installing the dense pack insulation so that they angle their way up the stairway from hole to hole, as shown in Figure 7-49.

This helps keep the wall more stable because the lath boards behind the plaster are not being "butchered" or weakened as much. After the dense pack cellulose insulation is installed in the walls, install plugs in the drilled holes. In some instances, if the client agrees, it may be cost-effective to install a permanently hinged, counter weighted, and well insulated hatch to avoid the additional labor and energy losses that come with insulating the door, the stairwell walls, and the underside of the stairs.

Whatever you do, do not leave exposed foam insulation, whether foam board or spray-on foam, on the underside

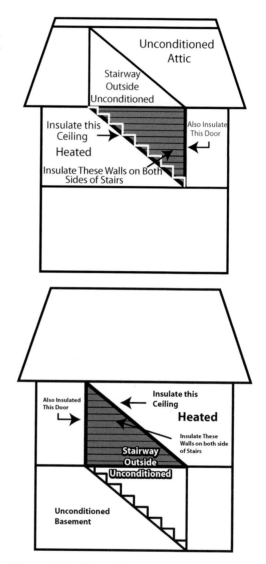

**Figure 7-48. (Upper) Example of how insulation is positioned if an attic stairway is considered "inside" the conditioned space. (Lower) A similar approach can be used for basements.**

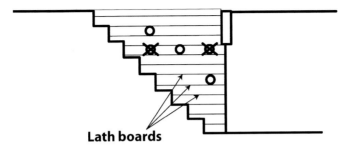

**Figure 7-49. Installers will often choose to drill holes angling up along a staircase in the different stud bays to or from attics or basements that need to be insulated. This avoids overly compromising the strength of just one horizontal wood lath board underneath the plaster.**

of the stairs or anywhere else on the inside of the home. This violates code in many areas. Cover any installed foam on the living space side of the stairwell with a thermal barrier of batt or spray-on mineral insulation (2″), drywall (1/2″), sheet metal (.016″), plywood (15/32″) or OSB (15/32″), or other material approved by the municipality.

## ATTIC VENTILATION

The two primary purposes of attic ventilation are to remove moisture vapor in cold weather and to remove solar heat in hot weather. Attic venting is particularly important in hotter climates where attic temperatures can easily exceed 140 °F for extended periods Wind effect can provide passive attic venting. Wind can create a positive pressure on the windward side and a negative pressure on the leeward side of a roof. A phenomenon known as the "Bernoulli Principle" provides a way for air to be exhausted from the attic when an airstream across an opening creates enough of a negative pressure to draw air out of the attic space. This is very similar to a jet pump that uses a higher velocity jet of air to entrain or draw air around it in a specific direction.

There seems to be a number of common misconceptions about attic ventilation, including that attic ventilation is always unnecessary, that the more attic ventilation the better, that attic ventilation will resolve an attic moisture issue, and that attic ventilation always eliminates moisture vapor in cold weather. These items will be discussed below.

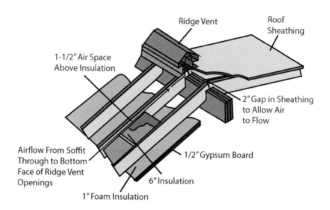

**Figure 7-50. Eave-to-Ridge Attic Ventilation**

**Moisture in Attics**

Sometimes retrofitting an attic can potentially cause problems in the home. For example, frost and mold might show up on the underside of the roof deck sheeting after roof vents and attic insulation are installed. Before these retrofits, when the attic was unsealed and uninsulated, the house could sufficiently warm the attic air to cause ventilation. When insulation is installed, attic air is not warmed, reducing the air's buoyancy and its ability to ventilate. The combination of high moisture on the interior of the home and poor attic air sealing is considered one of the primary causes of failure in a roof in heating climates. Attic flues actually have the potential to worsen an already existing attic moisture problem. Openings around flues can allow rising, warm, moist air from the house side through unsealed bypasses up into the attic. Here moisture can condense on the cold surfaces such as the backside of the roof deck, as shown in Figure 7-51.

Attic surfaces can be cooled even further on very still, cold days when the cold dense air sinks into the attic space from outside through roof deck vents. It is generally thought that, in most cases, providing adequate attic insulation and simple passive, non-mechanical ventilation is the best method to avoid these problems.

**Code Issues in the Attic**

With regard to providing passive attic ventilation, the code is specific. If the ceiling is the thermal boundary, then attic ventilation is required.

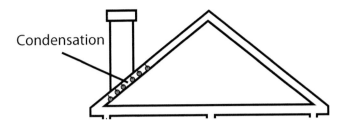

**Figure 7-51. Condensation can occur on the bottom of roof decks if the conditions are right**

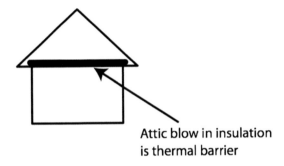

Attic blow in insulation
is thermal barrier

**Figure 7-52. Thermal Barrier in Attics**

On the other hand, if the roof deck is the thermal boundary, no ventilation of the attic is required, as in Figure 7-52. If ventilation is required, the code typically requires a 1/150 or 1/300 ventilation opening based on the square footage of the attic space and whether a vapor barrier is present. Thus, if there is a 300-ft² attic, under the 1/150 formula, only two square feet of ventilation opening would be needed. This is called the "net free area" or NFA. The rule states that if the ceiling does not have a vapor retarder, then use the 1/150 ratio. If the ceiling has a vapor retarder, use the 1/300 ratio. Since most code officials will recognize a painted ceiling as a vapor retarder, most homes can have the more liberal, and more easily provided, 1/300 rule applied.

NFA is a reference to the actual open area available for effective ventilation. If there is a grille or louver over the vent, you must correct for that restriction. For instance, if the vent opening is 12" x 12" and is covered by grilles or louvers, a correction must be made for the space occupied by the grilles or louvers. Thus, the true, corrected opening

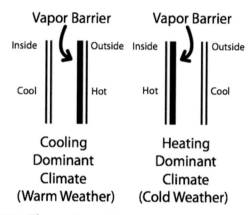

Figure 7-53. The position of the vapor barrier is a function of the climate in the area.

will be less than the 12" x 12" or 144 in² vent opening. While the louver or grille manufacturer is the best source for determining NFA of a particular louver or grille, if this number is not known, it is assumed that a metal grille reduces the opening by 25% and a wood grille by 75%. Thus, using the 12" x 12" example, reduce 144 in² down 75% (144 x .75) to get the true NFA, assuming a metal louver, which would be 108 in². The NFA rating for a vent is represented by an R-value. This is not the same as the R-value for insulation. (It should have been given a different name to avoid confusion.) The R-value for vents indicates the number of square inches of net free area provided by a given louver or grille. For instance, an R 61 roof vent has 61 in² of NFA.

If the roof deck, rather than the ceiling, is the pressure and thermal boundary, then attic ventilation is not required and not recommended.

Finished attics, as in Figure 7-54, are a good example of this. Another example is the recent development to use open or closed cell

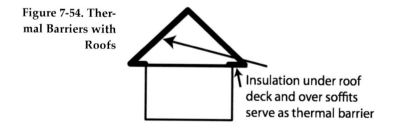

Figure 7-54. Thermal Barriers with Roofs

Insulation under roof deck and over soffits serve as thermal barrier

spray foam insulation to make attics airtight and insulated. To keep the attic space airtight, it cannot be vented. Another exception occurs when HVAC or ducts systems have been placed in the attic. Sometimes it is better to insulate the roof assembly rather than the floor of the attic and the ductwork in cases like this.

Here is another example on how to apply the ventilation formula. Assume there is an attic space where the attic floor is 60' x 30' and the ceiling below is painted with latex paint with a 1.0 perm rating which is the maximum perm rating to count as a vapor barrier. 60' x 30' is the square footage of the attic space to which the 1/300 ratio is applied. Since there is an adequately painted drywall ceiling below, assume that each of the vents being installed has an NFA of 60 in². The question is "how many of these vents need to be installed to provide adequate ventilation for the attic under the code?" First, calculate the square footage of the attic floor, 60' x 30' = 1800 ft². Then calculate the total net free area required for all the ventilation— 1800 ft²/300 = 6 ft². Convert 6 ft² to square inches by multiplying 6 ft² x 144 in²/ft² = 864 in². If the vents are known to provide 60 in²/vent, then divide 864 in² by 60 in² per vent to find out how many vents need to be installed to provide the requisite square inches required by code: 864 in²/60 in² per vent = 14.4 vents. Since the vents cannot be modified, it is best to install 15 vents to be sure that adequate ventilation is provided for this attic.

---

**NFA = Net Free Area**

---

It is important to understand "net free area" of attic or crawl-space openings/ventilation after deducting for metal louvers (25% reduction in size from measured opening) or wood louvers (75% reduction in size from measured opening). For instance, often the net free area after deducting for louvers must be 1/300 of the square footage of the area. So if there is a 300 square foot attic area, it will need a 1 square foot vent after taking into account for, say, a metal louver and the consequent 25% reduction in area. In this case, a 1 square foot area is needed, but you can start with a minimum of a 1.33 square foot opening if a metal louver is used because ¼th or 25% of 1.33 square feet is lost, which brings it down to exactly 1 square foot. If a wood louver was used, you would have to start with 4 square feet because it would have to be reduced by ¾ or 75% which would leave one square foot "net free area" when done.

**Practicality**

It is worth mentioning that even though the more liberal ratio 1/300th requirement for attic ventilation was used, there are still an awful lot of holes needed in the roof. It is not practical, nor cost-effective to be installing this many roof vents to ventilate an attic space. A better option might be to install larger vents to reduce the number of holes in the roof. At the same time, having this many square inches of openings to the attic space may actually promote air loss as a result of the stack effect - forcing warm, moist air from the house into the attic where it could condense on the underside of the roof assembly in cold weather. Unfortunately, you may still have to install vents to satisfy code. On roofs with difficult to work with shingles, such as those made of concrete or slate, install the vents on the gable ends rather than through the roof deck itself if possible. Placing vents through the soffits is also a possibility. In fact, vents should be placed at low and high points on the roof, soffit, or gable end. This provides a natural, rising ventilation of air in warm weather regardless of the type of passive attic vent used. For most homes, this may mean using primarily gable and roof vents to promote attic ventilation. Gable vents are installed on the gable end of gable roofs, while roof vents are installed on the roof deck itself. Figure 7-55 shows a gable vent cross section.

Soffit or eave vents have some drawbacks: too many are needed, they require sufficient soffit area and cross flow, and they are difficult to install. In addition, turbine vents tend to be subject to mechanical failure and can contribute to air leakage, and therefore, are not recommended for installation. If there is no other option, consider using an R-144 mushroom vent. It has more than twice the NFA of two R-61 roof vents and it can serve as an attic access from the roof deck side.

Perhaps the best con-

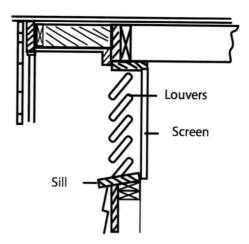

Outlet Ventilator

Louvers

Screen

Sill

**Figure 7-55. Gable Vent Cross Section**

figuration for attic ventilation is to have continuous soffit and ridge venting to allow for good upflow of circulating air from the low to high points in the roof and attic assembly. This provides air circulation without any energy cost based on the principle that hot air rises. This "passive ventilation" can work as well as a small fan powered mechanical ventilation system. Unfortunately, this is most easily only done during the original construction process. It is not practical and can be too costly to try to ventilate every retrofitted roof with this type of venting after the building has already been built.

**Installing Vents**

When installing a vent, a number of guidelines should be followed. Test fit the vent after cutting an appropriate size hole. The roof flange of the vent is designed to go under the roofing material on the slope of the roof above it. After establishing a good fit, caulk a bead of roof cement under the shingles near the edges where the flange will be. Then slide the vent into its proper position, caulk around where it meets the roof with roof cement, and then fasten the vent with galvanized roofing nails through the roof deck. Finally, cover the heads of the nails with roof cement so no nails will be exposed.

Typically, gable vents are installed near the peak of each gable end of the home by cutting an appropriately sized hole for each vent. Then caulk and fasten the flange of the vent to the gable wall with screws or galvanized nails driven into the wall material. Sometimes a gable window can be used in an unfinished attic as a location for a gable vent. If the window is too large for the expected vent opening, install a plywood panel into the window or window's spot and then cut an opening into the plywood to fit the gable vent. Installing gable vents in attic windows avoids unnecessary penetration through an outside wall into the home.

If soffit vents are to be used, chutes or insulation baffles should be installed above the soffit between each of the trusses or rafters in the attic, as shown in Figure 7-56. "Chutes" or insulation baffles are typically cardboard or thin pre-formed foam and should be installed when using soffit vents. They help maintain the airspace between the blown-in attic insulation and the underside of the roof sheathing to prevent the insulation from clogging the opening to the soffit vents. The baffles need to be installed before insulation is blown in.

Finally, all plumbing stacks and mechanical ventilation ducts, such as bathroom and kitchen vents, must be vented to the outside prior to

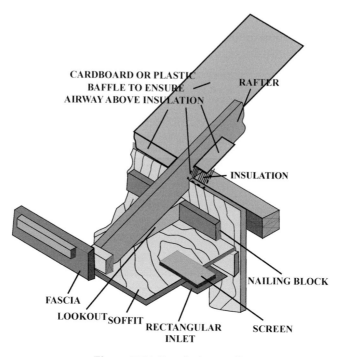

**Figure 7-56. Insulation Baffle**

installing insulation or attic vents. Be sure that the terminations of these vents prevent the intrusion of animals, materials, and moisture into the vents. How and where the vent opens to the outside and whether it has proper grilles in it can help eliminate these issues.

## ENCLOSED AND COVERED CEILINGS BELOW ROOFS

### Removable Tile Ceilings

If the house has a removable suspended or "T" ceiling, much like you see in commercial buildings, it is important to lift at least one tile in each room to check the condition of the plaster or drywall above it. These ceilings (Figure 7-57) are often installed to hide damaged plaster, etc. Suspended or dropped ceilings that are hung from the original ceiling to lower it can often cause problems above them, including damage to the drywall from which the metal ceiling grid hangs. Clearly this suspended ceiling grid with the "drop in" fiberboard squares cannot be counted as an air barrier itself.

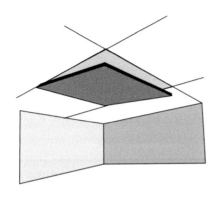

**Figure 7-57. Commercial style "T" ceilings often hide damaged or missing air barriers.**

If there are small holes in the original ceiling, you may be able to just use drywall tape and mud for repair. For holes that are wider than drywall tape, you can use the sheet metal repair patches up to about 8" x 8."

For larger holes, you may have to use drywall scraps, drywall tape, and drywall mud. If the original ceiling damage is extensive, consider meticulously installing a vapor barrier throughout the damaged area, appropriate to the climate, and then drywall to the underside of the joists.

Another related area of concern involves the smaller acoustical ceiling tiles that are about 1 foot square or less. These are prone to leakage and are difficult to seal, so it is probably easiest to just cover these tiles with drywall in their entirety. This can serve an additional useful purpose if you are having difficulty insulating a sloped or cathedral ceiling. The drywall would be a necessary fire barrier if you were to place foam board against these tiles as ceiling insulation.

Tongue and Groove Ceilings

**Figure 7-58. Tongue and groove slats are notoriously leaky and difficult to seal adequately or cost-effectively**

Another area of concern involves wooden plank tongue and groove ceiling applications, as shown in Figure 7-58. If there is no air barrier behind the planks, the treatment of choice is removing the planks and installing an air barrier behind them. If there is no special interest in keeping the tongue and groove plank system exposed, then drywall can be used to cover the planks. Once again, if drywall is chosen and it is a cathedral ceiling, it could be a great opportunity for installing rigid foam insulation over the planks to better insulate the ceiling. Be sure there is no air barrier behind the planks before you even consider conducting any work on the system. Look for openings in or around the planks to see if there is any evidence of drywall, polyethylene or other air barriers—foil or Kraft paper is not acceptable to some— that could serve as an air barrier.

### Sloped, Cathedral and Vaulted Ceilings

Sloped, cathedral and vaulted ceilings can represent a special concern for air leakage, especially above the top plate on the outside wall. If there is no air barrier extending up above the wall sheathing and across the bottom edge of the batt insulation in between the rafters or trusses, an open or vented soffit or eave can allow wind to "wash" up into the insulation (Figure 7-59). This weakens the insulation value and increases the risk of air leakage. This is especially true if there is a discontinuity in the drywall air barrier due to recessed lights, etc.

Perhaps the single most common defect in these sloped ceilings involves the lack of ventilation along the air space on the underside of the roof deck extending from the eave to a ridge vent. This missing ventilation can create condensation problems inside the roof structural system and insulation. It can also potentially damage the drywall, which acts as air barrier. Ventilating this space also creates a higher risk of air leakage if there is not an ad-

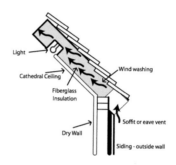

**Figure 7-59. Wind washing of batt insulation is an issue with cathedral ceilings.**

equate air barrier behind the drywall of the sloped ceiling or there are breaks in the continuity of the drywall air barrier due to incomplete drywall finish. There can also be a problem if there are recessed lights, inset ceiling beams, tongue and groove ceiling systems, intersecting interior walls, or other penetrations through the drywall of the sloped ceiling that compromise the air barrier from both the insulation or living side of the drywall.

Sloped ceilings can create additional problems such as not being able to seal off penetration openings for wires and pipes or the typically long thin openings between the drywall and the top plate of interior walls below the sloped ceiling. While it is typically emphasized to concentrate the insulation measures on the exterior walls, some make an exception here by dense packing any interior partition walls that may allow leakage through their top plates into a sloped ceiling in this manner, as in Figure 7-60.

Be aware that often scissor trusses may be used instead of true cathedral ceiling rafters in many homes. It is difficult to tell the difference unless you can see that the slope of the ceiling inside the house is at a significantly lower slope than the roof deck on the outside of the house. It may also be a challenge accessing some of these narrow attic spaces if they cannot be accessed from a more spacious attic area. Be careful in cutting through plywood in attic areas to access other attic spaces. Be sure that the plywood is not actually being used as roof deck. If it is a fully vaulted ceiling from the eaves up to the ridge of the roof, attempt to dense pack from the top down by opening a short width opening down each side of the roof ridge or from the bottom up by removing the soffit or facia at the eave. Then, blow in dense pack insulation by extending the insulation tube to the edge and making your way into the rafter bay space much like you would if filling a wall. If you are going to fill through the top, remove the shingles at the ridge of the roof, drill a hole or temporarily remove the upper planking on the roof deck. Both of these options are best utilized by attaching a longer muffler style pipe to the end of the flex tube rather than the typical flexible vinyl fill tube.

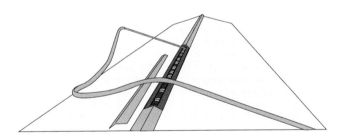

**Figure 7-60. Dense Packing a Cathedral Ceiling From the Ridge**

You can also use copper drainpipe, metal electrical conduit, or, as a last resort because of the static issues, a PVC pipe. If the access hole is large enough for the blower materials flex tube to fit through, you may even be able to just wire a smaller ½" to 1" diameter rigid pipe of one of these same materials to the end of the flex blower tube to push the end of the flex tube into the cavity. Another option is to drill holes in the ceiling drywall material and blow in insulation from these holes inside of the house. In any case, the roof is put at risk as a result of potential over-heating problems that could reduce the life of the shingles or moisture problems because the roof is unvented. Perhaps the best way to handle this is to be sure you have carefully air sealed the drywall etc. on the inside of the rafters, while making sure to avoid creating unnecessary moisture in the home.

If it is not a fully vaulted ceiling, but looks more like a partial sloped ceiling on the upper floor (as in a Cape Cod with a knee wall), you need to cap off at least one end of each rafter bay, either at the top or the bottom of the sloped ceiling portion. This can be done by tightly pinching in an adequately sized fiberglass batt into one end of the rafter bay, preferably the end where it would be the least comfortable to spend time blowing in insulation. That way you can blow in insulation from the other end that is more accessible. If the attic area above the upper floor is large enough, this is probably the preferred method of dense packing. In this situation, pinch fiberglass batts into the rafter bay from the lower knee wall end of the sloped ceiling before blowing in dense pack insulation from the upper attic opening.

When dense packing partially sloped ceiling rafter bays, you may be able to use the flex tube to blow in dense pack insulation, if you are blowing in the insulation from the top attic area down into the rafter bay. However, whether a sloped ceiling is a full vaulted ceiling or a partially sloped ceiling, do not install dense pack if there are recessed lighting fixtures in the ceiling that are not IC rated. Either replace the recessed lights with IC rated recessed lights or avoid insulating the sloped ceiling.

If the building is currently being built, perhaps the best treatment that allows for a sloped ceiling is one that uses an energy type scissor truss to allow for more insulation above the top plate of the exterior wall and free ventilation for the bottom of the roof deck. This is helpful because an energy truss raises the roof deck higher off the top plate to make more area for insulation near the exterior wall top plates. The scissor nature of the truss allows for thick, blown in insulation above the ceiling drywall, while it also allows an open attic area above the insulation to ventilate freely from the eaves up to roof vents. See the examples shown in Figure 7-61.

## Mansard and Gambrel Roof Ceilings and Walls

Mansard (Figures 7-62 and 7-63) and gambrel roofs with their steep slopes just outside the walls also represent an area that is difficult

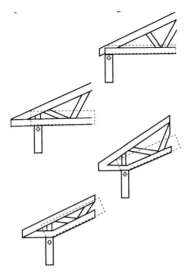

**Figure 7-62.
A Mansard Roof Style Home**

Upper: Regular Truss
Lower: Raised Heel or Energy Trusses

**Figure 7-61. The top traditional truss does not allow for adequate insulation above the exterior wall top plates. The lower three do. Dashed lines represent insulation. Note that the area for insulation is limited on the standard truss.**

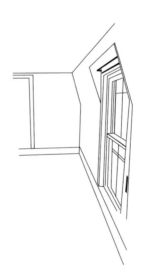

**Figure 7-63. The "Wall" of a Mansard Roof From the Inside**

to seal. While they are not much different than the Cape Cod style knee walls, there are often much narrower spaces behind the knee wall, and as a result, they do not allow access like the typical Cape Cod knee wall. Much like a roof overhang at middle height of a two-story wall, if you cannot access it by removing soffit material, you may simply have to fill the entire area of this narrow space behind the knee wall with dense packed cellulose, Figure 7-64.

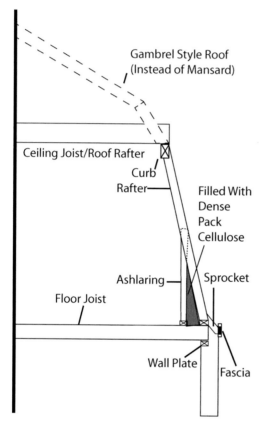

Gambrel Style Roof
(Instead of Mansard)

Ceiling Joist/Roof Rafter

Curb

Rafter

Filled With
Dense
Pack
Cellulose

Ashlaring          Sprocket

Floor Joist

Wall Plate

Fascia

Figure 7-64. This demon-strates placement of the insulation in a mansard or gambrel "wall" when access is not an option.

## ROOFS

There are two major concerns with roofs with regards to the energy arena: roof leaks and roof coatings. Roof leaks not only compromise the R-value of attic insulation by allowing it to get wet, they also allow the fire retardants in cellulose insulation to leach out, reducing the fire resistant nature of the cellulose. Therefore, it is a good idea to repair leaks before insulating the attic.

The color of the roof affects the surface temperature. For instance, a white strip on a black surface can be 10 to 15 °F cooler than the darker areas around it. Solar reflectance also significantly influences the surface temperature of any surface. If the roof receives full sun during the cooling season, consider applying a white or reflective roof coating to help keep attics from overheating as much and greatly reduce solar gain in the home. Reflective roof coatings can cut cooling loads 10% to 60% simply by reflecting solar radiation. This is in addition to color related temperature differences. Even higher numbers are typical if the home has no attic insulation. These coatings can be applied over common roofing materials, typically last 7 to 12 years, can protect the roof from water penetration if they are elastomeric, and can extend the life of the roof. A typical cost is between $0.75 and $1.50 per square foot (not including incidental repairs) if both labor and materials are included. Check actual local prices to make sure of cost-effectiveness. The solar reflectance values for several types of roofing are listed in Table 7-5.

**Table 7-5. Solar Reflectance of Materials**

| Material | Solar Reflectance, % |
|---|---|
| Metal Roof | 50-70 |
| Red Concrete Tile | 18 |
| White Concrete Tile | 74 |
| White Reflective Coating | 70-80 |
| Black EPDM | 6 |
| White EPDM | 69 |

To choose the right material, be sure that the solar reflectance of the coating is 65% or higher when it is new and that its thermal emittance is 80 to 90% or more. Thermal emittance is the ability of a material to release absorbed heat. A surface on a roof that has a high thermal emittance will absorb heat from the sun and then radiate it back into the atmosphere more readily than a surface with a low thermal emittance. Even just applying white industrial grade paint to the exterior ductwork, doors, window louvers, or awnings can help reduce heat gain on a home. Installing solar film, louvers, or shading on the south, west, or east facing windows can have the same effect. Appendix A gives information on what measures are allowed under the weatherization program. In the private arena, the clients decide what they will pay for.

Admittedly, reflective roof coatings carry a heating penalty because they reflect winter sun. Actual studies show that the cooling savings will outweigh the heating losses even in cold weather areas. One reason is that the sun's angle of incidence on the roof is much lower in the winter and so it is a less effective heat absorber than in the summer when the sun is almost directly overhead, as demonstrated in Figure 7-65.

There are some guidelines for applying reflective roof coatings. First, evaluate the cost-effectiveness. For instance, if the client does not have to use air conditioning, applying a reflective roof coating is not likely to be cost-effective. In the weatherization program, remember that a SIR of 1 or greater is necessary to justify a retrofit measure. Another consideration when applying reflective roof coatings is the fact that primer may be, and often is, required to prepare the substrate or

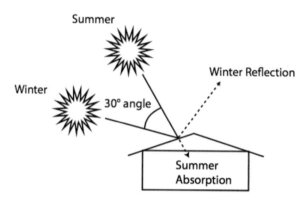

**Figure 7-65. Because of changes in the angle of the sun, solar rays tend to be absorbed in the summer and reflected in the winter.**

existing roof. For good adhesion of reflective roof coating and to reduce the progression of rust on metal roofs, the roof must be primed before applying a roof coating. In addition, for mobile homes manufactured before 1976, using scaffolding can help the installers avoid potentially damaging the roof by walking on it. The typical reflective roof coating installation involves one coat of primer and then two coats of the reflective or white roof coating. Finally, educate and encourage clients to periodically wash the roof to help maintain the benefits of the roof coating.


FLOORS

**Basement and Crawlspace Floors**

Installation of dense pack into the floors above basements and crawlspaces is much the same as for sloped ceilings except that fiberglass blown-in insulation is often used because of possible moisture issues. Covering an open cavity wall or floor (a wall or floor without any rigid cover, such as drywall over the face of the framing) with a vapor permeable air barrier or house wrap and then blowing insulation into the stud or joist spaces provides a far superior insulation than batt insulation. Covering a floor cavity above a basement with a rigid material such as drywall, plywood, or OSB and then dense packing the stud or joist space with insulation is even better. Some contractors even install sheathing this way, but leave a 4- to 6-inch gap between the sheathing boards at the edges to allow for inserting a fill tube. Once they have blown in the dense pack, they cut strips of OSB etc. to fill in the gaps they used to blow in the insulation. The same installation for dense pack applies to crawlspaces except drywall cannot be used as a rigid material due to the moisture concerns.

If there is already a rigid cover, such as drywall, over the bottom of the floor joists, drill holes in the drywall to access the space, and dense pack the joist space following the procedure for dense packing sloped ceilings. If there is not a rigid cover over the bottom of the floor joists, apply a vapor permeable air barrier or house wrap and then blow in the insulation. Seal the blow-in holes with a manufacturer-approved air barrier tape. Best practice suggests supporting the air barrier with twine or perforated foam board as described for fiberglass batt insulation if you want the insulation to last.

### Slab Floors

You can help keep the perimeter of a slab floor from being cold by installing foam board on the vertical edge of the slab on the outside, Figure 7-66. Typically, it is expected that the foam board extends at least 2 feet below the top edge of the slab. This is not often done unless some excavation work is already being done around the slab foundation since excavation work is typically very expensive.

Another option for a cold slab floor is to first cover it with a thick layer of polyethylene sheeting that is at least 6-mil thick. Then, cover the sheeting with 2-inch sheets of foam board and a series of 1 x 3 furred in wood "sleepers" (attached with appropriate concrete nails or screws into

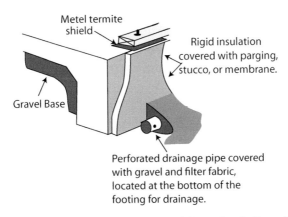

**Figure 7-66. Perimeter Insulation For Slab-on Grade Foundations**

the slab through the foam). Screw plywood on top of the sleepers as a sub floor. This configuration helps keep cold (in the winter) and condensation (in the summer) from getting into the house through the slab.

Never place carpet directly on uninsulated slab floors due to the risk of generating condensate and creating a mold problem.

### Cantilevered Floors

A cantilevered floor can represent a major concern for air infiltration and conductive heat loss due to inadequate insulation. If the space between the floor joists of the cantilever floor appears uninsulated, then the best approach is to dense pack insulation into the floor cavities to provide a good quality insulation and air seal at the same time.

Use zone pressure diagnostics to tell whether there is significant communication with the outside. Test while creating a -50 Pascal pressure in the home with a blower door, setting up a gauge outside so it reads "Pressure, Pressure" and placing a non-conducting plastic tube through an opening or joint in the bottom soffit cover of the cantilever or overhang (see ZPD testing procedure with the gauge outside). If the reading is close to 0, the cantilever is very leaky.

## WINDOWS AND DOORS

In the early days of energy retrofits, when there was not the same worry about cost-effectiveness, door and window replacements were common. However, this was before blower door testing became established and it was discovered that windows and doors typically do not represent extensive leakage areas. Now with experience and computerized modeling, window and door replacements are known to have a low savings to investment ratio and are therefore typically a last resort.

### Why Not Replace Windows

The primary reason it is often not cost-effective to replace rather than repair windows is that the R-values of windows are quite small (an R-value of 1 or less for a pane of glass). Even some of the best windows, like triple pane, wood framed, low-e windows, have an R-value just above 3. Typically, complete replacement of a window, rather than just repair, is not justifiable as a cost-effective measure in the weatherization program unless:

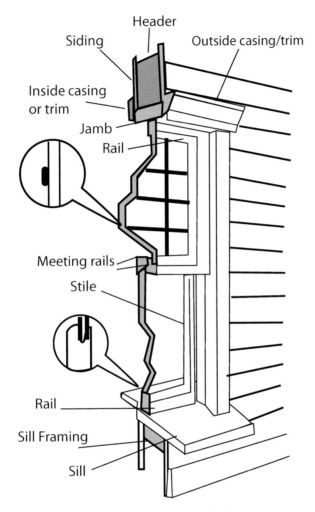

**Figure 7-67. Window terminology**

1) the window is either so weathered or damaged that replacing it is cheaper than repairing it. This could include major air leakage or comfort issues related to poor seals and operational problems with the window,

2) the cost of the lead safe weatherization involved in properly repairing a window exceeds the cost of its replacement (which may not involve extensive lead safe weatherization), or

3)    the window creates a compromise to building durability or repre-
      sents a hazard to safety or health that could not be corrected with
      repair. Window replacement should rarely be considered a mea-
      sure to reduce infiltration since that can typically be accomplished
      with repair.

If it has been determined replacement is the only option, make
sure to choose windows with a high solar heat gain coefficient or SHGC
and low-e coatings on interior panes in cold climates and a low SHGC
and low-e coatings on exterior panes in hot climates.

Table 7-6 shows the Building Performance Institute (BPI) default
U-values for a variety of windows. Use this table if you are unable to
identify the actual National Fenestration Rating Council (NFRC) U-values from the NFRC labels etched on the window glass corner or stamped
on the metal spacer on double glazed or better units. Remember the R-
values are the inverse of the U values. In other words $R = 1/U$.

### Table 7-6. Default Window Values

| Frame Type | Glazing Type | U-Value | SHGC | U-Value with low e | SHGC with low e |
|---|---|---|---|---|---|
| Wood | Single | .90 | .65 | NA | NA |
| | Single w/ Storm | .49 | .71 | NA | NA |
| | Double | .49 | .58 | .39 | .45 |
| | Triple | .39 | .53 | .30 | .45 |
| Vinyl | Double | .46 | .57 | .36 | .45 |
| | Triple | .36 | .52 | .36 | .45 |
| Metal | Single | 1.31 | .80 | NA | NA |
| | Double | .87 | .73 | NA | NA |
| Metal w/ Thermal Break | Double | .65 | .66 | .53 | .52 |
| | Triple | .53 | .60 | .43 | .52 |

## Window Repairs

Typical window repairs include caulking the interior trim around
the window, weather stripping the sliding surfaces and meeting rails
where the upper and lower sashes meet on double or single hung windows, replacing sash locks that are broken, and replacing broken glass
panes. If the windows are of the older variety, as in Figure 7-68, and
have a sash counterweight system that helps control the movement
of the lower sash with a pulley and counterweight, then pulley seals
should also be installed. A pulley seal fills in the gaps around the pulley
to help prevent air from entering the wall cavity where the sash weights

are located. Because sealing every possible joint around windows is not as cost-effective as sealing significant duct leaks to the outside or significant attic bypasses, limit this involved process to windows that are around areas where the clients seem to spend much of their time. Thus, do not spend a great deal of time sealing all windows in the home.

To replace glass, first remove the old broken pane, remove any old putty etc., obtain a new pane after measuring the opening and install it, along with glazier points and the glazing compound, with a putty knife. If you have the training and experience, you may be able to cut the glass yourself with a glasscutter using a wax crayon to identify where the cuts should be. Be sure to wear appropriate safety equipment such as safety glasses and gloves.

**Figure 7-68. Older style single and double hung windows often have pulley-supported counterweights integrated into the windows. These were intended to help the occupants raise the windows against gravity. In this display, the lower window has been pulled out on one side to see how the rope of the system was attached to the counterweight.**

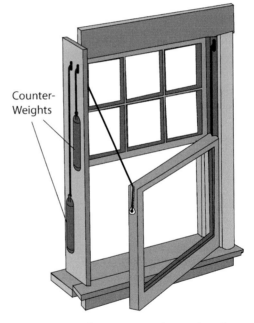

Counter-Weights

If old windows are removed, make sure to follow proper lead safe weatherization procedures, if applicable. This means protect the floor, wet the surfaces before removing them to avoid creating dust, vacuum and clean up, and properly dispose of the old windows.

Some retrofitters avoid having to deal with lead safe issues by simply taking out the old windows, but leaving on the unpainted metal trim and getting replacement windows that overlap the old window framing. This overlapping hides the old framing and avoids having to deal with any old paint that might be on it. This technique could be used for a variety of windows. Figure 7-70 shows various types of windows.

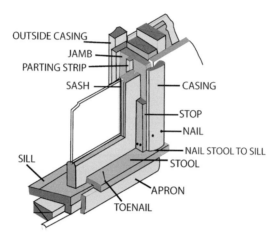

**Figure 7-69. Detailed Window Terminology**

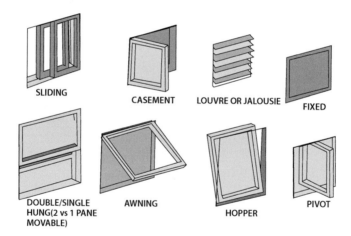

**Figure 7-70. Types of Windows**

If a double hung window is to be replaced, consider replacing the old frame and sash with a custom ordered vinyl replacement. When removing an entire window, first remove the old frame and trim using a thin bladed pry bar after scoring the painted joints along the trim and window with a sharp utility knife. Fit the new window inside the existing frame for the old window and then nail the window jamb in place. Inject nonexpanding foam in the gap between the window jamb in the frame and the wall of the home and then reinstall the trim and seal it with caulk. Follow the manufacturer's instructions for specifying, measuring, and installing these windows. Also, consider some other allowable window treatments such as movable insulation systems, interior storm windows, or simply educating the client on ways to provide high/low circulation of flow during warm periods or use of draperies during cold periods.

Some homes, especially mobile homes, have jalousie, awning, or hopper windows that are very leaky. To replace these, follow instructions similar to home window replacement except that putty tape and hex head screws are used to seal the flange of the window against the wall of a mobile home. Remember to use factory replacement windows to make sure that they fit properly.

**Doors**

As with windows, most door replacements have a high cost and relatively low impact on energy savings, and therefore are not cost-effective under the weatherization program. Solid wood core doors that are typically used as exterior doors in homes have an R-value of only 2.5. Energy Star qualified, foam-filled, steel doors have an R-value that is only close to 6. Thus, replacing doors is not cost-effective except in extreme circumstances such as where the door is weathered or damaged beyond repair, there is significant air leakage, there are operational problems, the door creates a compromise to building durability, the door represents a hazard to safety or health that cannot be corrected with repair, or the replacement cost, including material and labor, is less than the cost to repair. Thus, only in extreme circumstances is it cost-effective to replace the door. In the private retrofit arena, it is a function of whether the client wants a door replacement and is willing to pay for it.

If replacing the door is the best option, install a solid core wood or an energy-efficient pre-hung door that will fit the size of the rough, not finished, door opening. This process is typically much quicker than installing a solid core wood door blank unless you are sure you can buy an exact fit door replacement that will require little, if any cutting or trimming of the door. Once the old door and jamb are removed, you should be able to install and line up the new door and jamb, shim the gap between the jamb and the rough opening, and then nail or screw the doorjamb in place. If a solid or wood door is installed, trim or plane the edges of the door to fit the finished opening size, chisel slots for the hinges, install a lock, and weatherstrip the door. If needed, replace the hinges for the door.

If it is decided to repair the door, make sure the door has a threshold, a door sweep, a doorstop, and is properly weatherstripped. If the door does not have all of these items, retrofit the door so it has them. In addition, repair or replace damaged or broken hinges, locksets, and latches.

Perhaps one of the most significant leakage situations with doors is the cellar door accessed

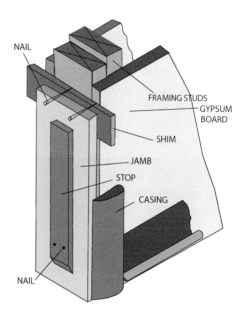

**Figure 7-71. Door Terminology**

through an angled bulkhead near grade level on the outside of a home (often found in the Northeastern U.S.). Often homeowners do not see the need to seal and weatherstrip these doors because they think the bulkhead door serves as a seal. Too commonly, neither the bulkhead door nor the basement door below it is typically sealed well enough to prevent significant air infiltration into the house in this area. Various types of door and window seals are shown in Figure 7-72. Because of the way bulkhead doors are designed, typically the best option is to tighten the basement door, especially between the jamb and the foundation, where non-expanding foam or blocking and caulking works well. The other option is to replace it with a pre-hung door that has treated- or decay resistant-wood for a jamb.

For mobile home doors, choose a prehung, factory replacement, single exterior door unit. Whether on a regular home or a mobile home, replacing deteriorated wood around the rough opening is important. If you are replacing a hot water heater closet door on a mobile home, for safety reasons make sure the replacement door is louvered if the hot water heater is fossil fueled.

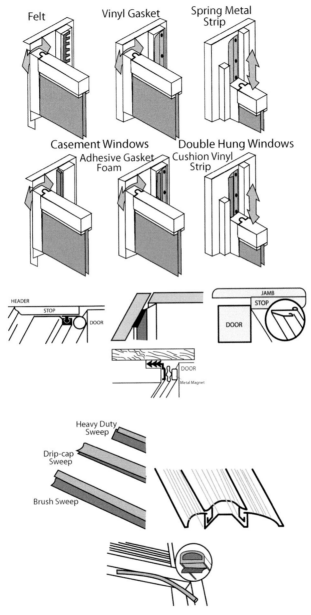

**Figure 7-72.**
**Various Types of Door and Window Seals**

## FURNACE DUCTS

### Energy Issues

Up to 30 to 40% of the home's energy can be wasted as a result of major leaks in ductwork. Thus, sealing duct leaks is a top priority. It is important to recognize that holes or other leaky spots in ductwork represent a more significant issue than leaks in the building shell because ductwork is pressurized to a much higher degree. Thus, if there are holes of the same size in a duct and in the building shell, far more air will rush out of the duct than through the building shell hole.

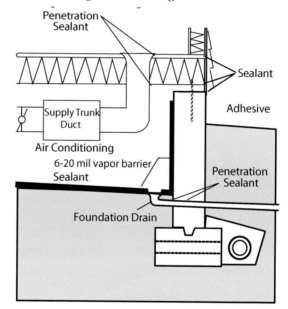

**Figure 7-73. Unvented Crawlspace with Insulated Floor Joists and Air Conditioning**

With furnace ducts, be primarily concerned with ducts located in unconditioned spaces because this is where the conditioned air will most easily leak to the outside of the thermal and air barriers. At the same time, houses that have virtually all of their ductwork located in a finished basement area may not be as cost-effective to test or repair as virtually all of the leaks are "inside" the conditioned space. Unconditioned spaces are also called "intermediate zones" because they are not inside or outside. Thus, duct sealing should be done when the ducts are located in unconditioned spaces such as crawlspaces (Figure 7-73), attics, or inside of exterior walls. Because duct leakage also significantly pressurizes or depressurizes these intermediate zones, it exaggerates the leakage through the building shell into and out of the home. If a return air leak draws in unconditioned air into the return air duct, or

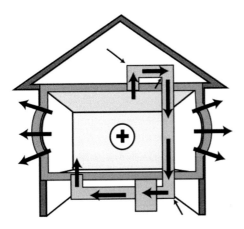

Figure 7-74. Having a return air leak can pressurize the house with respect to (WRT) the outside.

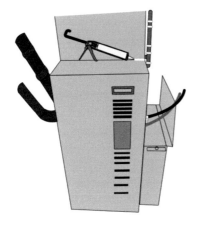

Figure 7-75. Caulking narrow joints closest to the furnace is very important.

a supply air leak pushes conditioned air out into the unconditioned space, energy has been wasted in the process, Figure 7-74.

Imbalanced pressures in rooms due to duct leaks in the home encourage leaks to the air barrier. These imbalances can cause air leakage to occur up to 2 to 5 times what would occur if the imbalances were not present.

Be most concerned about duct leaks that occur close to the furnace (Figure 7-75) or air conditioner air handler because this is where the higher positive and negative pressures are, depending upon whether it is the supply or return ducts, respectively. Some of the worst air leakage occurs near the air handler. If the air handler is located in an unconditioned space or outside, the effect is more severe. Some of the leakiest ducts are metal ducts that have not been sealed, most likely because of poor fit in their joints.

Any part of the distribution system that uses building framing for either supply or return air is not true ductwork (Figure 7-76) and must be well sealed. One of the more common examples of this is metal panning or sheet-metal on the underside of a basement or crawlspace ceiling joist cavity that serves as a return air duct or plenum. These are notorious for pulling cold air from unconditioned space. In cases where the duct is connected to a leaky exterior wall cavity, it can pull cold air directly from outside. See Figure 7-77.

Covered wall cavities or stud bays also should not be used as ducts since lumber and drywall are inappropriate duct materials. If there are moisture issues, condensation could deposit on the drywall and lumber and result in mildew or mold developing on these surfaces. It also could ultimately result in degradation of the drywall. Stud bay ducts are especially inappropriate on outside walls because of exposure to outdoor temperatures. In addition, "toe kicker" vents below the cabinets in bathroom and kitchen sinks can represent major leakage sites and can be very difficult to seal unless a direct duct connection is made between the register in the cabinet foot and the floor duct where it opens.

It should also be mentioned that return duct leakage in an unconditioned area of greater temperature differential from the house (such as attics that are hotter during the summer) could take a retrofit priority over other unconditioned areas that do not reach such a big temperature differential (such as crawlspaces).

### Health and Safety

There is also a safety reason for sealing ducts, especially return air ducts near combustion furnaces. Negative pressure can be created in a furnace closet by a leak in a return air duct that draws air into the return air duct. This results in the backdrafting of the flue gases down a furnace chimney or flue and into the living spaces. Similarly, pesticide treatments, mold that has become airborne, water vapor, and radon gas

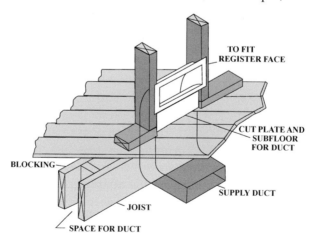

**Figure 7-76. True Ductwork**

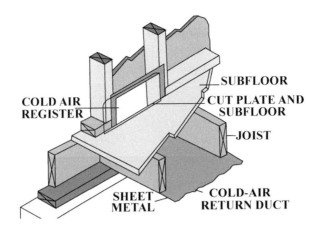

COLD AIR REGISTER

SUBFLOOR

CUT PLATE AND SUBFLOOR

JOIST

SHEET METAL

COLD-AIR RETURN DUCT

**Figure 7-77. Metal Pan "Duct" Between Floor Joists is Discouraged**

can be drawn up into these return air leaks to create other indoor air quality problems when the return air leaks are in a crawlspace or basement.

Dirty ducts typically do not have a major effect on airflow, but they can allow dust to become part of the indoor air. Asthma can be irritated by this indoor air pollution, so it may not hurt to clean dirty ducts to help with occupant health and safety.

**Connecting and Sealing Ductwork**

Before sealing or installing insulation, make sure the ductwork is properly attached, anchored, or supported. Metal to metal connections of ductwork must be fastened with a minimum of three equally spaced screws. This includes connections between the air handler cabinet and a metal plenum. Flex duct connections to metal must be fastened with a tie band using an appropriate tie band-tensioning tool. Duct board to duct board connections should be fastened with a clinch stapler before sealing. Duct board to flexible duct must be connected with an appropriate takeoff metal collar. When attaching duct board to a metal air handler cabinet, screw the duct board onto the termination bar on the air handler. Use screws or nails to attach a register boot to the wood around it to provide a better seal. If the register boot is supported by gypsum board, provide a boot hanger attached to some of the adjacent support framing with screws or nails, and connect the boot hanger to the boot with screws. In addition, provide snap boots to hold the boot in place next to the gypsum. All the different duct types must be supported at least every 4 feet with a minimum of a 1 1/2 inch wide material and attached in a way that it will not cramp the ductwork (see Figure 7-78). Specifically, metal ducts must be supported with metal rods, strapping or similar materials.

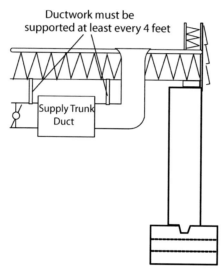

Figure 7-78. Ductwork must be supported at least every 4 feet.

### Retrofitting Ducts

Duct leaks outside the conditioned space are usually relatively easy to treat because they are typically more accessible than ducts in conditioned spaces. Make sure all the ductwork is properly sealed with a mastic compound and mesh. A properly sealed duct helps keep the conditioned air inside the home. Do not worry about the appearance of the mastic and mesh; it is more important that the duct is sealed properly. For the most part, ductwork is not exposed in the living space and so cosmetics are not an issue. Creating a tight seal on the ductwork however is very important. Seal all the joints on any part of the ducts that are accessible. You want to obtain a post-retrofit pressure pan reading of 1 Pascal or less on all the registers in the home if possible. Make sure all the connections or joints beginning at the main trunk coming off the furnace, any end caps, register boots, or elbows are carefully and properly sealed (Figure 7-79). In addition, where the register boot extends up to

**Figure 7-79. Caulking and mastic application should be done where the boots of the furnace ductwork connect to the floor or ceiling.**

the floor level, attach and seal the outside face of the boot to the edge of the hole cut out in the floor.

The closer you seal the metal boot to the hole, the easier you can fit the register back over the opening. The better you seal the boot to the edge of the floor or ceiling opening, the better you can keep cold air from basements, attics or other non-conditioned spaces on the other side from leaking into the living space. Perhaps the best way is to screw or nail the edge of the boot to the inside edge of the hole along its edge so there is a 1/4 inch or less gap and use mastic to seal this joint.

Flexible duct is usually used as duct in open attics and crawlspaces where a finish around the ductwork is not needed. Flexible duct should be stretched to its maximum length when installed and then supported in accordance with the manufacturer's specifications. When pulling it tight or supporting it, do not significantly pinch the flex duct. The interior lining of flex duct must be sealed with appropriate UL 181 BM listed tapes and mastic products. When connecting flex to a metal duct, the liner must be pulled up onto the metal duct as far as possible before using the tie band. Make sure the vapor barrier on neighboring duct insulation is taped to the flex duct using an appropriate tape recommended by the manufacturer. With metal ducts, the insulation must be attached to the duct by a rot-proof nylon twine or metal wire in a pattern that will adequately secure the insulation tight to the duct. A manufacturer-approved tape must seal separations in the duct insulation.

If you find a supply or return vent (Figure 7-80) in the garage, it must be appropriately removed.

**Figure 7-80. Vents like this should not be in garage walls.**

Cut the ductwork dedicated to the garage back as close as possible to the major supply plenum of the furnace. If there is a Y or a T in the branch system before the duct extends to the garage, the duct can be cut off at that Y or T. The hole remaining in the ductwork after the removal of the duct needs to be capped with sheet metal and fastened with at least three screws. The hole in the wall must be patched and taped using materials that meet the firewall requirements for garage/house walls in your area.

If a good portion of the ductwork is hidden from access because it is in outside walls, an inaccessible attic area, or cathedral ceilings, consider having these ducts sealed with an appropriate UL 181 aerosol. This involves spraying an aerosol into the ductwork under pressure so that the aerosol droplets congregate at the leaking holes and plug them. This can be a relatively expensive operation, but in some cases may be the only answer to leaky ducts exposed to these outside or unconditioned spaces.

Concentrate on insulating ducts in unconditioned spaces, but only after sealing them. R-10 is generally considered a minimum R-value for ductwork insulation. In colder areas, the insulation requirement may be higher. Use DOE approved audit software or a priority list to clarify how much insulation to put on ductwork in unconditioned spaces (according to the weatherization program in your area). In the private arena, check with local code officials for the minimum duct insulation requirements.

Vinyl faced fiberglass wrap works well for sealing. Make sure to use an appropriate tape to seal the joints of the insulation. Follow the manufacturer's instructions. If attic ducts lie close enough to the attic floor, you may be able to cover them with the loose fill insulation for the attic area. If they do not sit close enough to the floor, then some suggest that a minimum of R-10 insulation duct wrap be installed in attics.

Make a special point to seal all returns in closets or other spaces where combustion appliances are located, because the return air ductwork for combustion appliances can create a negative pressure in the appliance closet if it leaks. This can result in back draft of atmospheric appliances in the closet, a very serious health and safety threat.

Make a special effort to appropriately seal the duct materials coming immediately off the furnace that are nontraditional materials such as framed plywood, etc. This "platform" style return plenum often occurs where a furnace is in a smaller closet, as shown in Figure 7-81.

You may find that you have to remove some parts of these platforms to seal them off from the inside to better reduce the leakage. The closer these are to the air handler on the furnace, the more likely it is preferred to seal the joints from the inside, even if this means having to cut open an access hole to do so and then to seal the access hole from the outside.

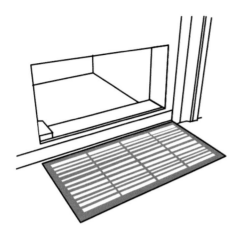

**Figure 7-81. A Framed-in Platform Style Return Plenum with the Grille Placed on a Hall Wall**

Platform style ducts are of special concern if they are in the garage because of the additional risk of indoor air quality issues if fumes from the garage can be drawn into the return air duct system for the living space in the home. All the ductwork from the furnace to the common wall between the house and the garage must be fire resistant to the same level as required for the garage wall.

The sealant of choice for ductwork is a latex-based mastic embedded in fiberglass mesh for holes or openings up to 1/4 inch. For openings greater than 1/4 inch but less than 3/4 inch, the best method is installing a backing, such as duct tape or foam backer rod, and then sealing over it with mastic embedded in fiberglass mesh. Some standards suggest that if a ¼ inch or less opening is more than 10 feet from the air handler, you need not use fiberglass mesh. For openings greater than 3/4 of an inch, attach a rigid duct material, such as metal, and then seal over the joints of the attached metal with mastic embedded in fiberglass mesh. The mastic must be UL 181 approved and you must overlap all repair joints by at least 1 inch on all edges with the fiberglass mesh and mastic combination. One example of an appropriate mastic would be RCD #6 Mastic. Mastic can be easily applied with gloves or an old uncontaminated brush. Some recommend that you wear three layers of vinyl gloves or one layer of vinyl gloves over your hands with a cotton work glove on the outside. If you are applying mastic to gypsum, spray water from a spray bottle before applying the mastic. Perhaps the best overall application involves first spreading a layer of mastic over the joint, placing fiberglass mesh tape over the top of that layer, and then applying yet another layer of mastic material over the tape.

## OTHER SPECIAL ISSUES IN AIR LEAKAGE

### Flues/Chimneys

The so-called "high-temperature" issues with flues, chimneys, or recessed light fixtures represent a special category of concern. As shown in Figure 7-82, there are minimum clearances for brick chimneys.

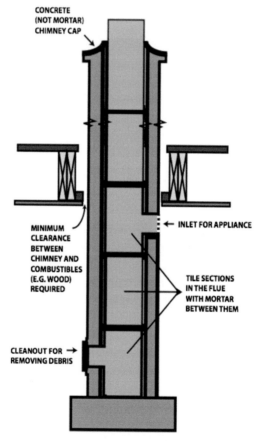

**Figure 7-82. There are minimum clearances for brick chimneys.**

To seal around the flues or chimneys use minimum 28-gauge sheet metal as a flange and high temperature caulk to seal the sheet metal joints, especially where they meet up with the chimney or flue. Where the metal needs attachment to the wood framing around the chimney, construction adhesive will provide a good seal if galvanized nails or screws are used to attach the sheet metal to this framing. Other noncombustible materials besides sheet metal may be used if they are rated for the higher temperatures of flues, etc. You will also need to build a metal dam at least 3 to 6 inches away from the flue or chimney (depending on the type of vent pipe material or chimney inside the dam) and high enough to keep the blown-in cellulose and fiberglass insulation safely away from the flue or chimney (20 or more inches high). Instead of a metal dam, you can also stack batts of rock wool carefully cut to fill in around the chimney or flue on top of the metal flange to hold the cellulose insulation back. This serves as a dam and better insulator than the open air inside of a metal dam. However, be sure the blow-in insulator knows not to blow insulation on top of the batts of rockwool.

As a special note, if you find a combustion or make-up air opening in the attic, do not seal it off. These openings provide oxygen to the burners of combustion appliances. They are typically made of solid round or square ductwork that is larger than the usual exhaust fan duct that simply opens up into the attic space. If possible, trace where the vent goes to verify it is a combustion air vent.

Wrapping brick or metal chimneys or flues with rockwool batt insulation to an adequate height can help avoid having to fit time-consuming metal insulation barriers around them, see Figures 7-83 and 7-84. Do not confuse these with exhaust fan ducts (like those for bathrooms and kitchens) that need to be extended to exhaust bad indoor air to the out-

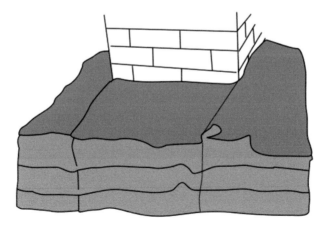

**Figure 7-83. Placing rock wool (not fiberglass) against high temperature flues, chimneys, etc. can make insulating much easier.**

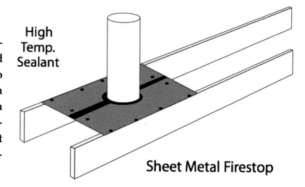

Figure 7-84. Bypasses around flues and chimneys must also be sealed underneath the insulation with high temperature materials such as sheet metal and high temperature sealant.

High Temp. Sealant

Sheet Metal Firestop

doors. Turning on all the fans in the house before going up into the attic can help you distinguish exhaust fan ducts from combustion air vents.

### Fireplaces

Since all the heated air from a fireplace escapes out the flue, use fireplaces to take advantage of radiation heat from the hot surfaces in the fire. Fireplaces that are not being used, or fireplaces without glass doors represent a significant breach in the air barrier. There are many ways to seal off a fireplace including closing the damper, using an inflatable balloon-like plastic pillow or using high temperature materials. If you are using the inflatable pillow, be sure to follow the manufacture's installation instructions. Make sure you provide some method of warning any future occupants that the fireplace has been sealed off to avoid anyone using it before realizing it will not properly draw. The warning should be conspicuous for anyone who would be in a position to ignite a fire in the fireplace. The current occupants should also be verbally warned as well. Take any precautions you feel are necessary to make sure no one will try to use the fireplace without removing the obstruction and opening the damper.

Glass doors are probably the best measure to significantly seal off, and yet later be able to easily use, a fireplace, Figure 7-85. They

Figure 7-85. Glass doors on fireplaces allow radiant heat to be provided from the fire without the loss of conditioned air from the home.

also keep a fireplace that is being used from becoming a large air leak in the home and still allow the fire to generate radiant heat.

### Vanities, Cabinets, Bathrooms

Vanities and other cabinets that have supply or drainage plumbing, electrical wiring, or other unsealed penetrations through the wall behind them, can represent compromises to the air barrier, especially if the wall is an exterior wall. Figure 7-86 shows a hole behind a sink.

In fact, even an interior wall can allow for significant air leakage if it communicates with an attic because there is no top plate to the wall or there are many electrical, plumbing, or mechanical penetrations through the top plate into the attic. Perhaps the best way for sealing these leaks behind cabinets, etc. is with drywall or other nonflammable rigid board. Foam board typically cannot be used here since it would be exposed to the living space, unless the foam is covered with a fire rated material like drywall.

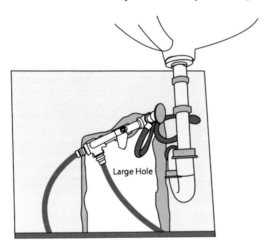

Built-in cabinets (including china cabinets) that are attached to a wall often have no air barrier between their backside and the studs. Thus, any intentional or unintentional slits in the wood or between wood pieces that are part of the backside of the cabinets will create an automatic air leakage site. A leak site can also be where the drywall or lath and plaster butts up against the outside edge of these cabinets. One of the best ways to air seal behind these cabinets may be to dense pack the wall behind the cabinet if it is an outside wall. If the wall is an interior wall, it is probably best to seal any penetrations through the top plate or install a top plate equivalent air barrier if there is none.

Recessed wall systems such as medicine cabinets, electrical panel boxes, recessed wall niches and shelves, etc. can represent significant

**Figure 7-86. Holes often are made into the wall behind the plumbing under sinks. These must be sealed to avoid bypasses in these areas.**

Large Hole

air leakage sites. The best treatment may be to dense pack around the recess if it is an outside wall, or sealing the top plate (or installing one if the wall lacks one) in the case of an interior wall. The only other option may be to remove the recessed wall item if it is feasible (if it is not an electrical panel, etc.)

A related issue has to do with cutouts in the flooring under showers and tubs to accommodate supply and drainage plumbing, including P traps, as shown in Figure 7-87. You may be able to use foam board and spray foam to seal off these areas. If the foam board is visible from a living space, such as the ceiling or basement area, you need to cover the bottom side of the floor (or at least the foam board and foam) with a fire resistant material such as drywall.

**Figure 7-87. Cutouts around plumbing penetrations in floors can represent major air leakage sites, otherwise known as "thermal bypasses."**

The wall area behind tubs, fiberglass showers, etc. is usually not covered with drywall because these fixtures are typically installed before the drywall is installed. In fact, if the plumbers do not install insulation with an air barrier, the shower or tub wall could be the coldest wall in the house. Perhaps the best option is to install dense pack cellulose or fiberglass insulation in the wall behind these fixtures using holes in the wall above the fixtures if possible.

**Porches and Additions**

Porches and additions can represent a special challenge for good air sealing and insulating. The areas where porch covers and additions are tied into an existing home can represent one of the largest hidden air leaks in a home. Perhaps the porch cover or addition was installed without first installing sheathing or siding on the wall where it attaches to the house or around the overhanging joists that serve as structural support for the porch cover.

Careful use of zone pressure diagnostics can tell whether there are interconnections between an addition wall and roof or porch cover and

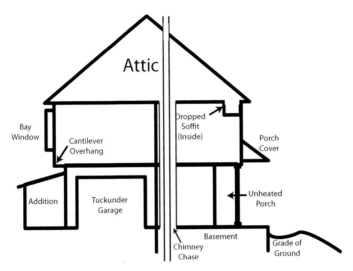

**Figure 7-88. The opportunities for air leakage are greater the more convoluted the home design. Each change in direction is an opportunity for poor sealing. This diagram shows some of the locations where air leakage commonly occurs.**

the inside of the original or primary walls of the house. If a blower door test shows a significant difference between the pressure outdoors and the pressure *inside the addition roof or porch cover,* such as more than 25 Pascals, then be concerned that the porch or addition attic space significantly communicates with the *inside of the house.* If there is communication between these areas, then determine the best place to put the air barrier and insulation. This often may be a function of accessibility. If you cannot adequately access and seal off the outside sheathing of the primary wall of the home from the porch or addition attic, then you probably need to air seal and insulate the entire porch or addition using dense pack insulation.

### Garages

The air barrier between the house and the garage (Figure 7-89) is very important because of the risk of pollutants such as chemical vapors and vehicle exhaust getting into the home and creating poor indoor air quality.

Garage doors tend to be very leaky and the floor and walls of the garage, including any ductwork, are essentially exposed to the outside. There can also be significant heat loss when the floor cavity of a conditioned floor

**Figure 7-89.** Assuring an intact air barrier between the garage and the living space is critical in protecting the owners from high utility bills, and more importantly from the risk of pollution in the garage. An often overlooked important component to reducing

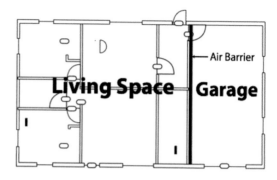

this risk is the conversion of ordinary hinges to auto-closing hinges on the house-to-garage door.

above the garage is inadequately insulated. An often overlooked important retrofit item to complete in this regard is the replacement of ordinary hinges with auto-closer hinges on the house-to-garage door.

Use zone pressure diagnostics to evaluate the quality of the air barrier between the garage and the house. With the house at -50 Pascals, compare the pressure in the garage with the pressure inside the home. You can attach a short plastic tube into the end of a tube on a digital gauge set to the "Pressure, Pressure" mode to find out if it is close to a 50 Pascal difference between garage and house. While inside the house, place the tube on the threshold of the door and close the garage/house door over it to read the pressure difference. (See ZPD section for more details). Air seal and insulate the wall using dense pack insulation in the wall and the enclosed floor cavity above a tuck under garage if applicable. Seal any openings or breaks in the wall. Usually it is easiest to seal from the garage side, and, since this is typically drywall, a sheet metal drywall patch may be the quickest and easiest method for sealing smaller holes. Holes larger than about 8" x 8" may have to be sealed with adequately backed drywall board. Cracks can be sealed with drywall tape and mud.

## BASEMENT AND CRAWLSPACES

### Inspect

When conducting a walk-through inspection of the basement, look for the following situations: Make sure stone foundations are in relatively good shape and they have been properly pointed with mortar. Check for debris in the perimeter drain, and see if the cellar runway door in the foundation wall is well sealed and in good condition. In addition, look for bypasses through the floor where gas pipe, water supply or drainage plumbing, electrical wires or cables, or mechanical vents or flues penetrate the floor. Another common bypass includes the rim or band joists area just above the foundation walls. Rim joists are particularly leaky in older homes built before foam seal was used under the sill plate during construction. Also, look for poorly weatherstripped exterior cellar doors, broken windows, and whether insulation exists on the hot water heater tank and piping. Remember, if there is at least limited use of a basement, such as the existence of a washer and dryer, then you may be more inclined to have the basement treated as conditioned space, especially if there is no duct insulation or any basement ceiling insulation.

Use the zonal pressure diagnostics technique for a crawlspace basement as described for the garage, but place the tube underneath the door or hatch cover to the basement or crawlspace. This will allow you to determine whether the basement or crawlspace is more connected to the house or the outside. See Figures 7-90 and 7-91 for different situations. Set the blower door to create a -50 Pascal pressure in the house (see ZPD section for more detail). If the pressure is basically the same in the basement as it is in the house, or if there is a differential of only 0 to 2 Pascals between the two, then there is probably a very leaky floor that would take a great deal of work to seal off to create an unconditioned space. In this case, you would probably make the basement or crawlspace conditioned space and create air and thermal barriers along the exterior foundation walls and floor of the basement or crawlspace.

If, on the other hand, the pressure is significantly different or there is a differential of closer to 50 Pascals between the two, then the floor may be relatively tight. There might also be a vented or relatively leaky foundation wall to the outside (such as a rubble stone wall). In this case, it would take a great deal more work to seal the foundation to make the basement or crawlspace conditioned space. In this case, you would probably make the basement or crawlspace unconditioned space and use the ceiling of the crawlspace or basement to create air and thermal barriers.

A concern with unconditioned basement and crawlspaces is ductwork leakage. Conduct a pressure pan test on all the vents on the lower floor of the house to determine how leaky the ductwork is in the base-

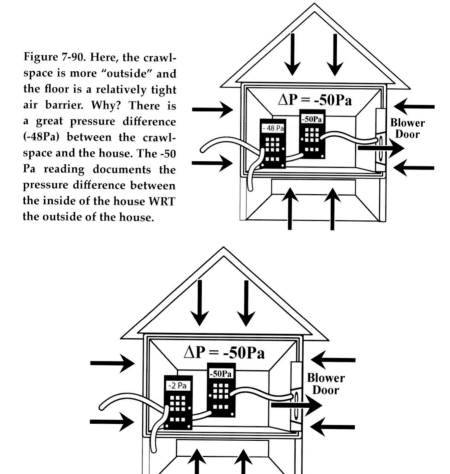

Figure 7-90. Here, the crawl-space is more "outside" and the floor is a relatively tight air barrier. Why? There is a great pressure difference (-48Pa) between the crawl-space and the house. The -50 Pa reading documents the pressure difference between the inside of the house WRT the outside of the house.

Figure 7-91. The crawlspace is more "inside" and the floor is a very leaky air barrier because there is very little difference in pressure (-2 Pa) between the house and the crawlspace. (The -50 Pa reading documents the pressure difference created by a blower door between the inside of the house WRT the outside of the house.)

ment or crawlspace (see Pressure Pan section for more detail). If it is decided to leave the basement or crawl unconditioned, the reading should be as close to zero Pascals as possible to feel confident that heat is not being lost to unconditioned space. In fact, even when the basement or crawlspace is conditioned, the pressure pan testing results should be close to zero Pascals. The tighter the basement or crawlspace, the less supply air will leak from the ductwork.

Since combustion appliances are often located in basement or crawlspaces, a worst-case combustion appliance zone/closet (CAZ) test needs to be performed to assure safe appliance operation. This test will better assure that there is no back draft on any atmospheric appliance flues.

### Basement and Crawlspace Strategies

Sealing off basements and crawlspaces can be almost as important as sealing attic leaks. One way to help prevent attic leaks due to the stack effect is to plug off the holes in the basement to reduce the flow of air into the house that becomes available to exit out the attic. Basement and crawlspace leaks are also sealed for indoor air quality reasons. A good seal helps keep pesticides, mold, radon, and moisture from getting into the house.

Whether to retrofit basement or crawlspaces to be conditioned or unconditioned depends on several factors. As shown in a previous paragraph, the diagnostics have probably shown whether the space is already acting as conditioned or unconditioned. Another factor is how the owner is using the space. If the owner only uses the space for storage that is accessed only occasionally during the year, then the basement or crawlspace probably does not have to be treated as a living space that needs to be conditioned. If on the other hand, the stored items are accessed on a more regular basis or if there is a washer and dryer in the area (Figure 7-92) that indicates weekly use, you would be more inclined to retrofit it as conditioned space.

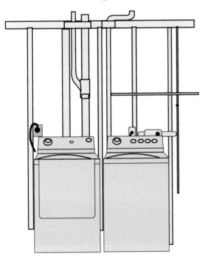

**Figure 7-92. Finding a washer and dryer in an unfinished basement would be a major factor in deciding whether to retrofit it as conditioned space.**

Some other factors to consider to determine whether the basement or crawlspace should be conditioned or unconditioned space include whether the outside walls of the basement or crawlspace are more than 50% below grade, whether the foundation is already relatively tight and unvented, whether water supply pipes or ducts exist in the space, whether there is a water heater or furnace in the space, and in the private arena, what the client prefers and can afford. If the basement (Figure 7-93) or crawlspace already feels like it is a conditioned space (it feels warm to you inside when it is cold outside, or vice versa), then retrofit it as conditioned space regardless of whether it appears to be purposefully or unintentionally conditioned. If 50% or more of the exterior foundation walls of the basement or crawlspace, from top to bottom, are below the grade of the ground outside, it is more likely to be treated as conditioned space because much of the ground effectively provides some degree of free insulation.

If there are exposed water supply pipes in the basement or crawlspace that cannot be adequately insulated to prevent or protect from freezing, bring the crawlspace or basement "inside" the house to protect the pipes from freezing. If the crawlspace or basement cannot be adequately conditioned to keep pipes from freezing, use pipe heating cables or heat tape that can plug into most any outlet in a home to protect the supply piping. Be sure that the full length of the piping is protected from freezing within the space. Some heat tape is available with a thermostat built in so that electricity is used only when it is cold enough to require it. This saves energy and avoids the risk of overheating if the heat tape is not disconnected before the weather warms up in the spring. It also avoids the risk of freezing if it is not turned on before the weather gets cold in the fall.

If the basement or crawlspace foundation is not vented,

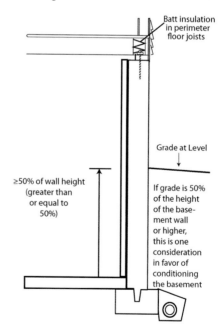

Figure 7-93. Conditioned Basement with Insulation on the Walls

Batt insulation in perimeter floor joists

Grade at Level

≥50% of wall height (greater than or equal to 50%)

If grade is 50% of the height of the basement wall or higher, this is one consideration in favor of conditioning the basement or, from its appearance or testing, it is a relatively tight foundation, treat it as conditioned space. On the other hand, if the basement or crawlspace foundation is vented or there is a relatively leaky foundation (such as often happens in rubble stone foundations), it is more likely to be treated as unconditioned space because of the additional work involved in sealing so many leaks in the foundation. An exception to this would be in homes that are in hot, humid climates where it is recommended to make crawlspaces and basements conditioned space. If there are ducts in the space, even if they end up being insulated, heat will be lost into the space and it would be preferred to keep the basement or crawlspace inside the house. Likewise, if a furnace or water heater is located in the space, then some of the heat generated by these appliances can provide some heat to the basement or crawlspace and encourage the client to treat it as a conditioned space. In the private arena, since you can usually tell your clients which energy retrofit measures will be more cost-effective, they end up deciding which measures they want to pay for. The owners could decide to have the retrofitter install measures that would not satisfy the SIR requirements of the weatherization program.

## Retrofitting Basements and Crawlspaces

If it is decided that the basement should be unconditioned, (Figure 7-94) then the retrofit work that should be done includes:

1) sealing all the bypasses between the house and the basement,

2) sealing all the supply and return ducts in the unconditioned space,

3) insulating all the ductwork in the unconditioned space to the recommended R-value for the area,

4) installing a vapor barrier to the heated side of the open floor joists (except in warm,

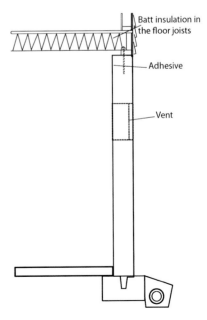

**Figure 7-94. Unconditioned Basement with Insulation in the Floor**

Batt insulation in the floor joists

Adhesive

Vent

humid climates or regions where no vapor barrier is recommended -see vapor barrier section),

5)    insulating all the open floor joists to the recommended R-value
      with batt insulation held up by wire or other supports,

6)    if the basement has "closed" floor cavities above it (for example,
      a drywall covered ceiling), insulate them with dense pack insula-
      tion,

7)    if the floor includes a plank sub floor, you may find that the only
      way to adequately air seal the floor is to caulk or foam in the gaps
      between the plank boards as well as the other bypasses, or to rely
      on a vapor barrier as an air barrier,

8)    with rim or band joists bypasses, seal the sill plate/foundation
      joint on the inside or outside with caulk, and, if the gap warrants
      it, foam. Otherwise, consider installing a 2-inch thick foam board
      insulation cut to fit against the band or rim joists and foam seal
      around the edges of the foam board in between the floor joists. Be
      sure to seal any gaps between the bottom edge of the floor joists
      and the foundation as well. An even tighter seal could result from
      using a spray in foam application in the rim joist area from the top
      of the foundation to the floor deck above. Seal all the penetrations
      all the way around them with foam or caulk.

Never seal off the opening of any vent or duct that appears to be
providing combustion air for a combustion appliance such as a furnace,
water heater, or boiler. When the retrofitting work is complete, conduct
pressure pan tests on all the ductwork registers to show a reading close
to zero Pascals.

If it is decided that the *basement* should be *conditioned*, then the
retrofit work should begin with air sealing the perimeter mudsill, band
joist, and all the air pathways between the house and the basement.
Also, seal vertical joints in stepped wall transitions, and seal and weatherstrip doorways to the basement or cellar from outside, including bulkhead doors. Seal where gas pipe, water supply or drainage plumbing,
electrical wires or cables, or mechanical vents or flues penetrate the
foundation wall. Also, insulate the band joist area to the recommended
R-value. Seal the sill plate/foundation joint on the inside or outside
with caulk, and if the gap warrants it, foam (Figure 7-95). Install a 2-inch
thick foam board insulation cut to fit against the band or rim joists and
foam seal around the edges of the foam board in between the floor joists.

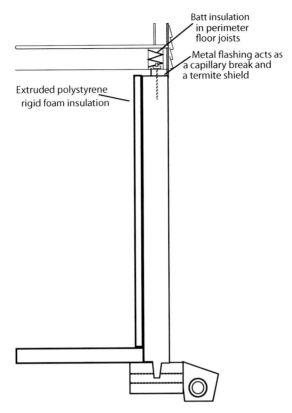

Batt insulation
in perimeter
floor joists

Metal flashing acts as
a capillary break and
a termite shield

Extruded polystyrene
rigid foam insulation

**Figure 7-95. Conditioned Basement with Foam Board on the Inside Wall**

Be sure to seal any gaps between the bottom edge of the floor joists and the foundation as well. An even tighter seal could result from using a spray-in foam application in the rim joist area from the top of the foundation to the floor deck above. Seal all the penetrations coming into the house, such as plumbing, telephone, etc. all the way around them with foam or caulk either on the outside or inside, or both if accessible.

If the penetration is below grade, the only option may be to seal on the inside. If the top of the foundation wall has two or more different levels, you may find gaps on the vertical edge of these stepped foundations so wide as to require insulation being stuffed into the gap before you foam or caulk the opening.

If a door from the basement to outside has major damage or major gaps around its edges, you may want to replace the door. If you replace any of the wood framing on the inside that even touches the foundation, be sure to replace it with treated wood, and, if possible, place some water repellent material between the wood and the foundation, such as ice shield. Also, make sure to seal any wood that already exists next to the foundation with caulk. Since there is also a need to insulate the band joist area to the recommended R-value, you can alternatively both air seal and insulate the band joist and the mudsill with two-part spray foam.

In addition, seal off any areas of the basement that will remain unconditioned such as coal bins or vented crawlspaces next to basements. Seal all return ducts and the return plenum for safety, seal cracks in the foundation wall, and weather strip any doors to the outside. Consider insulating basement walls inside with either two-part spray foam or a batt insulation down to 2 feet below where the grade sits outside.

Insulating the inside of basement walls just to 2 feet below the grade of the ground outside can be four to five times more cost-effective than insulating to 5 feet below grade. One concern with two-part form is that the temperature of the wall cannot be near or below freezing or the two-part foam will not adhere or cure properly. Another concern is that the air temperature within the crawlspace or basement must be within about 70 to 75°F for the foam to properly cure.

Make sure that the boot to register transitions in the ductwork are properly connected, and repair any branch ducts or broken trunk lines in the ductwork. It is not a bad idea to seal off the bypasses (Figure 7-96) between the basement and the living space above when conditioning

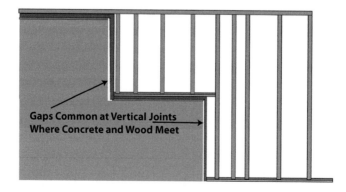

Figure 7-96. An important place to check for bypasses in on the vertical edges of stepped foundations.

the basement. At the same time, do not install insulation between the floor joists.

After completing the retrofit for the conditioned basement (Figure 7-97), use zone pressure diagnostics to confirm whether the basement has been well sealed from the outside.

Insulate at least two feet below the grade in basements. It is not necessary to go more than two feet below the elevation of the grade on the outside.

## Crawlspaces

Be on the lookout for possible moisture problems in crawlspace areas that are likely to put retrofit work at risk. Determine the underlying causes of existing or possible moisture damage or problems, or the

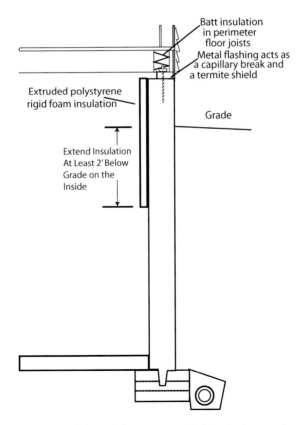

**Figure 7-97. Conditioned Basement with Insulation on the Walls**

purpose of insulating and retrofitting a home is being defeated. The risk of moisture problems goes up when a crawlspace rather than a basement is retrofitted. Often builders will install a crawlspace rather than a basement because of water table or other related issues.

Crawlspaces can be retrofitted for either moisture control (a health and safety reason), or energy savings (something that must be justified under the SIR of the weatherization program or under affordability in the private arena). Under the weatherization program the criteria for health and safety measures is different than it is for energy saving measures. It may be easier to justify placing the crawlspace inside under a health and safety concern rather than based on energy savings. If a crawlspace is overcome with moisture that is difficult to treat such as in flood prone areas or in poorly drained areas that have standing water or polluted soil, a conditioned crawlspace should not even be considered. Furthermore, if a crawlspace cannot be accessed because there is no opening, obviously one must be installed before retrofit work can begin (Figure 7-98). Whether the crawlspace is turned into a conditioned or unconditioned space, there will be a benefit to sealing the floor of the living space above from the crawlspace.

If it is decided that the crawlspace should be unconditioned (Figure 7-99), then the retrofit work will be the same as on unconditioned basements (see previous section) but with some additional points. Check local codes, especially about any requirement for ventilating crawlspaces. Make sure to seal all the supply and return ducts, etc. as suggested for unconditioned basements. Also, resolve moisture issues (including those caused by poor drainage around the foundation), remove debris, seal any foundation vents that exist (especially in hot hu-

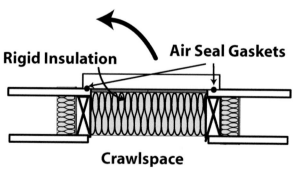

**Crawlspace**

Figure 7-98. This shows a simple crawlspace access proposal for insulating and sealing the crawlspace hatch.

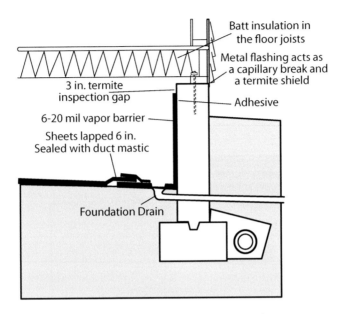

Batt insulation in the floor joists

Metal flashing acts as a capillary break and a termite shield

3 in. termite inspection gap

Adhesive

6-20 mil vapor barrier

Sheets lapped 6 in. Sealed with duct mastic

Foundation Drain

**Figure 7-99. An Unconditioned Crawlspace**

mid climates), and install a ground vapor retarder that is at least 6 mil plastic on the crawlspace floor to help prevent ground source moisture vapor from entering the crawlspace. Overlap this vapor barrier at least 12 to 24 inches and seal all the seams with an appropriate tape or sealant (sealant is preferred). Use spray foam around any penetrations in the vapor barrier, such as foundations for pier supports, the pier supports themselves, water supply or drainage pipes, etc. and make sure the vapor barrier is sealed up against the foundation wall. In some cases, even metal strapping and bolts are used at the edge to hold the vapor barrier tightly against the foundation wall.

If it is decided that the crawlspace should be conditioned, (Figure 7-100) then the retrofit work will be the same as on conditioned basements (see previous section) but with some additional points. Make sure all the return ducts, supply ducts, and return plenum, etc. are sealed. Resolve moisture issues including those caused by poor drainage around the foundation by removing debris, sealing any foundation vents that exist, insulating any access hatches that open to the exterior, and installing a ground vapor retarder that is at least 6 mil plastic on the crawlspace floor to help prevent ground source moisture vapor from

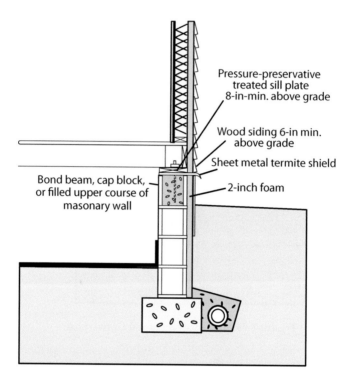

Pressure-preservative
treated sill plate
8-in-min. above grade

Wood siding 6-in min.
above grade

Sheet metal termite shield

Bond beam, cap block,
or filled upper course of
masonary wall

2-inch foam

**Figure 7-100. A Conditioned Crawlspace with all Insulation on the Exterior**

entering the crawlspace. Properly overlap and seal the vapor retarder at overlapping seams and attach the vapor retarder onto the foundation wall. This can potentially overlap the vapor retarder with the wall insulation depending upon how far down on the wall the insulation goes, which is a function of where the ground grade level is outside.

If you decide to use foam board insulation on the inside of the wall of the crawlspace, make sure to tape or seal the vapor retarder coming up the wall to the board insulation as well as the joints in the foam board. Use a minimum of 2-inch foam board for this application.

### Foundations

With regard to the foundation, note in detail what type of foundation exists, whether the basement or crawlspace is currently conditioned or unconditioned, whether there are moisture problems including bulk water intrusion or other indoor air quality issues, whether combustion appliances or a hot-water tank exist within the space, whether special zone issues exist such as a tuck under garage or coal bin, how the dryer venting is arranged and installed, whether there seem to be direct penetrations to the outside or major air bypasses in the area, what the components of the thermal boundary are (especially if the basement is unconditioned and next to a conditioned space above), what the condition is of the duct system and its insulation, whether there are hazards to individuals in the area such as electrical hazards, and what the band joist, wall, and ceiling insulation levels are.

Other things that can be important in evaluating foundations include identifying and recording the location on the sketch of any ceiling wiring and plumbing chases in the first-floor ceiling, crawlspaces without vapor barriers over a dirt floor, a wet crawlspace, water in the basement, open sump pumps, and too small of a clearance between the ground and the floor joists. Remember that the presence of moisture in a home represents a concern for the health and safety of weatherization workers as well as occupants.

Fiberglass batt insulation that has been designed for metal buildings in R-11, 4-foot-wide batts can be hung horizontally from the sub floor and draped against the inside of the crawlspace or basement wall to quickly and easily insulate foundation walls. If two-part foam is applied to the foundation wall, not only does the foam air seal and insulate the foundation, it also is more dependable in keeping the foundation walls above the dew point. This keeps water vapor from condensing between the insulation and the cool foundation walls, which in the case of fiberglass batt insulation, would wet the insulation and deteriorate its R-value.

### Pier and Beam or Post and Beam Structures

A pier and beam structure has piers supporting the house throughout but without a foundation wall around the perimeter. A typical mobile home is a type of pier and beam system except that the supporting piers are not permanent, but removable. Because there is no perimeter foundation on permanent pier and beam structures, the retrofit measures are limited to: sealing the bypasses in the floor above the under area, covering and attaching a strong rigid air barrier to the underside of the floor, and installing loose fill dense pack insulation inside this air barrier. If batt insulation is preferred, install a layer of fiberglass batts underneath between the floor joists and cover the batts with their own air barrier.

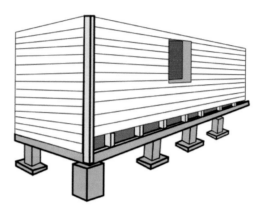

**Figure 7-101. Pier and Beam Support Structure for the House**
Notice the heavy duty built up joist that helps keep the wall from sagging between posts where the siding has been removed.

### Slab Foundations

With slab foundations, retrofit options are limited. Since it is not cost-effective to dig down to check for or install insulation around the outside perimeter of a slab foundation except in very, very cold climates, this measure is not allowed, at least not under the weatherization program. It is an available option in the private arena only if the client is willing to pay for it. If is it attempted, a rigid foam board insulation must be used, and the house siding or a flashing above it must cover its top. To make matters more complicated, many municipalities require all the foam board to be completely covered to protect it from deterioration from the elements, such as by a hardboard, stucco, parging, or other durable material.

You can also attempt to seal the joints between the wall mudsill and the slab. Use foam (Figure 7-102) to seal the outside where the wall meets the slab. If there is carpet inside the house, you may be able to caulk at the bottom edge of the baseboard at the joint with the slab.

SPECIAL TOPICS

### Special BPI Standards for Insulation and Air Sealing

The BPI standards require that you either verify an effective air barrier between the attic living space or specify attic air sealing as the

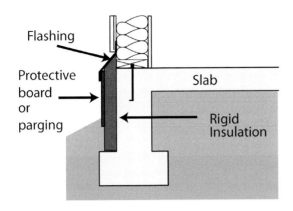

**Figure 7-102. Proper Installation of Rigid Foam Insulation on the Outside of a Slab Home**

workscope before you recommend attic insulation or ventilation. Until air sealing is complete, if it is necessary, no insulation should be installed. Likewise, attic ventilation cannot be recommended or installed until a thermal barrier has been verified as well.

Three types of techniques can be used to determine the effectiveness of an air barrier including:

1) depressurization using a blower door and "smoke stick" in the attic;
2) visually inspecting the attic for cavities (thermal bypasses) underneath the insulation, blackened insulation, and evidence of moisture (including a requirement for determining the source of the moisture); or
3) differential pressure diagnosis such as series leakage tests or "add a hole" type methods.

Attic leakage paths are given a priority over other air sealing measures. In addition, the minimum insulation levels required are based on local codes.

Both pre-installation blower door testing and post-installation blower door testing must be conducted whenever any one of the three following recommendations or conditions arise (Table 7-7):

1) Any air sealing work is done,
2) Enclosed cavity insulation representing 15% or more of the total building shell area is done, or
3) Any duct sealing of ducts outside of the thermal envelope is done

**Table 7-7. BPI—Situations When Pre-
and Post-blower Door Testing is Performed**

| |
|---|
| 1. Any air sealing work |
| 2. The enclosed cavity insulation represents 15% or more of the total building shell area |
| 3. Any duct sealing outside of the thermal envelope |

Any connections in heating ducts outside of the building envelope and cooling ducts in attics must be sealed with duct mastic followed by a minimum of R-5 insulation underneath the duct wrap.

**Electrical Issues**

Figure 7-103 shows how electricity gets from the source to the outlet. Perhaps electricity can best be understood by analogizing it to flow and pressure. Current, amperage, or electrical flow rate is analogized to water flow in a water supply line while voltage is analogized to water pressure in a water supply line. Thus, the higher the amperage flowing through a wire, the more electricity there is "flowing" along that wire. Voltage has more to do with the "power" or "pressure" behind the electrical flow such that the higher the voltage, the more power there may be available. When the electrical flow or amperage is too high for a given size wire, the wire is likely to overheat, possibly destroying the insulation around the wire. In a properly protected electrical system, if the flow through a given wire is too high for its size, the fuse will burn out or the breaker will trip so that no electrical flow can continue through that wire until the fuse is replaced or the breaker reset.

Electrical systems for small residential buildings are usually simple in concept and layout. The primary components are the service entry, panel box, and branch circuits. In unaltered buildings built since about 1940, the electrical system is likely to be intact and safe, although it may not provide the capacity required for a planned retrofit of the building. Adding a larger panel box between the service entry and the existing box can increase electrical capacity. Existing circuits can continue to use the existing box, and new circuits can be fed through the new box. The electrical systems of buildings built prior to about 1940 may require overhaul or replacement if any significant upgrades of equipment using electricity are expected such as upgrading from window air conditioning to central air conditioning. Parts of these older systems

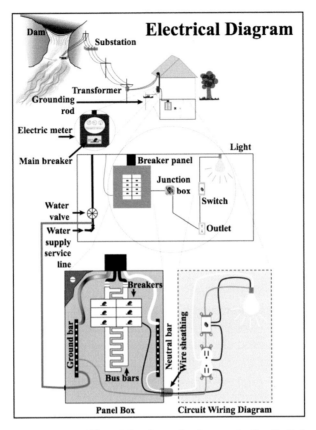

**Figure 7-103. Electricity from the Source to the Outlet**

may function very adequately, and can often be retained if the proposed retrofit is not extensive.

When inspecting wiring, look for missing or deteriorated wiring, a separate electrical circuit for the heating system, and proper wiring connection and enclosure of wires. Look for any charring or evidence of overheating on wiring, fixtures, etc. All wire connections should be enclosed in a code compliant junction box. In addition, all wires passing through any metal enclosure should be protected with an appropriate clamp or other protector. Furthermore, tape should never be used to repair or connect wires. As a general rule, if you find any taped wires, know that these are unsafe and cannot be covered with insulation unless the taped portion is turned into a wire connection and placed in an appropriate electrical box. Damaged wires such as those that have deteriorated or are missing protective insulation should be replaced. All wires should be connected to other wires by an appropriate wire nut, etc., (Figure 7-104) and the connection should be confined to the inside of an appropriately covered junction box.

Improperly modified wiring can draw more amperage than expected, which can cause a higher resistance and generate heat. If you insulate over a problem like this, heat will not be able to dissipate into the air surrounding the wiring because it will be covered with insulation. Any wiring repairs should be conducted in accordance with state and local electrical code. If it is not, there is a risk that if it contacts insulation, it could cause a fire.

Electrical boxes on exterior walls, whether for outlets, switches, or fixtures, should not pose a significant air leakage problem if you have properly sealed penetrations and chases in the top plate or opening above and the sill plate below, and have dense packed or air sealed the wall the box is in.

One of the biggest concerns with electrical in older homes has to do with knob and tube wiring, Figure 7-105. If local or state code allows adding insulation over the top of knob and tube wiring in the attic, be sure to have a licensed electrician inspect it and certify that it is safe. If local or state code allows adding wall insulation to wall cavities that contain knob and tube wiring, be sure to measure the voltage drop of the appropriate circuits while under load. To conduct a voltage drop test, plug the circuit analyzer into an outlet that lies along the electrical circuit being tested. The circuit analyzer tells the percent voltage drop. The voltage drop is represented by the percent difference between a

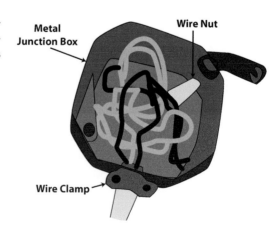

**Figure 7-104. Plastic boxes have their own built-in, self-closing clamps (not shown).**

house voltage of 120 V and the possible reduced voltage in a specific circuit that is caused by long wiring runs, bad connections, or overloaded circuits. If the voltage drop is greater than 3% to 5% or up to about 6 V (the voltage drops down to below 114 V), the knob and tube wiring may be in questionable condition.

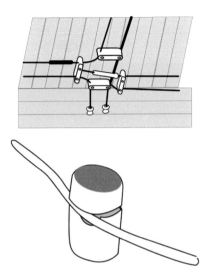

**Figure 7-105. Knob and Tube Wiring Examples**

Do not insulate in the attic or walls if you find a significant voltage drop like this. Instead, contact a qualified electrician to investigate the possible wiring problems. Replacement of wiring may be an option in order to insulate if enough safety funding is available under the weatherization program, or in the private arena, if the client is willing to pay for it.

Check to make sure all the breakers or fuses are appropriately amped— 20 amps for 12 gauge copper wire and 15 amps for 14 gauge copper wire. 12-gauge aluminum wire can only be connected to 15 amp breakers. If the house has any fuse-type electrical panels, install color-coded S type fuses (Figure 7-106) into the fuse holders. These fuses are tamperproof because a special "s" fuse holder allows only an appropriate amperage fuse to be screwed into it. A barb in the S type fuse holder prevents the "s" holder from being removed from the original fuse holder in the fuse panel, thus preventing anyone from putting in a fuse with too high an amperage rating.

**Figure 7-106. S type fuses provide an extra level of protection with the tamper proof bases that only the correct amperage fuse will fit.**

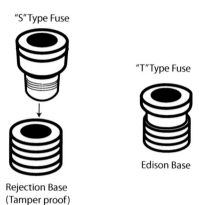

"S" Type Fuse

"T" Type Fuse

Edison Base

Rejection Base
(Tamper proof)

## PLUMBING AND MECHANICAL

Plumbing penetrations may be so large they do not fit inside the wall or up through the top plate of a wall. In these cases, a chase way or other larger area is provided for the plumbing components to pass up from floor to floor. The main soil stack for the drainage plumbing is often a good example of this in most homes. Since this is usually a 3- or 4-inch diameter drainage pipe and does not fit inside a regular stud wall, the builder frames in around it and covers the framing with drywall to keep the pipe from being exposed. If the chase involves a flue or chimney, special provisions must be provided. Foam board, cardboard, etc. is used to seal off the larger area around major plumbing bypasses in attics, crawlspaces or basements. Seal around the foam board or cardboard to make sure this bypass is well sealed. In the case of a chimney or flue, there is a need for a metal flange connecting the high temperature heating components to the framing or support around them as they pass up through floors, etc. Remember different types of flues, etc. require different clearances to combustibles, such as wood framing. In addition, there should be a dam around the flue or chimney to keep insulation away from the high temperature material.

## STRATEGIES FOR DIFFERENT TYPES OF HOMES

### Cape Cod House

The distinction with the Cape Cod home (Figure 7-107) from other platform-framed homes is in the attic space (see above diagram). This half story attic above the first floor has unconditioned spaces surrounding it, including knee walls (C), essentially vaulted ceiling rafter cavities (D), and a small collar beam attic at the peak (E). Note that there needs to be continuity in the thermal barrier and air barrier along the CDE continuum. In addition, when you insulate the first floor attic space designated as A, you must air seal roughly at about the point where F is, and then dense pack insulation under the floor of the upstairs from the bottom edge of the C wall to the air seal at point F. The best fit for insulation in this case is to dense pack insulation in the roof vaulted ceiling rafter bays at D, and loose fill insulation on the attic floor and collar beam floor at A and E. The backside of the

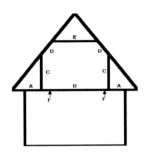

**Figure 7-107.**
**Cape Cod House**

knee walls C can be covered with batt insulation and house wrap, spray foam, or other similar material. Whatever type of insulation is installed, make sure that it is installed to the appropriate R-value.

**Balloon Frame House**

When retrofitting a balloon frame home, use ZPD testing to find unwanted connections between the wall framing and the floor. Air sealing and insulating should only be done around the conditioned space. To do so effectively, seal off the tops, and if open, see off the bottoms of all interior walls before insulating the walls, or you could end up with mounds of insulation where you do not want it.

As a rule, it is expected that the attic and basement will be left unconditioned. If there are unheated additions or unconditioned porch areas, make sure to create an adequate air barrier in the wall between the addition and the house, and the wall and floor between the porch, porch attic, and the house (see ZPD section). Be sure to adequately air seal the house using ZPD testing to check the pressure differences between the inside of the house and the attic, basement or crawlspace, porch attic, and any unheated addition. Insulate the attic spaces, the basement ceiling, and all the outside walls to complete the air sealing and insulation retrofit.

**Row Homes**

Row homes (Figure 7-108) are a special type of housing more often found in the larger cities of the northeastern United States such as New York and Philadelphia. They are typically two or three story masonry construction with a narrow front but long floors that extend relatively deep into the building.

Because these row houses are built side-by-side with their walls right next to each other and no gap between them, (except for end units), the only exterior sidewall exposure is at the front and rear of these buildings. These buildings are typically very leaky owing primarily to their furred-out exterior masonry walls and their large upper floor bypasses that open into a very shallow, if only slightly sloped roof cavity. The wood furring strips used on the furred out walls are typically rough sawn 1" x 3" strips that support the lath and plaster wall coverings. The furring strips are attached to the masonry walls. Unfortunately, too often

**Figure 7-108. The Facade of Row Homes in the Northeastern US**

the space between these furring strips on the wall opens up into the unconditioned attic space above the conditioned upper floor. These gaps between the furring strips, much like a miniature stud bay without a top plate, represent major air leakage and thermal barrier gaps much like a balloon frame constructed home without top plates in the attic (Figure 7-109).

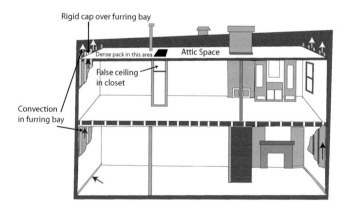

**Figure 7-109. Capping off the furring bays from the attic side on the front and rear of a row home is important in helping block the circulation of air up through these bays. Also, because of height restrictions on providing adequate loose fill insulation in the lower height end of the attic area, you may find the need to dense pack on that end of the attic between the bottom side of the roof deck and the attic floor.**

There are concerns with other major bypasses like chimney chase ways, dropped ceilings in closets, etc. To seal many of these bypasses in a row house, the roof retrofit typically involves a lot of crawling in the limited space of the attic. If there is not already an access opening into the attic, create one through the roof deck. Seal off all the openings between furring strips at the top, dense pack blown-in insulation into the low-height, more restricted attic space at the lowest point of the roof, and install loose fill insulation to the appropriate depth in the rest of the attic space that does not have as limited a height restriction. When the work is finished, it is best to cover the roof access hole with a single large roof vent that does not allow rain or water in.

**Insulating Framed Sidewalls That Have Existing Insulation**

Uninsulated wood framed sidewalls on any home could be dense packed to the appropriate density, such as 3.5 pounds per cubic foot for cellulose. If they have any type of existing fiberglass insulation in them, they are typically not cost-effective to re-insulate in the weatherization program. In the private arena whether to add insulation is a function of whether the client decides to absorb the cost.

## AFTER COMPLETING THE RETROFIT

You want to seal the house as tightly as cost-effectively possible to best achieve the benefits of energy savings, without causing any health and safety issues for the occupants in terms of inadequate oxygen supply, mold, etc. Thus, if the house is tightened so much that health and safety issues are created, the best option is to continue cost-effective air sealing until complete, and then create a powered house ventilation system that brings in outside air in a controlled fashion, such that energy is not lost in the process. This usually means putting in a cross ventilation system that both brings outside air in and pushes inside air out, while extracting the heating or cooling from the inside air before exhausting it to the outside. For many people this seems counterintuitive. They ask "Why not just leave the house leaky enough that it doesn't create the health and safety issues?" The answer—The energy savings from tightening the house as cost-effectively as possible is greater than the cost of a powered ventilation system with a heat exchanger to extract the heating or cooling from the air exiting the home.

How do you determine if venting is needed? If the post-retrofit blower door reading is below the minimum ventilation rate (MVR), you must ventilate the home appropriately. Consider requiring that ventilation be installed that meets the requirements of one of the two standards. In the case of the weatherization program, if the post-retrofit blower door reading falls between 70% and 100% of the MVR, then you have to determine if there is a need for mechanical ventilation. Mechanical ventilation is powered, not passive, ventilation such as fans. If you find no special circumstances to require a fan in this case, you still need to document the reason why you did not require the installation of mechanical ventilation.

Under BPI standards, when the post-retrofit blower door reading falls between 70% and 100% of the MVR under ASHRAE 62.1-1989, you only recommend to the homeowner that they have continuous mechanical ventilation installed to make up for the difference between the CFM50 and the Building Airflow Standard (BAS)—the equivalent of the MVR but at CFM50.

If the post-retrofit blower door reading is 70% of the MVR or less, whether under BPI or weatherization standards, you must specify and require the installation of mechanical ventilation that meets one of the two MVR standards, regardless of any special circumstances. Fans are the key to satisfying the MVR in a home when needed. Using an energy star rated fan can help for a number of reasons. Energy star qualified ventilating fans must meet certain efficiency levels (use at least 60% less energy), maximum sound levels in sones, and must pump a minimum percent of their rated flow under specific pressure conditions. For instance, an energy star qualified bathroom fan that is rated for an airflow of between 10 and 89 CFM must have a minimum efficiency of 1.4 CFM/Watt, a maximum sound level of 2.0 sones, and have fan power that still performs at the originally rated airflow when exposed to a static pressure of 0.25 inches of water gauge (0.25 in wg). Sones are a unit of perceived loudness. Unlike decibels, sones involve a linear relationship – a 2 sone fan is twice as loud as a 1 sone fan and a 4 sone fan is twice as loud as a 2 sone fan.

HOUSE VENTILATION

There are basically three types of ventilation that can be installed in homes to satisfy the MVR: supply ventilation that draws air from the outside, exhaust ventilation that pushes air to the outside, or an equilibrated or balanced system that brings the same amount of air inside the home as it separately removes from the home. A typical existing bathroom fan cannot usually satisfy the MVR because the ventilation system must be left on continuously or on a specified intermittent basis. For instance, under ASHRAE 62.2 bathroom fans must provide 50 CFM intermittently or 20 CFM continuously. Thus, if a 50CFM fan is selected to serve intermittently, the fan would have to be set to create the same flow as a fan that runs continuously 24 hours at 20 CFM. As a result, the 50 CFM fan would have to run 40% of a 24-hour day or 9.6 hours per day. This fan will work at a 40% duty cycle.

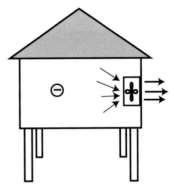

**Figure 7-110.**
**An Exhaust Ventilation System**
This creates a more negative pressure in the house with respect to (WRT) the outside.

## Supply Ventilation

Supply ventilation systems pressurize the home by pulling air into the house through a duct system, Figure 7-110).

This in turn pressurizes the house and pushes inside air out of the home through leaks in the shell. Use supply ventilation systems only in hot or warmer mixed climates. When the cooler air inside the home leaks through the walls and shell to the outside, it is heated and water vapor will not condense in the walls. If a home in a hot climate is depressurized, the reverse would happen and warm air would be cooled down as it passes from the outside of the home through leaks in the shell, to the inside of the walls in the home. When this warm air cools down, it cannot hold as much moisture, and condensate tends to form.

One easy method of installing a supply ventilation system (Figure 7-111) is to provide air through a duct directly from outside into the return air duct of

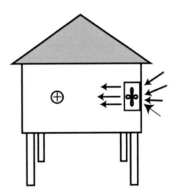

**Figure 7-111.**
**A Supply Ventillation System**
This creates a more positive pressure in the house with respect to (WRT) the outside.

the furnace system. Since this takes advantage of the larger furnace circulation fan there is a need to control both the frequency and degree of air being brought in from outside. Control systems are available to satisfy the ventilation requirements for a home without wasting energy by over ventilating.

### Exhaust Ventilation

Standard exhaust ventilation systems depressurize the home by pushing air out of the house through a duct system. This in turn depressurizes the home (creates a vacuum in the home) and brings outside air into the home through leaks in the shell.

Use this type of system in cold-weather areas only. When the cold outside air leaks through the shell into the walls of the home, it is heated and thereby can hold more moisture, which does not condense in the walls of the home. If a home in a cold weather area is pressurized, the reverse happens and warm air is cooled as it passes through the walls. When the warm air cools, it cannot hold as much moisture, and condensation tends to form, even inside the walls of the home.

Never install an exhaust ventilation system in a home with a fireplace or similar device because of the risk of backdrafting the fireplace. In fact, if an exhaust fan is installed in a home with combustion appliances, conduct an appropriate CAZ test in the home to make sure the fan is not causing any of the combustion appliances to backflow.

One of the simplest exhaust systems to install in the home is to replace a bathroom fan with a special fan that runs continuously on low-speed to provide the required minimum ventilation. It can be switched to a higher speed to serve the purpose of a normal bathroom fan to remove odors and moisture.

### Balanced Ventilation

A carefully installed balanced ventilation system should pull in the same amount of air as it pushes out of the house. It should neither pressurize nor depressurize the home so there is no worry about warm air being cooled in either direction through the walls, thus avoiding the risk of condensation. The exception is the vent itself. Thus, the balanced vent system should be able to collect and drain the condensate.

While the basic balanced ventilation system simply brings fresh air inside and pushes inside air out, balanced systems can be purchased with a heat exchanger to help avoid the energy losses that typically come with simple supply or exhaust systems. This can allow for higher flow rates and better ventilation than supply or exhaust systems without major losses of energy. However, if the intake and exhaust on these systems is not carefully balanced, then some of the same problems that exist with exhaust and supply systems can occur such as moisture forming inside the walls.

If the balanced system uses a heat exchanger, it is simply a system similar to a heat exchanger in a furnace often represented by a thin wall sheet of metal or plastic. On one side of the sheet, cold outside air can be drawn into the house while on the other side of the sheet, warm air can be pushed out of the house. With this process, the colder air is warmed as it passes along the heat exchanger, avoiding as much energy loss as possible in the process. The high purchase cost that typically comes with an air-to-air heat exchanger like this often keeps it from being cost effective to install, including not satisfying a good savings to investment ratio in the weatherization program. The savings to investment ratio or SIR is generally lower for air-to-air heat exchangers than it is for an exhaust only vent system.

There are two different types of balanced air-to-air heat exchangers. A heat recovery ventilator (HRV) uses only a heat exchanger to avoid energy losses. An energy recovery ventilator (ERV), Figure 7-112) uses both a heat exchanger and a process of recovering latent heat contained in moisture as well. This is sometimes referred to as an enthalpy exchange system. Thus, a good ERV will reduce energy losses better than an HRV. Good ERVs and HRVs that use a counter flow or countercurrent heat exchanger can have heat exchanger efficiencies of up to 95%—in other words, these ventilation systems can recover up to 95% of the heat that would otherwise be lost if the air were simply exhausted to the outside. In some advanced systems like this, using an energy efficient motor can save 8 to 15 times the energy used by the fan motors.

## ADDITIONAL VENTILATION CONSIDERATIONS

Upgrading to ASHRAE 62.2 allows air sealing the home to the greatest degree possible, cuts the air driven heat loss to a minimum, provides proper ventilation under all conditions all the time, and cuts the code required CFM from 15 CFM per person with a five person minimum down to 7.5 CFM per person actual without any minimum. At the

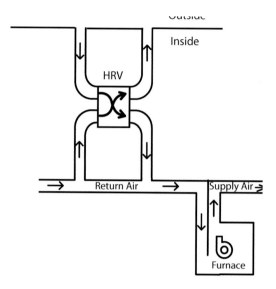

**Figure 7-112. Energy Recovery Ventilator**

same time, upgrading from ASHRAE 62.1 to 62.2 requires measuring the actual fan flow, having automatic controls, and has an incremental cost of the fan control switch and because air inlets are needed.

Spot ventilation typically refers to a fan system that exists in only one room, such as a bathroom or kitchen. As a rule, bathroom fans should exhaust at least 50 CFM as tested, not as calculated. Seeing if the fan will hold up a piece of tissue or toilet paper is a crude but often used test to see if a fan is drawing as least 50 cfm. Kitchen exhaust fans should probably be delivering at least 200 or more CFM. All homes should at least have these so-called "spot" ventilation systems as bathroom fans that exhaust to the outside unless they have an openable window. If the home needs ventilation, it must be added to the extent required whether bathroom vents exist or not.

### Static Pressure and Equivalent Duct Length

Ventilation fans are rated for a certain CFM. Several factors can alter the CFM rating such as static pressure. Static pressure is defined as resistance to airflow in the interior grille, ductwork, transitions in the ductwork such as elbows and 45° turns, and the termination of the ductwork outside. This resistance keeps fans from performing at their rated CFM. Static pressure is affected by the type and opening size of end caps and grilles, the number and degree of bends and elbows in the ductwork, and the material and length of the ductwork. The smoother the inside of the duct, the more easily air will flow past it.

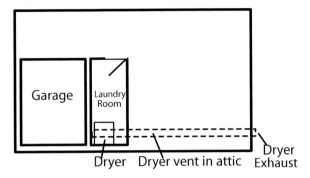

**Figure 7-113. Top or floor plan view showing why the dryer vent is sometimes much longer than it could be. It is done to avoid exhausting the dyer vent out the front of the house for cosmetic reasons and for code issues, it cannot go through the wall to the garage. In addition, when you place elbows (turns) in the vent, grille covers where the duct opens to the outside etc., you reduce the fan's operating CFM or its "static pressure." Corrections can be calculated to make sure you have an adequately sized fan to produce the proper CFM given the static pressure resistance created in the vent by adding elbows, grilles, etc.**

Generally, it is better to have a long broad turn rather than a short abrupt turn in the ductwork. The broader or wider the radius of the turn, the less resistance there is to airflow. Obviously, a kink in ductwork represents a restriction to airflow. Furthermore, if the opening on either end of a duct is not completely open to air, for example if there is a grille or cover on it, the grille or cover will add resistance to the flow.

All fan manufacturers have information tables to help determine the true fan flow in CFM given the particular ductwork configuration and the static pressure losses that occur as a result. The "equivalent duct length" or EDL method of calculation is one way to determine the static pressure to make sure you are using the properly sized fan for the situation. (See Appendix D for a sample calculation).

### Installing Fans
Suggested guidelines for installing fans include the following:
1) Seal the fan housing,
2) Seal the ducts,

3) Run the duct to the gable end if possible,
4) Slope the duct to the exterior,
5) Block soffit vent openings for 2 feet to either side of the vent hood if it runs to a soffit,
6) Make sure exhaust air flows all the way to the outside of the house,
7) Use ductwork that is smooth on the inside. This not only allows for easier flow, it is also easier to clean, and once cleaned, it typically stays clean for a longer period of time.
8) Provide a grille or end cap to keep animals out,
9) Use as short and straight a length of ductwork as possible,
10) Insulate the ducts to at least R5 when they are in unconditioned space. This helps reduce the risk of condensation in the ductwork as well as reducing the risk of corrosion and damage to the building.
11) Make sure the ducts are well sealed at the junctions or connections,
12) Use broad or long radius elbows to help reduce the resistance caused by turbulence, etc.
13) Provide 2 to 3 feet of straight pipe before installing an elbow. This allows the air to pick up speed before it meets the resistance of the elbow,
14) Always follow the manufacturer's installation instructions and specifications,
15) Remember, the larger the diameter of the ductwork, the less resistance. However, when the size of the ductwork is enlarged, it takes more work and thus more expense cutting the openings. Thus, be careful not to oversize the ductwork.
16) Remember that rigid ductwork results in less resistance to flow.

**Fan Noise**

There are noise maximums for fans based on whether they are continuous or on-demand. Continuous fans must be rated at one sone or less, while on-demand fans must be rated at three sones or less. Under this standard, all dryers must be vented to the outside of the building as well. Furthermore, if there is an attached garage, its design must prevent migration of contaminants from the garage into the house. One oft overlooked important consideration here is requiring the replacement of ordinary hinges with auto-closer hinges on the house-to-garage door to reduce the risk that a car warming up in the garage could expose the occupants inside the house to the combustion byproducts of the car.

As a rule, fans that have greater than a 1.5 sone rating are considered too noisy for installation in a home. Some experts believe that an hourly timer, a 24-hour dial timer, or a programmable timer (Figure 7-114) are suitable to be considered for use in a home to make sure adequate ventilation is provided.

**Figure 7-114. A manual ventilation timer either runs continuously or cycles on/off in 20, 40, or 60-minute intervals.**

# Chapter 8

# Work Order Development by the Auditor

After conducting the audit for the weatherization program, the auditor produces a work order to indicate what work is to be done to retrofit the home. Every work order should include a rough sketch of the footprint of the home including an arrow indicating north, identification of chimney locations, and the exterior dimensions of the building (length, width, height, etc.—see Appendix B for how to do these calculations). In addition, the work order should provide some rough elevation drawings that show each of the wall sections of the house from the outside. These should include an indication of the amount and type of exposed foundation, the dimensions and the number of windows and doors, and the distance from the ground to the highest point in the home represented by conditioned space. With this information, the minimum ventilation guideline can be calculated.

The framed wall square footage can be calculated for purposes of determining the amount of insulation to be installed. Here is an example of such a wall calculation, Figures 8-1 and 8-2. Assume a Cape Cod style house in the shape of an "L" with 8-ft walls on the first floor, and three-ft knee walls on the second floor.

Assume that there are eight 4 ft x 3 ft windows (12 ft² each) and eight 3 ft x 3 ft windows (9 ft² each) along with two 20-ft² doors (36" wide). Taking into account only the wall without the windows and doors, add all the perimeters (26' + 26' - 10' + 16' + 16' + 12' + 12' + 10' = 108') and multiply by the first floor wall height of 8 feet to get 864 ft², the area of the wall on the first floor. Subtract 10 feet for the common wall between the two portions of the building by subtracting the 10-ft length from one of the 26-ft walls on the larger portion of the building. Do not add a second 10-ft length for the smaller part of the building. As for the

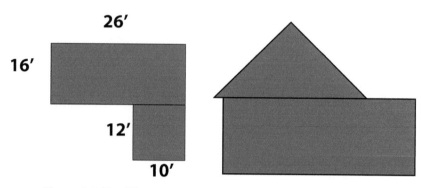

**Figure 8-1. Top View**                    **Figure 8-2. End View**

second floor that sits only over the larger part of the building, the knee wall perimeter wall dimensions are 26 ft x 8 ft with 3-ft high knee walls. Calculate this knee wall area by adding the perimeter of the knee walls (26′ + 26′ + 8′ + 8′ = 68′) and multiplying by 3 ft to get 204 ft². Do not take out any perimeter wall length for having a common wall with the smaller portion of the building. There is no common wall because there is no second-floor on the smaller part of the building. Thus, the total wall square footage without subtracting framing, windows, or doors is 864 ft² + 204 ft² = 1268 ft².

As a rule, the "framing factor" for the 2 x 4 frame construction in this older home is 0.23 or 23%. This means that roughly 23% of the 1268 ft² does not need to be filled because framing takes up that space. Therefore, 1268 ft² x 0.23 = 290 ft² must be subtracted from 1260 ft² to get the true wall square footage: 1268 ft² - 290 ft² = 978 ft².

Next, total the area of all the windows and doors. Assume there are eight 4 ft x 3 ft and eight 3 ft x 3 ft windows along with two 20 ft² exterior doors (6.66′ x 3′ doors). The calculation for the total area for windows and doors is:

$$(8 \times 4 \times 3) + (8 \times 3 \times 3) + (2 \times 20) =$$
$$96 + 72 + 40 = 208 \text{ ft}^2$$

Take the square footage of the walls (after correcting for the framing factor), 978 ft² and subtract the 208 ft² of windows and doors to end up with 770 ft² of wall area that needs to be insulated.

## FIRST STEP IN WORK ORDER DEVELOPMENT

The first step in creating the work order is using the on-site inspection to evaluate possible "deal breakers," an issue that keeps the home from qualifying for weatherization. An example would be if the client information does not match what was seen in the audit. Other deal breakers include a client who has moved or is no longer living, or if the client seems to be living at a standard much higher than allowed in a low-income weatherization program. In this case, contact the outreach department to see about the possibility of having the home recertified. In the private arena, the only deal breaker is anything the client cannot afford or does not want to pay for.

Another example of a deal breaker is unvented combustion appliances or any other major health or safety issue that is unresolvable. This may require sending the client a deferral of services letter. A home that has an unvented combustion appliance can be weatherized as long as it is not the primary heat source. Another deal breaker is health and safety issues or costs going beyond what is available under the weatherization program. Two other cases that represent deal breakers and involve a deferral of services letter include when the building is in such poor condition that weatherization work is virtually impossible, and when the building has been tagged for demolition.

A deferral of services letter typically lays out the conditions that must be satisfied before the home will be weatherized. It will tell the client that weatherization work cannot begin until action is taken on

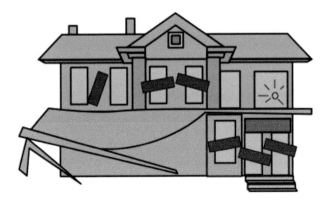

**Figure 8-3. Dilapidated homes cannot be weatherized under the program.**

the items noted in the letter, such as signing an agreement to not use an unvented space heater, or making a major non-weatherization repair like cleaning a basement perimeter drain before work can begin. If the heating system is found to be dangerous, the client can be required to repair or replace it before retrofit activities can begin.

## SECOND STEP IN WORK ORDER DEVELOPMENT

The second step in creating a work order is to designate on the rough drawings where the building envelope is and develop a list of the measures considered appropriate for the home. A rough drawing is shown in Figure 8-4.

Put health and safety issues first including heating system problems, moisture problems, and the isolation of the garage. Examples of heating system problems to be noted include insulation barriers that need to be at least two inches from masonry chimneys, a missing masonry chimney liner, inadequate ductwork sealing. It should also be noted if the furnace needs to be evaluated, cleaned, or tuned. Examples of issues to be addressed for moisture problems include the need for polyethylene vapor retarder in the crawlspace ("300 ft² 6 mil poly"), adequate rain gutters if there has been cellar flooding, or cleaning the perimeter drain in the cellar.

A good example of work needed to isolate a garage includes requiring any wall penetrations or openings between the house and the garage to be air sealed including weather stripping and adding auto-closer hinges the house-to-garage door. This would be specified in the work order as "1 Porte-Seal weatherstrip, 1 flexsteel door sweep, two tubes silicone caulk, two auto-closer hinges." To fill and seal large openings, the work order might say "Fill with scrap fiberglass and coat with spray foam—two cans expanding one part foam." Specify that zone pressure diagnostics be used by the retrofitters to evaluate the quality of this sealing.

After addressing the health and safety issues, and designating where the building envelope is located on the drawings, the next step is to specify the building envelope retrofit work that needs to be done. A good example of this might be to specify air sealing of the exterior walls ("dense pack all walls with cellulose as per the State Technical Manual (STM). Material: 770 ft² x 4 inches"), the crawlspace perimeter,

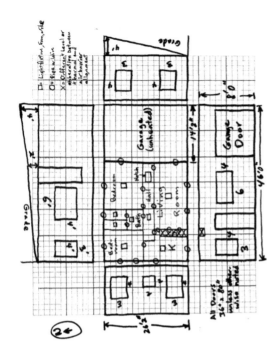

Figure 8-4. Providing a diagram of the location of mismatches of thermal and air barriers can help installers bring these two together as needed. Notice that this auditor identified different levels in the attic with an "X" pattern over the area where there is a dropped soffit in the kitchen. Notice that no combustion appliance vents or chimneys have been indicated, and that the type of foundation has not been recorded.

basement, the walls between the house and unconditioned spaces connected to the house, the attic floor ("Material: one can one part foam for plumbing vent stack and around new bath fan—two tubes caulk for all wire penetrations—4 ft² galvanized steel and 1 tube high temperature caulk for chimney chase air seal and barrier"), and furnace ductwork ("Material: 1 gallon RDC #6 mastic, two rolls web type drywall tape").

The work order should require the measures necessary to provide a continuous heat, air, and moisture barrier all along the building envelope surfaces. Specifically, this might involve insulating the furnace ductwork in crawlspace/basements ("Material: 100 ft² ductwork insulation"), weatherizing the attic hatch by replacing it with a new insulated hatch ("Materials: 4 ft² 3/4" AC plywood, 20 ft number two pine 1" x 4," 8 ft weatherstrip, two 3" hook and eyes latches, 8 ft² 2- in rigid foam insulation—two layers, miscellaneous fasteners."), blowing in insulation onto attic floors/flats ("Material: Main house = 204 ft², rear addition = 128 ft²"), dense packing vaulted ceilings ("Material 416 ft² x 6 in. cellulose") and uninsulated exterior walls, installing window locks, putting door sweeps on and weatherstripping front and rear exterior doors ("Materials: two Q-Ion™ weatherstrips, two Flexsteel™ door sweeps"), and installing two-part foam along the rim joist/box seals between floor joists in the crawlspace and basement ("be sure that two-part foam seals to the first-floor subflooring and to the six mil poly groundcover. Materials: one 200 board foot two-part foam kit"). Designate the need for incidental repairs such as sealing a roof leak ("1 quart plastic roof cement").

Identify health and safety tasks such as installing a low sone bathroom fan that vents to the gable end of the house, using an insulated metal duct that is sloped towards the exterior after sealing the joints ("install Panasonic 1.5 sone bath fan vented to the gable end as per State Technical Manual (STM) using 4-in metal duct sloped slightly to exterior. Insulate duct. Seal all joints. Materials: one bath fan, 8' of 4" metal duct, one 4" elbow, 1 quart RDC #6, one exterior vent hood, 10' of 4" duct insulation."). Require putting identifying tags on the rafters above the location of any live knob and tube wiring in the attic that could be buried by attic insulation, Figure 8-5.

The list of measures can either be entered into a weatherization audit program or compared with a priority list for your area to determine the overall priority for all of these measures. The weatherization audit software will exclude any measures that do not have an SIR of 1.0 or higher as do the priority lists (see savings to investment ratio section).

Ultimately, make sure that the total projected weatherization cost does not go over any dollar limit in the program or client dollar limit in the private arena. As you become more experienced with the types

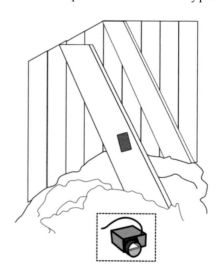

**Figure 8-5. Placing tags on the wood supports above items otherwise hidden underneath the insulation allows others to later more easily locate them if repair, etc. is necessary**

and brands of materials retrofitters use, you become more adept at being specific and drafting your work orders as well. Moreover, you will understand the cost implications of your work order proposals.

## THIRD STEP IN WORK ORDER DEVELOPMENT

The third step to creating a work order involves tying up loose ends and providing additional information. Provide the client's contact information and travel directions to the home, note special issues for the client such as the best contact times, a photo of the house from the front with its street number visible, accurate material quantities, specifics of insulation installation and products, pinpoints on the diagrams or other pictures to give the location of each retrofit measure, notes about unusual items, and the auditor's contact phone number and name.

Once the work order is completed, it is up to the retrofitters to install the measures indicated and re-test the home at the end of the work, and if the work takes more than one day, retest from day-to-day as well.

# Chapter 9

# Heating and Cooling

Heating and cooling are particularly important in the energy arena. The first part of this chapter will describe heating units that can be found around the country, how they operate, how to inspect them to identify and note problems, and in some cases, how to make repairs. (Making repairs on heating systems can require a license in some states). Because fuel-burning appliances have potential to cause serious problems in the home, the second part of the chapter will have extensive discussion of how to check for these types of problems. The final part of this chapter will describe cooling units, how they operate and how to inspect and repair them. (Making repairs on cooling systems requires certification if you are working on the refrigeration gas system).

In general, whenever possible, ask building occupants about the HVAC system's performance history. However, always try to observe equipment in actual operation. In most cases, you can assume the system has sufficient capacity since it has a history of use, but this is not always true.

## VARIOUS TYPES OF HEATING UNITS

### Gas Burning Furnaces

Warm air heating systems distribute warm air to the living areas of the house through two methods: forced air and gravity. Gravity systems are occasionally still found in older single-family houses, but most gravity systems have been either replaced or converted to forced air. Gravity systems are big, bulky, and easily recognizable (they are floor installations, they hang just below the floor level, or they look like an "octopus" if they rest on the ground in the basement). They provide heat to rooms in the house based only on the fact that heat rises. Because of this, they must be placed below the area(s) they are intended to heat. Lacking a mechanical means of moving air, such systems are inefficient, heat unevenly, and are generally considered archaic.

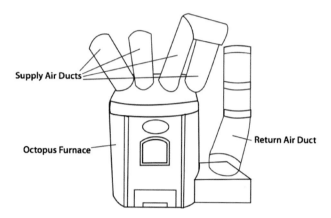

**Figure 9-1. "Octopus" Furnace**

Gravity system supply and return air ductwork is oversized compared with modern furnaces. Many have been converted to gas-fired from coal-fired (such as the "octopus" type, Figure 9-1). Some have had a fan (called an "air handler") installed into the ductwork to convert the furnace to a forced air system. Owners should install a blower or fan box in the return air duct if there is not one there already, or completely replace the furnace unless there are overriding reasons for doing otherwise.

Among forced air systems, low, medium and high efficiency furnaces exist. Low efficiency units are no longer manufactured, but are recognized in existing homes by the fact that they have no fan serving their flue exhaust. Medium efficiency units typically have an induced draft fan (a fan just below the flue to force the combustion air out of the flue), and a metal flue. This flue typically should be two walled. High efficiency furnaces have forced air flues made of plastic PVC-type pipe. Because they remove so much heat from the combustion air, the combustion air is cool enough to be handled with plastic.

In some cases, medium efficiency furnaces share a common flue with a standard water heater with a gravity-fed flue, Figure 9-2 (no fan in the water heater flue). Sometimes the forced air furnace flue "wins" in the battle with the common flue, and backflow occurs down the gravity-fed water heater flue. This spills combustion air out around the bell or draft mixer on top of the water heater. You can feel near the draft diverter with the back of your hand, or use a draft gauge, CO tester, tissue, mirror, artificial "smoke" (a Wizard Stick, smoke puffer tube, etc.) or match/incense that has been lit and then blown out to generate a smoke to detect any outward flow of hot exhaust or combustion gas out of the bell housing. This indicates a hazardous draft problem that must be corrected.

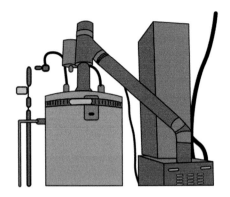

Figure 9-2. Having a "gravity fed" water heater flue tied in with a fan-assisted medium efficiency flue may have backflow of flue gases designed into it

Older gas burners typically have a series of holes or ports in a "pipe" that allow gas to burn along the length of the "pipe," much like the burners on a gas fireplace or barbecue. This type of burner is often called a "ribbon burner." Oxygen is provided in an optimum ratio by adjusting air slots or shutters typically positioned just in front of the burners that allow air to be drawn in with the gas (typically based on a Venturi effect) for a hopefully efficient air-to-fuel ratio.

Jet burners, which are becoming more common in gas burning furnaces, also use a Venturi effect to mix gas and air, but have only one hole or orifice out of which the mixed air and gas pass before igniting.

A thermocouple is a heat sensor type safety device that detects if the pilot is on in the furnace. If the pilot is not on, the thermocouple prevents the gas valve from opening and supplying gas to the burners. In addition, if the thermocouple is defective, it typically sends a message to the valve so the pilot light cannot stay on by itself. If you try to light a pilot light that has thermocouple protec-

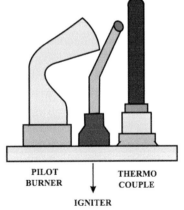

PILOT BURNER    THERMO COUPLE

IGNITER

Figure 9-3. Intermittent Ignition Device (Igniter)

tion, the pilot light typically must be left on for about a minute to give the thermocouple time to send the message to the valve that the burners can operate. This is typical in most modern furnaces except some of the medium and high efficiency furnaces that use an electronic ignition such as a "glow plug" or spark generator type igniter, Figure 9-3. These ignite the burners or pilot light only after the thermostat demands heat. This avoids having the pilot light on all the time and avoids some of the subsequent damage that occurs to furnaces from the condensation that results from pilot lights.

Other safety devices that exist on more modern furnaces include a flame roll-out sensor that shuts down the furnace if flame roll-out is detected and a backdraft sensor that shuts down the furnace if it detects that backflow is occurring on the furnace.

Gas supply lines are normally made of black iron or steel pipe (some jurisdictions allow copper lines with brazed connections). There should be a drip or sediment trap leg installed in the pipeline. Shut-off valves should be easily accessible and all piping well supported and protected.

### Oil-fired Systems

Oil-fired furnaces need an oil tank. Oil tanks should be maintained in accordance with local code or the recommendations of the National Fire Protection Association. All tanks must be vented to the outside and have an outside fill pipe. Buried tanks normally have a 550, 1000, or 1500 gallon capacity. Basement tanks are usually restricted to a 275-gallon capacity, with no more than two tanks allowed.

Tanks must be located at a minimum of seven feet from the furnace or boiler, and should be adequately supported and free of interior rust. Outside above ground tanks should have an adequate supporting base. Warn your client about possible existing or abandoned underground oil storage tanks because they represent a possible serious environmental concern. If you notice abandoned supply and return oil lines coming into the house from outside, this is good evidence of a possible oil tank.

Metal oil tanks often begin to leak after about 20 years when the bottom of the tank corrodes from moisture that has condensed inside the tank and settled to the bottom. Feel along the undersides if exposed. You can also look for an oil level gauge and see if it works. Decide if the tanks should be replaced. Check the oil supply line to the furnace; it should be equipped with a filter and protected from accidental damage

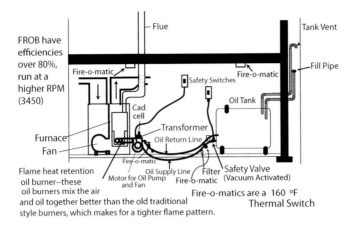

**Figure 9-4. A Typical Oil-Fired System Set Up**

and rupture. Oil burning units should have fire-o-matics (a safeguard fire protection system) above or around high fire risk areas such as the fuel tank (if inside), the burner, or the heating plant.

Oil-fired burners operate by an oil pump (near the heating system, not the tank) pumping the liquid fuel oil through a nozzle, which "atomizes" it (turns it into a spray of small droplets) while an air blower or fan blows air into the burner/nozzle combination to provide oxygen to the oil spray so that it can burn efficiently. Oil-fired units, Figure 9-4, have a cadmium cell sensor to detect whether there is a flame when the fuel oil is being supplied to the burner and nozzle. If the cad cell does not detect a flame, the oil supply is shut off to the nozzle to avoid build-up of unburned oil in the fire chamber. In addition, a "primary control" or oil burner combustion control typically is mounted to a stack, flue or smoke pipe to detect if the exhaust gas temperature is high enough to indicate that oil is burning. This can be reset, but care should be taken. Several non-igniting attempted starts of the heating unit can leave significant amounts of un-ignited oil in the chamber.

Proper draft is critical to the efficient operation of an oil-fired unit. A barometric draft regulator is required above the unit. This regulator is flue-mounted with a weighted horizontally hinged circular "door" that allows air to mix with the combustion air as the unit warms up. All such openings should be sealed. Check the draft regulator by observing its motion when the heating unit is in operation. It should open as the heating unit warms up. If a tissue drifts away from the damper, backflow

may be occurring. If a tissue is drawn into
the damper, the damper is probably properly mixing air into the flue, Figure 9-5.

## ITEMS COMMON TO BOTH
## OIL- AND GAS-FIRED UNITS

### Heat Exchanger

The heat exchanger is located above
the burners in gas- and oil-fired furnaces,
and separates the products of combustion from the air to be heated. In warm air
furnaces, it is typically nothing more than

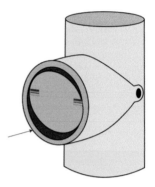

**Figure 9-5.**
**Barometric Draft Damper**

a piece of sheet metal. Exhaust air combusted by the burners flows on
one side of the sheet metal; on the other side, the fan blown air becomes
heated from the hot sheet metal surface and travels through the ductwork to the rooms. Heat sources that use a heat exchanger are referred
to as indirect fired heating systems. In fuel-fired furnaces, it is critical
that the heat exchanger is intact and contains no cracks or other openings that could allow combustion products into the warm air distribution system. Visual detection of cracks, even by heating experts, is a
difficult and sometimes unreliable process without extensive testing
equipment. Look for signs of soot at supply registers and smell for oil or
gas fumes. Look for rust on the exchanger. A major cause of premature
heat exchanger failure is water leakage from humidifiers or blocked air
conditioner condensate lines. Check for other signs of water leakage.

The durability of the heat exchanger determines the service life of
the furnace. Furnaces installed since the 1950s normally have a useful
life of 20 to 25 years or less. Older furnaces with cast iron heat exchangers may last much longer.

### Thermostatic Controls

Residential HVAC controls consist of one or more thermostats and
a master switch. Thermostats are temperature sensitive switches, such
as in Figure 9-6, that automatically control the heating or cooling system.
They normally operate at 24 volts (low voltage). Thermostats should be
located in areas with average temperature conditions, and away from
heat sources such as windows, water pipes, or ducts. Programmable

or "set back" thermostats can also be used to save energy by programming them to turn on only during hours when the client will be home or when they are getting up in the morning. You can check the thermostat (or a random sample of a few thermostats if there is one in every room) by adjusting it to activate the HVAC equipment. Thermostats in each room can help keep from cooling or heat-

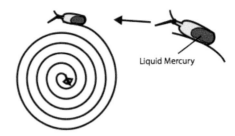

Figure 9-6. A mercury switch on a thermostat relies primarily on the expansion and contraction characteristics of the metal in the coil for serving as a thermostat.

ing rooms that are not being used. Sometimes two thermostats separately control the heating and cooling system, and sometimes the living unit is divided into zones, each with its own thermostat. Buildings with electric baseboard heat may have a thermostat in every room or on every heating unit. You can check the functioning of multizone systems by operating the HVAC system and noting if distribution is adequate in a few randomly selected zones. You should not activate an air conditioner system if the temperature is less than 60°F outside.

### Master Switch

Every fuel-burning system should have a master switch that serves as an emergency shut-off for the burner. Master switches are usually located near the burner unit (gas burners), or preferably near the top of the stairs leading to the basement (oil burners). Cooling system controls on a thermostat may also include a master switch, which in the "off" position will not allow the compressor to start, as well as a switch allowing only the circulating fan to operate. You can operate all master or emergency shut-off switches when the burner is in operation to see if they deactivate the unit. In hot water heating systems that are used to generate domestic hot water, the thermostat controls the circulating pump and/or the burner.

### Internal Furnace Controls

Gas- and oil-fired furnaces have two internal controls: a fan control or low temperature limit and a high-temperature limit control, Figures 9-7 and 9-8.

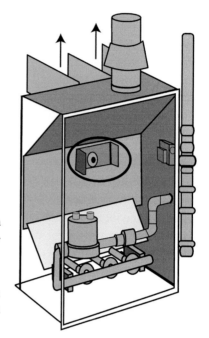

**Figure 9-7. A High Limit Switch (Aquastat) for a Boiler with Sensor Rod Showing Behind**

**Figure 9-8. Typical Location for a Low Limit/High Limit Switch on a Furnace**

The fan control prevents cold air from being circulated into the rooms without having first been warmed by the furnace. This prevents the blower from blowing cold air after the heat exchanger has cooled off (well after the thermostat has quit demanding heat). The fan control is a temperature-sensitive switch, completely independent of the thermostat, that turns the furnace blower on and off at preset temperatures. When the thermostat calls for heat, the furnace burner turns on. After the heat exchanger warms to a preset temperature (usually 110-120°F), the fan control activates the blower. The thermostat will shut off the burner when the building warms to the thermostat setting. After a while, when the heat exchanger cools to about 85°F, the fan control will switch off the blower. If, as you observe the above sequence, it is faulty, the fan control should be adjusted or replaced.

The high-temperature limit control is a safety device that shuts the burner off if the heat exchanger gets too hot. The control is usually set at about 175°F. Should the burner automatically turn off before the blower is activated, the blower or the fan control is faulty, or the high-temperature limit control needs adjustment. In addition, if the burners turn off well before the house seems to reach the temperature demanded at the thermostat, a problem may exist with the high-temperature limit switch, or the anticipator on the thermostat may be out of adjustment. In either case, the furnace may "short cycle" (the burners turn off before the demanded temperature is reached even though the heat exchanger temperature has not become too hot). If the high-temperature limit switch or anticipator allows the heat exchanger to get too hot, or the furnace to run too long, "overshooting" (a temperature higher than that demanded at the thermostat) can occur in the house.

### Circulation Blower (Fan)

Direct driven blowers (Figure 9-9) have the motor attached directly to the fan blower and do not have a V belt. The motor on direct drive blowers usually sits "inside" the blower and is not easily seen. Most modern furnaces made after the early 1970s are typically direct drive.

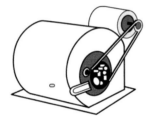 

**Figure 9-9. Indirect Drive With Pulley (left) and Direct Drive (Right) Air Circulation Blowers for Furnaces**

Some furnaces are set to allow for fan-only operation. This can be useful if the cooler air in the basement of a home can be circulated upstairs for a more even temperature distribution in the house. In addition, persons with allergies who have an electronic air cleaner can also benefit from a fan only setting. Such a setting is usually at the thermostat and can be selected instead of the automatic setting.

To check the blower fan and motor, remove the blower cover and look for proper maintenance and oiling. Check for wear or misalignment of the fan belt on blowers driven by a pulley and for dirt buildup on the motor or fan. When the system is operating, listen for unwarranted blower noise and try to determine its cause.

## DISTRIBUTION SYSTEM AND CONTROLS

The distribution system consists of supply and return ducts, filters, dampers, and registers. Supply and return ducts may be made of sheet metal, glass fiber, or other materials. Supply ducts provide warm air to the rooms, while return-air ducts return cool air back to the furnace to be heated. Glass fiber ducts are self-insulated. However, sheet metal ducts are usually not insulated except where they pass through unheated (or uncooled) spaces. Sheet metal ducts are occasionally insulated on the inside; you can determine the presence of insulation by tapping on the duct and listening for a dull sound. Check ducts for open joints wherever the ducts are exposed. Look for possible asbestos ductwork wrap. This is especially common in homes built before the mid 1960s.

**Types of Ductwork**
There are many types of ductwork but some are more common in homes than others.

*Galvanized Steel Sheet Metal Ductwork (Figure 9-10)*
This type of duct is nothing more than shaped 28-gauge sheet metal to serve as a conduit for airflow. The rectangular cross-section variety can be pre-fabricated or bent, joined, and crimp sealed on site. The round variety is usually pre-fabricated. It is often insulated with foil-backed fiberglass insulation or spray foam. This is by far the most common ductwork that has been used in homes. Perhaps its biggest advantage is its better durability over other types of ductwork, with the exception of high velocity ductwork.

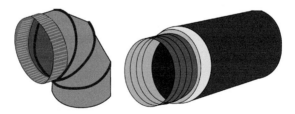

**Figure 9-10. Galvanized Steel Sheet Metal Ductwork**

*Flexible Duct*
This is typically a flexible plastic duct pre-insulated at the factory with a plastic-backed, fiberglass insulation. It typically has some plastic wire-like spiral support embedded within the flexible duct to help keep it from collapsing or pinching. The most common problem with this type of duct is that it is often stretched too much in extending it to its endpoint such that it is pinched off at some point turning a corner, going over a

rafter, being supported by a tie, etc. One of the advantages of this duct-work is that it is less labor intensive in putting together and installing in a home. It also comes pre-insulated, unlike the sheet metal ductwork.

*Fiberboard Duct (Figure 9-11)*

This type of duct is fiberglass board that is compressed with enough pressure that it forms a rigid board. It is often fabricated on site by taking a 4 ft by 10 ft 1" thick sheet of pressed fiberglass duct board and cutting it with special cutting tools so that it can be folded and taped at the final joint into a rectangular cross section of ductwork. The board is typically covered with a reinforced metal foil on the outside of the duct while joints are sealed with a special reinforced foil tape for strength. One advantage of this type of ductwork is that it tends to be a quieter system when providing airflow.

**Figure 9-11. Fiberboard Duct**

*Pre-insulated Duct (Figure 9-12)*

Typically this is a foam board type duct, often manufactured in the same way as fiberboard duct, but with greater insulation value. Typically, a thicker metal foil tape is used in sealing joints in the duct material than is used in the fiberboard variety. This type is rare to find in homes but more common in the commercial setting. This is more frequently pre-fabricated than fiberboard. Special connections or joints are used between sections to provide the potential for a better seal.

**Figure 9-12. High Velocity Pre-Insulated Duct**

*Mini or High Velocity Duct*

This is perhaps the least common type of ductwork to find in homes but is being used more frequently in modern homes. One advantage of the high velocity variety of ductwork is its small size so it takes up far less room in installation than any of the other varieties. Its primary disadvantage is that it, and the equipment that services it, are often far more expensive than the other varieties of ductwork. It is however a very durable ductwork.

*Other Components*

Air filters are usually located on the return side of the furnace next to the blower or just inside the return air grate in the hall or ceiling, Figure 9-13. However, they may be found anywhere in the distribution system. Check for their presence and examine them for correct placement and condition (for example, see if they are plugged with dust). Some filters are one-time use disposable (such as fiberglass) while others are washable and can be used again (the "mat" type and metal type). Most filters are rectangular and fit in the wall of the return grille, while some on more modern furnaces "hang" in the blower compartment.

Electronic air cleaners generally work much better than standard filters and can even remove pollen and cigarette smoke particles from the air. Many persons with allergies can be helped significantly by the higher quality filtering devices. In general, if you hear a popping sound while the furnace blower is running, the electronic air cleaner is working. Taking apart the electronic air cleaner involves dealing with potentially high voltage electricity and is not

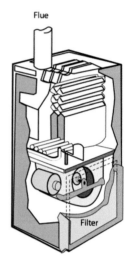

**Figure 9-13. The filter on most furnaces is placed to one side or the other on edge in the blower compartment.**

recommended. However, use of electronic air cleaners can increase the level of radon exposure.

Supply ductwork is often provided with manual dampers inside the ducts to balance airflow in the distribution system. You can locate them by looking for small damper handles extending below or to the side of the ductwork on the basement side. In zoned systems with more than one zone (multi-zone heating systems using more than one thermostat that allow each thermostat to control a different part of the home), automatically controlled dampers may be located in the ductwork, usually near the furnace. Most often movable louvers built into the heat registers/vents are used to balance the heating in individual rooms.

These allow for thermostatic control of each room or region of the house where heat flow is controlled by the damper, Figure 9-14. You can check the operation of a few randomly selected automatically controlled dampers by activating the thermostats one at a time. If the dampers are working properly, air should begin to circulate in each zone immediately after its thermostat is activated.

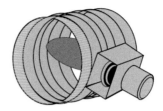

**Figure 9-14. Duct Damper to Thermostatically Control Flow of Warm Air to a Room**

Check the location of supply and return grilles in each room. Warm air registers are most effective when positioned on the floor or low on the exterior wall; cold air registers are most effective when high on the walls or in the ceiling. Return grilles should be on the opposite side of the room from the supply registers. If return grilles are located in a hallway or in a different room, make sure intervening doors are undercut by about one inch to allow return air to flow out of closed rooms. Return air ducts should never be located in garages. Only by special applications (which are very rare) are vents and dryer vents allowed in garages. This is due to the firewall breech that results, regardless of whether their location is higher than 18" off the floor. Another serious concern is that heavier than air gasoline fumes could potentially enter the house.

**Vent Types for Combustion Appliances**

Tables 9-1 and 9-2 shows the appropriate vent types to use for venting various combustion appliances. Combustion appliance vent sizing is typically established by the manufacturer. B-vent is typically a double wall flue for use with propane or gas-fired combustion appliances. The outer wall is galvanized steel and the inner wall is aluminum. The air gap between the two metal walls serves as an insulator to allow for smaller clearances to combustibles. A BW-vent is also a double wall vent in an oval shape to allow it to fit inside studded wall frames and still provide a large cross-sectional area. The BW-vent is used primarily for venting wall furnaces. Another type of B-vent is a corrugated single wall vent. The corrugated wall allows it to be bent and shaped align the vent to serve its purpose and be properly pitched. An L-vent is a double wall vent for oil-fired, propane, or gas appliances. It is built like the B-vent except that the inner sleeve is made of stainless steel rather than aluminum. It must be used for oil-fired equipment and some propane and gas-fired appliances if required by the manufacturer.

A single wall galvanized metal vent is for open interior use only because of its significant clearance requirements to combustibles. Direct vent space heaters and mobile home furnaces are sealed combustion appliances that must use a manufacturer approved venting system. Typically, this double wall concentric vent system exhausts combustion gases out of the inner pipe and draws in combustion air for the burners through the outer pipe. 90%+ high-efficiency condensing furnaces typically have extracted enough heat from the combustion gases that they can use a schedule 40 PVC pipe for a vent.

Several factors affect vent operation. The higher the gas temperature of the combustion gases in the vent, the better the flue or vent will draw the combustion

**Table 9-1. Appropriate Vent Types**

| Combustion Appliance | Minimum Allowable Vent Type | |
|---|---|---|
| | Conditioned Space | Unconditioned Space |
| Oil – Furnace, Boiler, or Water Heater | Single Wall 24 or 26 Gauge Steel or L Vent | L Vent |
| Gas -Forced Air Furnace - Gas Fired (propane or natural gas) | Single Wall Galvanized, or B-Vent | B-Vent |
| Gas -Vented Space Heater – Gas Fired (propane or natural gas) | | |
| Gas - Water Heater – Gas Fired (propane or natural gas) | | |
| Mobile Home Furnace | | |
| Direct Vent Space Heater | Special (Manufacturer's Instructions) | Special (Manufacturer's Instructions) |
| 90+ AFUE Condensing Forced-Air Furnace | | |
| Gas – Wall Furnace | Single Wall Galvanized – Almost always requires BW because inside of wall | BW-Vent |
| Floor Furnace | Not applicable – always in unconditioned space | Gas – B-Vent |
| | | Oil – L Vent |
| Woodstove | Single Wall 24 Gauge Black Steel, or Shielded Single Wall (such as Duravent or Super Six) | Double Wall Stainless Steel with insulation between walls (such as All-Fuel, or GSW) or Triple Wall |

**Table 9-2. Vent Clearances**

| VENT CONNECTOR TYPE | CLEARANCES | | | |
|---|---|---|---|---|
| | Combustion Appliance Type | Without Protection | With Heat Shield | |
| | | | **Wall** | **Ceiling** |
| Single-Wall Galvanized | Vented gas space heaters Atmospheric gas furnaces and water heaters Fan-assisted-draft gas furnaces (80+ AFUE) | 6" | 2" | 3" |
| | Oil furnaces, vented space heaters, boilers, and water heaters | 18" | 6" | 9" |
| 24 Gauge Steel (Black or Stainless) | Wood stoves | 18" | 6" | 9" |
| B Vent (including BW) | Atmospheric gas furnaces and water heaters, Fan-assisted-draft gas furnaces (80+ AFUE) | Install per manufacturer's instructions (usually 1") | | |
| L Vent | Oil furnaces, boilers, or water heaters | Install per manufacturer's instructions (usually 3") | | |

**Table 9-3. Chimney Type**

| CHIMNEY TYPE | Minimum Clearance from Combustible Surfaces |
|---|---|
| Interior chimney masonry with fireclay liner | 2" |
| Exterior masonry chimney with fireclay liner | 1" |
| All-fuel metal vent: insulated double-wall or triple-wall pipe | Install per manufacturer's instructions (usually 2" or less) |

products up and out of the house. In addition, the colder the air inside and outside of the house, the better the vent will draw. If heat is lost from the vent because it is not very well insulated as it extends up through the house, it can reduce the draw in the vent. The greater the negative pressure in the CAZ, the more difficult it is for a vent to draw properly. The capacity or size of the venting system can affect flow through it, but there is a good range of cross-sectional area for venting most appliances.

A vent can be too small or a chimney too large, to provide for adequate draw from a combustion appliance.

Any restrictions in the vent, such as a pinch or a corroded inner pipe can cause the draw on the vent to be reduced. Good cross flow of air across the top opening of the venting system can also help it draw better. This is why there are minimum requirements for chimney heights and vents above roof decks, Figure 9-15. Specific requirements for oil burning equipment state that the vent or chimney must be at least 2 feet higher than any portion of the building that rest within 10 feet of it.

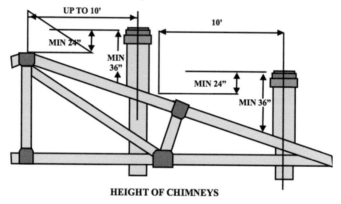

**Figure 9-15. Minimum Flue Heights on Roofs**

## OTHER HEATING UNITS

### Forced Hot Water (Hydronic Heating) Systems

Hot water heating systems, like warm air systems, are of two types, forced or "hydronic," and gravity. Gravity or open systems are sometimes found in older single-family houses, but in most cases such systems have been replaced or converted to a forced hot water system. Gravity systems have no water pump and use bulky 3-inch piping for hydronic distribution of the hot water. They tend to heat unevenly, are slow to respond, and can only heat spaces above the level of the boiler. Like gravity warm air systems, they are considered inefficient.

Forced hot water systems are usually heated by gas- or oil-fired boilers. The term boilers is a misnomer since the water in hot water boiler systems does not actually boil, unlike steam boilers in which the water is boiled. Occasionally they may be immersion-heated by electric resistance heating coils. These coils replace the burner found in gas- and oil-fired boilers. The hot water pump and distribution piping for electrically heated systems are similar to those of gas- and oil-fired systems.

*Boiler*

Most hot water and steam heating systems have steel boilers with a service life of about 20 years. Cast iron boilers, which are less common, have a service life of about 30 years. Old cast iron boilers converted from coal-fired units may last much longer but are usually quite inefficient. All boilers (except electrical resistance boilers) have both a heat exchanger and a burner. The heat exchanger works much like a heat exchanger in a furnace except it is usually thicker metal and is a pressure vessel quality material. The burners operate on one side of the exchanger while the water to be heated occupies the other side. Check boilers for signs of corrosion and leakage that could represent a damaged or deteriorating heat exchanger. Also, check for water stains. Occasionally a boiler fitting will leak slightly before it warms up, expands and returns to watertight fit. Do not confuse condensation droplets on a cold boiler with water leaks.

*Expansion Tank, Figure 9-16*

The expansion tank for hot water boilers is usually located above the boiler (although it may be in the attic) and is connected to the hot water distribution piping. As boilers heat up, the water expands. Whether the hot water system is a closed system (fully contained within a piping system) or an open system, it must accommodate water expansion. Expansion tanks serve this purpose. Most expansion tanks are

**Figure 9-16. The large amount of air in a normal expansion tank allows it to serve as a compression buffer for the expansion of water in the hot water heating system as it heats (upper diagram). If the tank loses enough air, it loses its buffering characteristics and becomes "waterlogged" (lower diagram).**

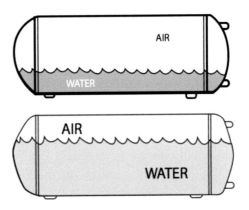

compression-type tanks that are designed to permit heated water to expand against a cushion of pressurized air within the tank. When the tank loses this air cushion, it becomes "waterlogged" and expansion of the water cannot be accommodated. Instead, water discharges from the boiler's pressure relief valve each time the system heats up. Some of the newer expansion tanks are of the diaphragm variety that separates the water from the air cushion and thus keeps the air from dissolving into the water and waterlogging the tank. However, the diaphragm on these tanks can sometimes fail. Expansion tanks may have to be emptied (standard expansion tanks) or recharged (diaphragm type expansion tanks) to continue to serve their purpose.

On open system boilers (the water supply to the boiler opens to atmospheric pressure), the system is not under direct pressure. The expansion tank is usually located above the highest radiator in the house (for example, the upper half of a closet on the uppermost floor of the house) and is open to air above it by a pipe that usually continues up through the attic and the roof and drains out onto the roof if overflow occurs. Some of these have a glass gauge on them that allows determination of the need to add more release water in the system. Since open systems are not under pressure at their uppermost points, their expansion tanks are not considered pressure vessels. In addition, since they are not pressurized, they do not have circulating pumps, pressure relief valves, or a pressure-reducing valve and they must be manually fed make-up water.

### Boiler Controls

All boilers should typically be equipped with a pressure gauge, a pressure relief valve, a backflow preventer, and a pressure-reducing valve. See the example in Figure 9-18.

The pressure gauge indicates the water pressure within the boiler, which should normally be between 12 psi (when cold since this is the typical setting for the pressure reducing valve) and 22 psi (when warm because of the expansion of the water and the subsequent increase in pressure).

A temperature gauge may be included in the pressure gauge.

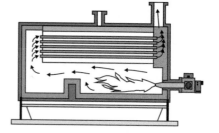

**Figure 9-17. Fire Tube Boiler System**

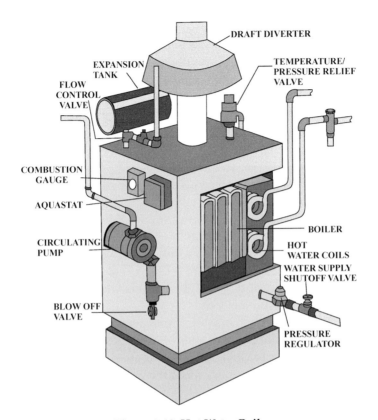

Figure 9-18. Hot Water Boiler

The pressure-reducing valve automatically adds water to the system from the domestic water supply when the boiler pressure drops below 12 psi. Pressure readings lower than 12 psi indicate a bad gauge or a faulty or maladjusted pressure reducing valve and can be identified by little if any air relief when opening a radiator bleed valve above.

The pressure relief valve (Figure 9-19) should discharge water from the system when the boiler

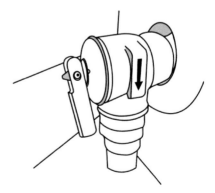

Figure 9-19. Close Up of a Temperature and Pressure Relief Valve

pressure reaches 30 psi. High-pressure conditions are usually due to a waterlogged expansion tank. If the boiler also generates domestic hot water, high pressure may be caused by cracks in the coils of the water heater, since the domestic water supply pressure usually exceeds 30 psi. The pressure relief valve should be mounted on the boiler and have a drain line much like hot water heaters. The backflow preventer eliminates the possibility of contaminated boiler water reversing into the supply plumbing in the house in the event the pressure is lower in the supply plumbing.

Hot water boilers should have a high-temperature limit control or aquastat that shuts off the burner if the boiler gets too hot (greater than 200°F). A low water cutoff can control turns off the burners when the level of the water in a boiler gets too low.

These devices typically have a "blow off" valve flush out water from time to time. The blow off valve is used to empty the low water cut off and thereby check if it shuts off the burners when the water level is low. This test is not recommended.

Aquastats are simply thermostats that evaluate and help with controlling the temperature of water instead of the air. A modulating aquastat keeps the water in the boiler at a higher temperature as the outside temperature gets lower, generally when there is a demand for heat from the thermostat.

*Circulating Pump and Controls (Figure 9-20)*

The circulating pump forces hot water through the system at a constant pressure and speed. It must be a closed system and is much more efficient than a gravity fed system. It should be located adjacent to the boiler on the hot water distribution piping.

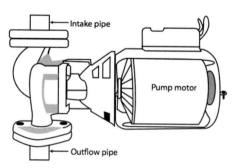

**Figure 9-20. Typical Circulating Pump for a Hot Water Boiler**

The pump may be operated in one of the following four ways:

1.  Constant-running circulator, where a manual switch controls the pump. The pump is usually turned on at the beginning of the heating season and runs constantly until it is turned off at the end of the heating season. The thermostat independently activates the boiler as heat is required.

2.  Aquastat-controlled circulator, which turns the pump on and off only when the boiler water reaches a preset boiler temperature (normally 120°F). Like the constantly running circulator, the thermostat independently activates the burner as heat is required.

3.  Thermostat-controlled circulator, in which water is maintained at a constant temperature in the boiler by an aquastat and the pump only turns on when there is a demand for heat.

4.  Relay-controlled circulator, in which the pump is activated (via a relay switch) whenever the boiler is activated by the thermostat. This is probably the most efficient method of boiler operation—neither the boiler nor the pump operates unless there is demand for heat.

Some boilers have a safety interlock which keeps the boiler from operating unless the pump is running. You can determine what kind of device controls the pump and check its operation. You can also inspect the condition and operation of the pump. A loud pump may have bad bearings or a faulty motor. You can inspect the seal between the motor and the pump housing for signs of leakage. Examine the condition of all electrical wiring and connections. Feel the return line after the system has been operating for a short time, it should be warm. If it is not, the pump may be faulty. In heating systems used for generating domestic hot water, the thermostat will control the circulating pump and an aquastat will control the burner.

*Distribution Piping*

The forced hot water distribution system consists of distribution piping, radiators, and control valves, as seen in Figure 9-21. Distribution piping may be one of three types: series-loop, one-pipe, and two-pipe.

In a series loop, radiators are connected by one pipe directly in a series. Since the last radiator will receive cooler water than the first,

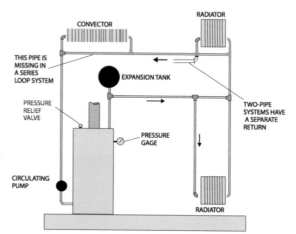

THIS PIPE IS MISSING IN A SERIES LOOP SYSTEM

CONVECTOR

RADIATOR

EXPANSION TANK

PRESSURE RELIEF VALVE

PRESSURE GAGE

TWO-PIPE SYSTEMS HAVE A SEPARATE RETURN

CIRCULATING PUMP

RADIATOR

**Figure 9-21. A Typical Hot Water Boiler Distribution System— Heat Sources can be Finned Convectors or Iron Radiators**

downstream radiators should be progressively larger. Alternatively, series-loop systems may be divided in small zones to overcome this problem.

One-pipe systems differ from series-loop systems in that their radiators are not connected in series. Instead, each radiator is separately attached to the water distribution pipe via diverter or Venturi fittings, which are used to regulate the amount of entering hot water. This type of system, however, may suffer from the same downstream cooling problems that affect the series system. Two-pipe systems use separate pipes for supply and return water, which assures a smaller temperature differential between radiators regardless of their location. Individual room control is possible with both one-pipe and two-pipe systems, although a change in the valve adjustment on one radiator will affect the performance of others downstream.

Distribution piping can be checked for leaks at valves and connections. Make sure pipes are properly insulated in unheated basements, attics and crawl spaces. When the distribution piping is divided in zones (a system with more than one zone is a multi-zone system), each zone will have either a separate circulation pump or a separate electrically operated (for instance through a separate thermostat for the zone) or manually operated valve, see Figure 9-22. Most valves on any boiler system are not normally tested, including manually operated zone and radiator valves.

You can check the operation of a few randomly selected zone valves by activating one thermostat at a time. If hot water is being

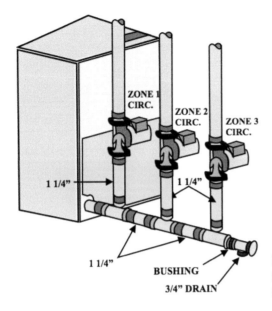

Figure 9-22. Separate Hot Water Circulators for Separate Zones

distributed properly to each zone, the radiators in that zone should be warm to the touch within several minutes. Locate all valves, inspect their electrical wiring and connections, and look for signs of leakage.

**Radiators and Control Valves**

Radiators (Figures 9-23 and 9-24) are of three types: cast iron (which in most cases are free-standing but are sometimes hung from the ceiling or wall), convector (Figure 9-25) (which may have a circulating fan), and baseboard.

Older residential buildings usually have cast iron radiators that are extremely durable and can normally be reused, although they are

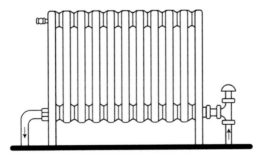

Figure 9-23. Typical Hot Water Radiator

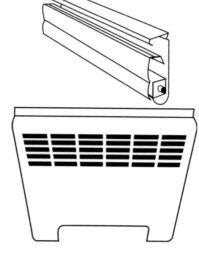

**Figure 9-24. Zone Valves for a Hot Water Heating System where Separate Pumps/Circulators Not Used**

**Figure 9-25. These two types of convectors look like and work much like an aluminum finned radiator, although the fins are not always readily visible.**

less efficient than convectors. Baseboard radiators are considered the most desirable for residential use because they are the least conspicuous and distribute heat most evenly throughout the room. Radiators should be located on outside walls whenever possible.

Most radiators have a small air bleed valve, such as shown in Figure 9-26a, located near the top of the radiator that allows trapped air to be released for better circulation of the hot water. You can activate the system and look for signs of water seepage. Feel the surfaces of a few randomly selected radiators to ensure that they are heating uniformly. Examine the fins of all baseboard units and convectors for dirt and damage. These fins can be "combed" straight. Check the condition of radiator control, safety, and bleed valves to see if they are operational. Often the valve stem cap needs tightening or the stem packing needs replacing.

Hot water distribution piping may be embedded in floors (as shown in Figures 9-27 and 9-28), walls, and ceilings to provide radiant heating. Because the piping is embedded, it can only be examined by looking for signs of water leakage or rust on the floor or wall surfaces. Such heating is normally trouble-free, unless there are major structural problems that damage distribution piping and joints.

The following represents how one hydronic boiler might operate

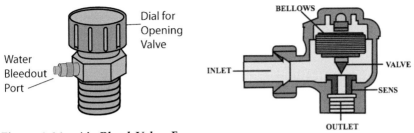

**Figure 9-26a. Air Bleed Valve For Hot Water Boiler System**

**Figure 9-26b. Steam Trap**

in sequence to heat one of the rooms in a house that is on its own thermostat and zone in a multi-zone (more than one zone) heating system:

A.   When the room temperature drops, the room thermostat switches "on" (the switch closes).

B.   The zone valve opens and the circulator starts. The boiler burner flame starts after the primary control has turned on the oil burner (if the boiler water temperature is too low, having been detected via the low limit switch). This does not apply in Canada where circulators are often set to run continuously and where the thermostat directly turns on the oil burner. For the burner flame to start, the burner typically pumps oil from the oil tank through a copper fuel line, during which the transformer creates a high voltage spark near the burner nozzle to ignite the oil.

C.   Hot gases pass through tubes of steel or between sections in a cast iron boiler, sending heat into the water through the heat exchanger

D.   The hot water leaves the boiler by being drawn by the circulator pump

E.   The water passes through the distribution piping and the baseboards, which warm the rooms.

F.   The room thermostat senses the heat increase while less warm water is flowing back to the boiler, having lost some of its heat to the radiators.

G.   The boiler temperature eventually rises to the high limit temperature at which time the sensor indicates such to the primary control, and the primary control turns off the boiler burner.

H.    When the room becomes warm enough relative to the thermostat setting, the thermostat opens its switch, the circulator stops pumping boiler water, and the zone valve closes.

## Radiant Panel Heating

Hot water distribution piping may be embedded in floors (as shown in Figures 9-27 and 9-28), walls, and ceilings to provide radiant heating. Because the piping is embedded, it can only be examined by looking for signs of water leakage or rust on the floor or wall surfaces. Such heating is normally trouble-free, unless there are major structural problems that damage distribution piping and joints.

You can operate radiant heating systems to determine functional adequacy. If boilers have already warmed up, radiant surfaces should be warm to the touch within several minutes. You can check the condition of the shut-off valves for each distribution zone and the main balancing or tempering valves near the boiler, and look for signs of leakage. You can also inspect the expansion tank and air vent (if any).

## STEAM HEATING SYSTEMS

Steam heating systems are no longer installed in small residential buildings, but are still found in many older ones. They are simple in design and operation, but require a higher level of maintenance than modern residential heating systems. They are gravity type systems since they rely on the hot steam to rise to the radiators without the use

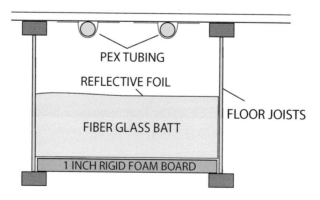

PEX TUBING

REFLECTIVE FOIL

FLOOR JOISTS

FIBER GLASS BATT

1 INCH RIGID FOAM BOARD

**Figure 9-27. Under-floor Hot Water Heating System**

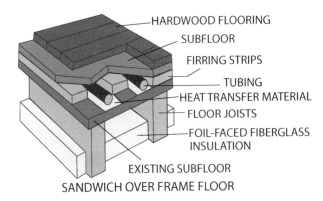

HARDWOOD FLOORING
SUBFLOOR
FIRRING STRIPS
TUBING
HEAT TRANSFER MATERIAL
FLOOR JOISTS
FOIL-FACED FIBERGLASS
INSULATION
EXISTING SUBFLOOR
SANDWICH OVER FRAME FLOOR

Figure 9-28. Radiant flooring can be placed under the floor (above) or above the floor (lower)

of circulating pumps. In addition, the radiators may become extremely hot and as such may represent a hazard to children.

### Boiler and Controls

Steam boilers are physically similar to hot water boilers. Figure 9-29 shows the components of steam boilers. Unlike hot water boilers, steam boilers operate only about three-fourths full of water and at much lower pressures, usually 0.2-5.0 psi. Steam boilers should be equipped with a water level gauge, a pressure gauge, a high-pressure limit switch, a low water cut off, and a safety valve.

You can activate a steam boiler and observe the water level gauge, which indicates the level of the water in the boiler. The gauge should normally read about half to three-quarters full, but the actual level of the water is not critical as long as the level is showing. If the gauge is full of water, the boiler is flooded and water must be drained from the system.

Check to see if it has been converted to a hot water boiler, if it is a two-pipe system or if there is a circulating pump on some of the distribution lines. If the gauge is empty, the boiler water level is too low and must be filled (either manually through the fill valve or automatically through the automatic water feed valve, if the boiler has one). Unsteady, up and down motion of water in the gauge means the boiler is clogged with sediment or is otherwise operating incorrectly and must be repaired. You can see the clarity of the boiler water when checking the gauge.

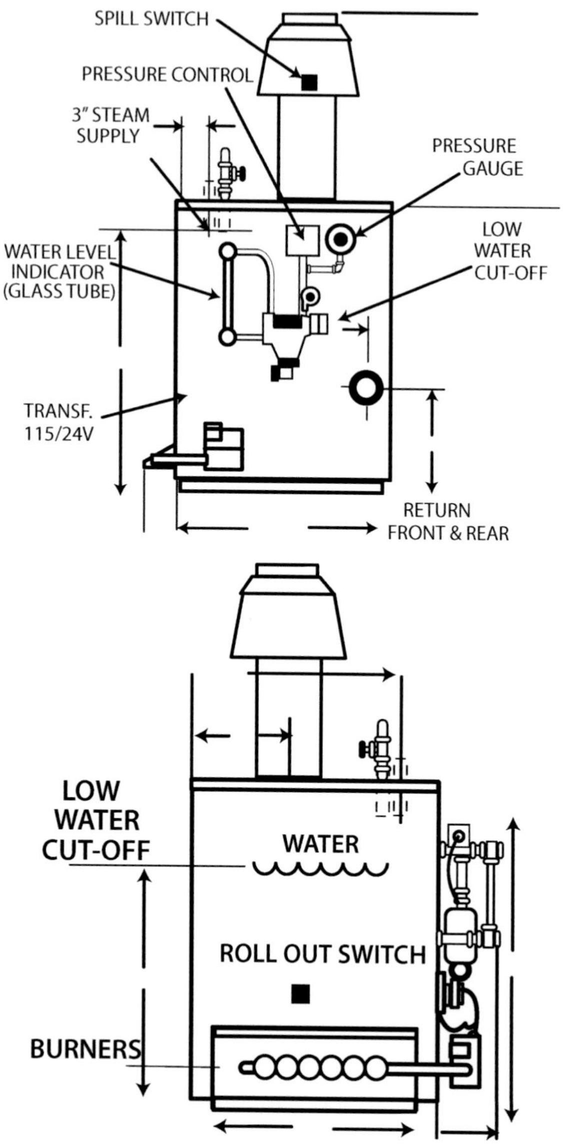

The high-pressure limit switch turns off the burner when the boiler pressure exceeds a preset level, usually 5-7 psi. It is usually connected to the boiler by a pigtail-shaped pipe. The low water cut off (Figure 9-30) shuts down the burner when the boiler water level is too low. The pressure relief valve is designed to discharge when the boiler pressure exceeds 15 psi.

The sequence of operation of a steam boiler is similar to a hot water boiler described previously. However, there is no circulator pump involved or needed because the steam rises in the pipes by gravity feed to the radiators and warm the rooms. As the steam gives off its warmth through the radiators, it condenses to water which drains back to the boiler, typically through the same distribution piping (one-pipe system). In addition, when the room becomes warm enough relative to the thermostat setting, the thermostat opens its switch and the burners are turned off.

**Figure 9-29. Steam Boiler Components**

SPILL SWITCH

PRESSURE CONTROL

3" STEAM SUPPLY

PRESSURE GAUGE

WATER LEVEL INDICATOR (GLASS TUBE)

LOW WATER CUT-OFF

TRANSF. 115/24V

RETURN FRONT & REAR

LOW WATER CUT-OFF

WATER

ROLL OUT SWITCH

BURNERS

## Distribution Piping

The steam distribution system consists of distribution piping, radiators, and control valves. Distribution piping may have either a one-pipe, as shown in Figure 9-30 or two-pipe configuration. Two-pipe systems are seldom seen in the single-family setting. In a one-pipe system, steam from the boiler rises under pressure through the pipes to the radiators. In the system, "quick vents" allow the cooler air to escape so the steam can rise more quickly through the pipes. At the radiators, they displace air. The air is evacuated through the radiator vent valves or "air

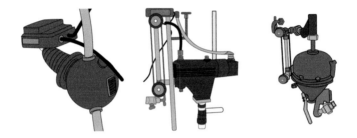

**Figure 9-30. Some Varieties of Low Water Cutoffs Used for Steam Boilers**

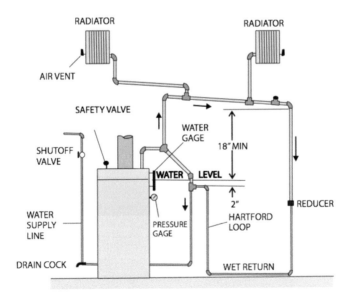

**Figure 9-31. One-Pipe Steam Heating System**

vents" that automatically shut off when the hotter steam reaches them.

The steam condenses on the radiator's inner surface, and gives up heat. Steam condensate flows by gravity back through the same pipes to the boiler for reheating. The pipes must therefore be pitched no less than one-inch in ten feet in the direction of the boiler to ensure that the condensate does not block the steam in any part of the system.

All piping and radiators must be located above the boiler in a one-pipe system unless there is a pump in the condensate return line. If the condensate return piping is located below the level of the boiler water (a "wet return" system), it should be brought back up to at least the level of the boiler water and vented to the supply piping in a "Hartford Loop." This prevents a leak in the condensate return from emptying the boiler.

In a two-pipe system, steam flows to the radiators in one-pipe and condensate returns in another, usually smaller pipe. A steam trap on the condensate return line can release air displaced by the incoming steam. Two-pipe systems can be balanced by regulating the supply valves on each radiator, and may be converted for use in a hot water heating system (although new, larger-size return piping usually may have to be installed). All radiators must be located above the boiler in a two-pipe system unless there is a pump in the condensate return line.

Distribution piping can be checked for leaks at all visible valves and connections. Make sure all piping is properly pitched to drain toward the boiler. "Pounding" may occur when oncoming steam meets water trapped in the system by improperly pitched distribution piping, or by shut-off valves that are not fully closed or open. Inspect the physical condition of all piping for domestic water supply piping. Make sure pipes are properly insulated in unheated basements, attics, and crawl spaces.

### Radiators and Control Valves

Steam radiators consist of cast iron and are usually freestanding. They are quite durable and in most cases can be reused. They can also be replaced by convectors. Radiators should be located on outside walls whenever possible.

You can activate the system and inspect the physical condition of the radiators. Look for signs of water leakage. Feel the surfaces to make sure they are heating uniformly. If they are not, check the radiator air vents and supply valves on a one-pipe system and the radiator supply

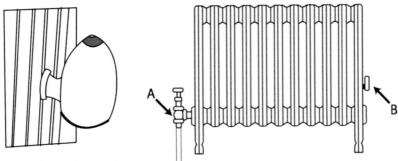

**Figure 9-32. Steam Ra-
diator Air-Bleed Valve/** **Figure 9-33. Typical One-Pipe Steam Radiator**
Vent

valves and the steam trap on the condensate return on a two-pipe system. Often air vents need cleaning, supply valves need tightening, and
valve packing needs to be replaced. "Pounding" near the radiator is
often cured by lifting one edge of the radiator slightly; this reduces condensate blocking in the pipes. If air vents are defective and are bleeding
after the steam has reached the radiator, some staining on the radiator
may be evident or you may actually see steam emitting from the air vent
while the boiler is operating.

ELECTRIC RESISTANCE HEATING

Electrical resistance heating elements are commonly used in heat
pump systems, wall heaters, radiant wall or ceiling panels, and baseboard heaters. They are less frequently used as a heat source for central
warm air or hot water systems. Such heating devices usually require
little maintenance but may be expensive to operate. One of the advantages of these systems, in some cases (such as with wall heaters and
radiant panel heaters), is that they can potentially be set up so that each
room may be placed on its own thermostat. This way there are multiple
zones in the system (a multi-zone system), thus allowing for savings in
heating by not heating rooms that are not being used.

**Electric Resistance Heaters**
Electric resistance heaters are used in warm air and hot water
systems, and in heat pumps. They incorporate one or more heavy duty

heating elements that are actuated by sequence relays on demand from the thermostat. The relays start the fan. Each heating element is then started at 30-second intervals, which eliminates surges on the electrical power system. In warm air and heat pump systems, electric heating elements are normally located in the furnace or heat pump enclosure, but they may be located anywhere in the ductwork as primary or secondary heating devices. Once the thermostats in the room(s) detect a temperature equal to the settings of the thermostats, the electrical resistance heating elements automatically shut off and the fan is then automatically shut off. You can inspect the condition of all electric resistance heaters, including their wiring and connections.

Electric resistance heating systems have no moving parts (except the blower) and require no adjustment. The circulation blower and air distribution ductwork for electric resistance heating systems (and heat pumps) are identical to those of gas- and oil-fired warm air systems.

### Electric Wall Heaters

Electric wall heaters are compact and generally trouble-free. They are often used as supplementary heating units, or as sole heat sources in houses where heating is only occasionally required. They may have one or more electric heating elements, depending on their size. Wall heaters often have a small circulation fan. You can check its condition and operation, and look for dirt build up on the fan blade and motor housing. Inspect all electrical wiring and connections that are visible without removing any housing.

### Radiant Wall and Ceiling Panels

These wires are stapled to the ceiling and covered with another layer of drywall to provide an electrical heating system in the ceiling. This type of system often allows for a separate thermostat for each room to avoid unnecessary heating of rooms not being used. As you might imagine, the drywaller needs to be careful not to short these wires by nailing into them when adding the last layer of drywall.

Electric heating panels that are embedded in wall or ceiling surfaces, Figure 9-34, cannot be directly inspected. However, you can check all radiant surfaces for signs of surface or structural damage. Finding the problem wire in these systems can be as difficult as finding a needle in a haystack blindfolded. When problems occur, it is not unusual for a new heating system to be installed to replace the electric heating panel system. If the panels do not provide heat when the thermostat is activated, check the circuit breaker. If it is not the circuit breaker, the system is not operable.

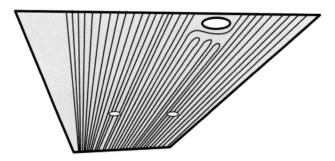

**Figure 9-34. Radiant Ceiling Panel**

SPECIAL TOPICS

**Baseboard Heaters**

Baseboard heaters rarely malfunction, but their heating fins can be damaged and become clogged with dust. You may decide to remove heater covers on a randomly selected set of baseboards and check for such problems.

**Space Heaters**

There are special rules for unvented space heaters because they release their combustion products, including $CO_2$, water vapor, $SO_2$, and CO compounds into the breathing environment in a home. The amount of water vapor released by a 25,000 Btu unvented space heater is about one quart of water per hour. Over time during the cold weather season, this moisture will condense on cold surfaces—including windows, walls, and even inside of walls—promoting mold growth.

With regard to the rules in the weatherization program, WP in 08-4: "Space Heater Policy" prohibits any weatherization work to proceed when an unvented gas or liquid fuel space heater represents the primary heat source for the home. The owner must agree to allow installation of a vented, code-compliant heating system before the home is weatherized. In addition, space heaters that are not used as the primary heat source cannot have an input rating over 40,000 Btu per hour. Furthermore, weatherization funds cannot be used to replace secondary unvented space heaters if they need replacing. In addition, a mobile home is prohibited from having an unvented space heater; any space heater in this case must be vented to the outside. Finally, minimum ventilation requirement guidelines apply when the home is heated by any space heaters as well.

## Humidifiers

Humidifiers are sometimes added to warm air heating systems to reduce interior dryness and reduce fuel bills during the heating season. It is generally believed that a higher humidity can allow comfortable living at three degrees lower than would be required in a dryer environment. The principle behind this is the same one that makes 90°F in a humid summer environment (such as in Florida) feel much hotter than 90°F in a dry summer environment (such as in Las Vegas). Humidifiers are installed in the air distribution system and are controlled by a humidistat (usually located on or near the humidifier housing). A humidistat is an automatic control that switches a fan, humidifier, or dehumidifier on and off based on the relative humidity at the control. Humidifiers can be of several types:

• Pan type (Figure 9-38), where an open pan of water is used to provide water vapor to the circulating air stream.

• Stationary pad (Figure 9-36), where air is drawn from the furnace plenum or supply air duct by a fan, blown over an evaporator pad, and returned to the air distribution system.

• Revolving drum (Figure 9-35), where water from a small reservoir is picked up by a revolving pad and exposed to an air stream from the furnace plenum or supply ductwork.

• Atomizer (Figure 9-37), where water is broken into small particles by an atomizing device and released into the supply or return air ductwork.

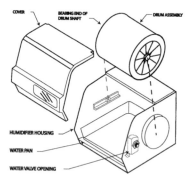

**Figure 9-35. Revolving Drum**

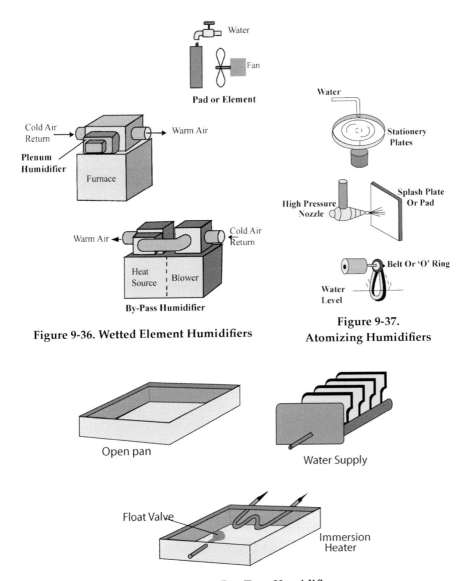

Figure 9-36. Wetted Element Humidifiers

Figure 9-37.
Atomizing Humidifiers

Figure 9-38. Pan Type Humidifiers

Turning up the humidistat activates the humidifier. (This may not be possible in summer if the unit is off). It should only operate when: 1) the furnace fan is on; 2) the system is in the heating mode; and 3) the indoor humidity is lower than the humidistat setting.

## COMBUSTION APPLIANCES—BACKGROUND, INSPECTION AND TESTING

### Background

Some important definitions and concepts with combustion appliances are as follows:

*CAZ—(Combustion Appliance Zone)*

The CAZ is any area of the building that contains a vented combustion appliance regardless of whether it is in the main living area or in an isolated area that is separated from the main living area by a door. Typically, this is the furnace or water heater closet, or the room where the combustion appliance is located.

*Direct Vent Heating Appliance*

A heating appliance that draws air directly from the outside, such has a high efficiency furnace.

*Atmospheric Heating Appliance*

A heating appliance that draws air from inside the house to provide oxygen for its fuel to burn, such as a standard (low or medium) efficiency furnace. Standard efficiency furnaces are no longer allowed to be manufactured, but they may still be in use in older homes. They rely strictly on the buoyancy to exhaust flue gases. A medium efficiency furnace relies partly on this same buoyancy and partly on a fan to exhaust its flue gases. An atmospheric appliance relies on the temperature and buoyancy of combusted gases for the gases (to at least some degree) to rise up a chimney or vent. Low efficiency atmospheric draft appliances have intentional openings called draft diverters or draft hoods for cooler air in the CAZ to mix with the hot combustion gases before they exit the flue. This helps provide a consistent draft and to counter the effects of outside wind. If the flue is not drafting properly to adequately exhaust the low efficiency combustion gases, the combustion gases can spill out into the CAZ.

*Combustion Process*

Air is an important part of the combustion process. An adequate supply of oxygenated combustion air or make up air needs to be provided to any combustion appliance. On standard and medium efficiency furnaces in older homes, this can be provided by adequate air exchange or communication with other areas of the house, or by outside air provided by ducts feeding air directly into the furnace room. If outside air is used, the furnace room should be insulated from the inside walls shared with other heated living spaces to avoid excessive loss of heat and freezing of pipes, etc.

On virtually all high efficiency furnaces the combustion air must come from close proximity to the exiting exhaust air outside and cannot come from the inside of the house. Thus, two plastic pipes are typically used on high efficiency furnaces, one to bring combustion air in and one to take combusted air out, as shown in Figure 9-39.

In addition, the highly caustic condensate that high efficiency furnaces typically generate usually drains to the nearest floor drain and often disintegrates the concrete it flows over to get to the drain. Sometimes, especially on houses with no lower level floor drain, a condensate pump (usually not much bigger than a large matchbox or shoe box) is used to divert high efficiency furnace and/or air-conditioning condensate water to a sewer line.

With atmospheric appliances, if the room is too small to provide an adequate volume of supply air, the vent or chimney may not draft properly. To get bad gases out of the house, NFPA 54 "National Fuel Gas Code" requires that combustion air must be provided for any combustion appliance zone/closet where the input exceeds 1000 Btu per 50 ft³ of volume in the house. Virtually no combustion appliance zone—typically the furnace or boiler closet—satisfies this volume requirement by itself. In this case,

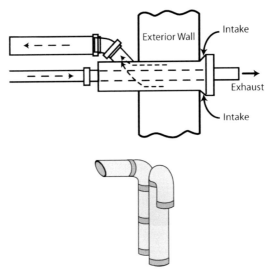

**Figure 9-39. A Concentric (upper) and Dual Port (lower) High Efficiency Furnace PVC Air Intake/ Exhaust System**

as seen in Figure 9-40 combustion air volume can be added by cutting openings and installing grilles to establish vents near the bottom and tops of all doors between the furnace and any other rooms—except bedrooms.

If there is still not enough cubic footage by allowing air exchange in this manner, then a permanent opening of an adequate size between the outside and the furnace room or CAZ must be provided.

Venting combusted or flue exhaust air from appliances is always ventilated to the exterior. The vent is the connection between the combustion appliance and a vertical metal or masonry chimney. The vent is also sometimes referred to as the smoke pipe. A good venting system should carry all the combustion byproducts to the outside, establish a draft or draw of the combustion gases up the chimney or vent quickly, have minimal restrictions and a properly sized flue, be provided with adequate makeup air to replace the air from the CAZ used to burn and mix with the combustion byproducts exiting out the flue, and not overheat the materials around it. There should not be too many elbows on a vent because this can restrict flow. The smoke pipe (flue) between the unit and the chimney should have a slight upward pitch with no sags. Sometimes if the horizontal run of the flue is too long, it will cause backflow or spillage to occur.

Figure 9-40. A furnace closet or CAZ door with grilles at the top and bottom provides adequate combustion air volume of supply air.

## Inspecting Combustion Appliances

You need to inspect combustion appliances to ensure safe operation. These appliances include furnaces, ovens, stoves, boilers, water heaters, and even space heaters. Although some of these items were mentioned previously, the following is a more

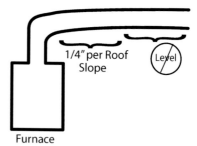

Figure 9-41. The flue on a combustion appliance should always have a minimum slope of ¼ in./foot

extensive checklist, with follow up detail:
1) overall safety and condition,
2) vent pipe condition and location,
3) adequacy of combustion air,
4) heat exchanger condition,
5) condition of the heat distribution system such as ductwork,
6) the condition of wiring that is tied into the combustion appliance,
7) fuel leaks,
8) fire and electrical hazards,
9) combustion efficiency,
10) clearances,
11) proper combustion safety,
12) identify the potential for carbon monoxide and check for CO,
13) backflow or spillage of carbon monoxide and other flue gases into the house,
14) evidence of flame rollout,
15) worse case draft, and
16) testing efficiency of combustion appliances.

Obviously, a visual inspection alone will not find all these problems; therefore, frequent use of specialized diagnostic tools is common. The following sections detail how to inspect and test the items mentioned above.

### Overall Safety and Condition

Check that the unit meets local fire safety regulations. No fuel-burning unit should be located directly off sleeping areas or close to combustible materials (see "Clearances" section). Locate the data plate on each unit and at least note its date of manufacture and its rated heating capacity in Btus. If you find a manufactured date, the ANSI code year tells you it was manufactured within three years of that date. Make sure the fuel-burning unit is easily accessible for servicing.

Open all access panels and examine the external and internal condition of each unit. On hot air furnaces, look for signs of rust from basement dampness or flooding, as seen in Figure 9-42. If an air-conditioning evaporator coil is located over the furnace, look for rust caused by condensate overflow.

On hot water boilers, look for rust caused by dampness and by leaking water lines and fittings. If possible, check the condition of the

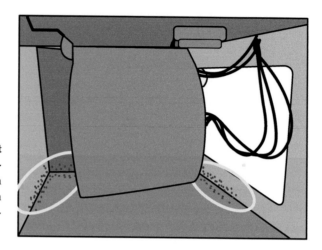

**Figure 9-42. Rust marks at the bottom of the fan compartment can indicate condensate overflow.**

interior high temperature refractory lining on all oil-fired units.

Check the furnace unit by turning off the emergency switch. After turning up the thermostat, return to the unit and turn the emergency switch back on while standing clear of the unit. Look for a puff-back in oil-fired units or flames licking under the cover plate of a gas-fired unit; both indicate potential hazards that must be corrected. If the unit does not light, check the master switch or emergency cut off to make sure it is on. Press the reset button (if applicable), and try again.

Once the unit is activated, closely observe the combustion process. This may be limited by the arrangement of the unit (for example, no access to view the flame). In oil-fired units, the flame should be clear and clean, and have minimal orange-yellow color. Flame height should be uniform. Gas-fired units should have a flame that is primarily blue in color. Note if the flame lifts off the burner head—this indicates that too much air is being introduced into the mixture. Check gas burners for rust and clogged ports. Turn the thermostat down after you complete the heating unit inspection.

Change the furnace filter and educate the client how to change the filter. Airflow in a furnace is drastically reduced if the air filters are clogged. Clogged air filters can cause heat exchangers to overheat and ultimately, fail over time. So, educate the client that dirty filters can not only affect their comfort and energy costs, but also reduce the life of the furnace.

If dirt accumulates on the fans of a furnace blower fan, it works less efficiently. These blower fans can be cleaned with a stiff brush or even compressed air. Often, in mobile homes, the blower can be easily removed, cleaned and then reinstalled.

## Vent Pipes

With vent pipes, make sure the venting is going to the outside, not the attic. Chimneys and vents should be unblocked and properly connected from one section to the next. They also should: a) rise at least 1/4 inch per linear foot in the direction of the outgoing air, b) extend at least 3 feet above the highest point where they pass through a roof deck, c) at least be the same diameter as the combustion appliance exhaust port, Figure 9-43, d) have a sound liner inside of them, e) be free of any holes or cracks, and f) be the appropriate size and type for the combustion appliance that they service (see "Flue Type" section).

**Figure 9-43. All manufacturers prohibit narrowing the flue on a combustion appliance to a smaller size than the connector on the combustion appliance. Such a narrowing is referred to as a "reducer."**

Inspect the vent pipe for corrosion holes, the tightness of its fittings, and the tightness of its connection to the chimney. Check for signs of soot build-up in the smoke pipe. Consider consulting local code requirements about the minimum size, required clearance from combustible material, and number of smoke pipes entering the chimney. The key is whether the flue seems to adequately remove the combusted air from the unit and the building without spillage (without backflow at the mixer, etc.) or significant leaks. Checking for backflow under the most challenging conditions can give you a good idea of any problems.

## Backflow and Spillage

If there is spillage anytime the appliance is on, this spillage is known as draft reversal or backdrafting. Backdrafting is a serious defect. Backdrafting can be caused by things such as a) physical problems with the vent system, b) a missing door on the furnace blower compartment, c) an undersized or oversized chimney or vent, d) blockage of the chimney or vent, or e) significant depressurization or vacuum in the CAZ. The fan inside the blower compartment is creating a negative pressure making the combustion appliance zone/closet negatively pressured. As a result of the vacuum, combusted air can be pulled back down out of the flue rather than exiting the flue by naturally rising. Most manufacturers admit it is normal for some initial spillage when a combustion appliance first turns on. However, spillage should not last more than 60 seconds while the chimney or vent heats up.

### Flame Rollout

Flame rollout occurs when flames flow back out from where the burners go into the heat exchanger. Flame rollout is caused by flue gas spillage or insufficient draft or draw. This issue typically manifests itself with burn or soot marks above the burner compartment. Flame rollout can be due to delayed ignition, letting in a larger than normal portion of unburned natural gas into the burner compartment before ignition (causing a sudden mini explosion), obstruction of the flue, a flue running up through the center of the tank, dirty burner ports, or improper mixing of fuel and oxygen.

### Fuel Leaks

Use a calibrated electronic gas leak detector (Figure 9-44) to help locate fuel leaks in the fuel supply lines of natural gas or propane lines.

To calibrate the instrument, adjust the dial until you hear a slow but steady clicking sound. Run the probe along all the exposed gas fittings and pipes. If you are required to follow BPI Standards, you cannot check for gas leaks at a rate faster than one inch per second. In addition, you must check all along each accessible gas pipe, but you can move faster than one inch per second along the pipe. If a gas leak is present, the intervals between the clicks will shorten and the noise level will rise to a shrill, virtually continuous sound. Test all the valves, fittings, and joints of the fuel lines. Natural gas leaks are more common above fittings, joints, and pipes because natural gas is lighter than air. Propane leaks are more likely to be detected below fittings, joints, and pipes because propane is heavier than air. If a gas leak is detected using the gas

**Figure 9-44. Checking for Gas Leaks at a Pipe Joint With a Gas Detector**

detector, confirm that it is a leak using a soapy water mixture dabbed around the suspected area. Some types of pipe joint sealant can set off the calibrated gas detector. If it is a real leak, the soapy water will begin to generate bubbles.

For evaluating an oil-fired combustion appliance for fuel oil leaks, no equipment is used. Do a visual inspection looking for signs of oil deposits on the ground and use a sniff test to confirm that it seems to be oil.

## Clearances

Clearance is the required minimum distance between chimneys, vent systems, or heat producing appliances and any combustible surfaces like wood. Sometimes the clearance required is determined through testing and is listed as a manufacturer's instruction. The industry standard for clearances can be found in NFPA 211 entitled "Standard for Chimneys, Fireplaces, Vents, and Solid Fuel Burning Appliances."

Be concerned about the clearances between the vents in the home and combustible surfaces. Tables 9-4 and 9-5 show a sampling of some of the clearances required for different types of vents depending upon which fuel is used.

## Proper Combustion Safety

Look for combustion safety problems. Some examples include rust on the draft hood of a natural gas furnace, which could represent spillage of flue gases into the CAZ, or soot stains at the inspection port on an oil-fired boiler, which could represent a need for a tune-up.

Depressurization or vacuum is typically a result of the interaction between the tightness of the home and the flow capacity of exhausting devices such as fireplaces, forced air systems, or dryers. Any exhaust device will tend to create more vacuum when the house is tight than when the house is leaky. Think of all the exhaust fans in a home as if they were mini blower doors. If the home is "loosened" by opening a window, the negative pressure or vacuum is relieved.

## Adequacy of Combustion Air

Soot buildup is a sign of inefficient combustion. In oil-fired units, look for soot below the draft regulator (barometric draft damper), on top of the unit housing, and around the burner. The odor of smoke near the unit is another sign of poor combustion.

| Oil and Wood Fired | Gas |
|---|---|
| Single Wall – **18"** | Single Wall – **6"** |
| Double Wall, Super Six, or Dura Vent – **6"** | |
| Manufactured Chimney – **2"**<br><br>Or Per Manufacturers Instructions | Double Wall, B-Vent, or B-W Vent – **1"** |

**Table 9-4. Clearances to Combustibles for Combustion Appliance Vents by Fuel Type**

**Table 9-5. Clearances to Combustibles for Combustion Appliance Vents by Fuel Type**

| Vent Connector Type | Clearances | | | |
|---|---|---|---|---|
| | Combustion Appliance Type | Without Shield | With Heat Shield | |
| | | | Wall | Ceiling |
| Single Wall Galvanized | Vented Gas Space Heaters,<br><br>Atmospheric gas furnaces and Water Heaters,<br><br>Fan-assisted draft gas furnaces (med. Effic. 80+AFUE) | 6" | 2" | 3" |
| | Oil Furnaces, vented space heaters, boilers, and water heaters | 18" | 6" | 9" |
| 24 gauge Steel – Black or Stainless | Wood Stoves | 18" | 6" | 9" |
| B Vent – incl. BW (wall) | Atmospheric Gas Furnaces and Water Heaters, Fan-Assisted draft gas furnaces (med. effic. or 80+AFUE) | Install as per manufacturer's instructions (usually 1") | | |
| L Vent | Oil Furnaces, boilers, or water heaters | Install as per manufacturer's instructions (usually 3") | | |

### Checking Heat Exchangers

Both visual checks and combustion analysis help assess the condition of the heat exchanger. The visual check involves using a strong light and an inspection mirror to look for cracks in the heat exchanger, rusty heat exchangers, and rust on the burner ports. Checking for cracked heat exchangers with a combustion analyzer involves checking the flue gas concentration before and after the blower starts on the furnace.

If there is more than a 1% oxygen change when the blower starts or if you see an enduring change in draft pressure, CO, or flame when the blower turns on, that could indicate evidence of a cracked heat exchanger. Note that often a short-term change in flame color can occur at the time the blower turns on just due to vibration, etc. If the flame does not shortly return to the color it was before the fan turned on, it could indicate a cracked heat exchanger.

Cracked or leaking heat exchangers are a major concern because combustion byproducts from the appliance can cross the heat exchanger to mix with the conditioned house air that is breathed. If there is a cracked heat exchanger, the furnace must be replaced. Replacement heat exchangers are usually not available for any significant period of time

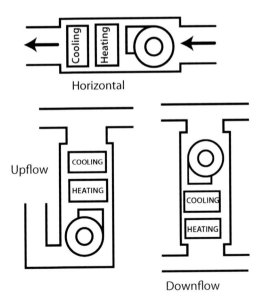

**Figure 9-45. Types of Circulation Patterns in Furnaces and General Location of Heat Exchangers and Evaporator Cooling Coils**

after a specific furnace model has been manufactured. Thus, not only is it important from the safety standpoint that a heat exchanger not be cracked, but if it is cracked, it becomes expensive because the entire furnace will have to be replaced.

The airflow pattern of HVAC units determines the location of the heat exchanger and air conditioning coils. Heat exchangers and coils are always downstream of the fan with the cooling coils always being furthest from the fan.

### Analyzing for Carbon Monoxide in Furnaces

Use a combustion analyzer to identify the flue gas $CO_2$, CO, and $O_2$ content. The analyzer also provides a flue gas temperature, an ambient or room air temperature around the combustion appliance, and combustion efficiency. In undiluted flue gas, CO content can either be "air free" or "as measured." "Air free" essentially means free of uncombusted air. You can only get a true "air free" sample if there is just the right amount of oxygen that is completely burned in the combustion process; otherwise, you get "excess air" (oxygen that is taken up into the burner but is not used in the combustion process). If you check the CO level at an even later stage where the flue gases have been diluted even more as a result of a draft hood, mixer, etc. that adds room air to the flue gases, you make it even harder to get an "air free" CO level. Thus, taking a sample of the gases before the mixer is the most you can do to get an "air free" reading. Some combustion analyzers can determine the oxygen content and thereby calculate what the CO reading would be if the CO were undiluted by the oxygen to get the "air free" CO. If the combustion analyzer does not have that capability, you will be getting the "as measured" CO level even if you take the sample before the draft hood, etc. dilution. Check with the manufacturer. Take the sample before a draft diverter of any kind has a chance to dilute the flue gases. To get a CO level on an appliance with a draft diverter or mixer, place the analyzer probe into the upper section of a heat exchanger on a 70%+ (low efficiency) furnace or on a water heater, down into the throat of the interior water heater vent before the gases reach the diverter or bell housing.

The "as measured" CO level is the actual CO level measured even if it has been diluted by the excess oxygen or air contained in the combustion byproducts or dilution in a mixer draft hood or bell housing. Because this air has diluted the CO, it will show a lower level than it really is. If it is known how much the excess oxygen or air has diluted the CO, then the true CO level produced by the appliance without the dilution from the excess oxygen can be calculated. Obviously, "air free" CO levels give a more accurate idea of how well the combustion process is going by checking the CO level. If the combustion process is burning perfectly, these should be no CO, only $CO_2$. Being able to measure the true CO level can better tell the extent, and potentially the reason for, the appearance of CO in the combustion byproducts. Most combustion analyzers give printed results of the tests so there is always hard copy documentation.

Anytime you are working around a home, you should wear your personal CO monitor, as shown in Figure 9-46.

**Figure 9-46. A personal CO monitor is typically kept near the upper part of the body so that it samples near the air that is being breathed**

This device constantly tests the CO content in the air around you. It is often placed in a shirt pocket or worn around the neck with a lanyard and has an alarm on it that can be set for the appropriate CO ppm. While in any combustion appliance zone/closet, you should check the ambient CO levels on your personal CO monitor to make sure that they do not exceed 20 ppm in the weatherization program or 35 ppm under BPI standards. In addition, if a house or a combustion appliance zone has CO levels exceeding 9 ppm without having conducted any testing, this could indicate that the standing CO level in the house, 24 hours a day, goes beyond the long-term exposure standard. If this situation is encountered, the combustion appliance zone could be ventilated, and perhaps the entire house could be ventilated until the CO problem is repaired.

### CO Measurements in Gas Cook Stoves and Ovens

Gas cook stoves or ovens must be checked for elevated levels of CO. Elevated CO in this situation is far more common than expected and this is especially disturbing since CO can be deadly. To make matters worse, many people use gas cook stoves as a heat source without understanding the risks. Each range top burner must be tested by holding the combustion analyzer probe about 6 inches above the flame and measuring for CO. To get a more accurate reading on the range top CO level, a short, modified, double wall section of flue with a hole drilled about 6 inches from the end of the flue section is used. This is called a CO "Hot Pot." This flue is placed on end, vertically over the top of the burner. The combustion analyzer probe is placed in the interior of the "flue" section to measure the combustion products. Under the weatherization standards, natural gas or propane-fired cook stoves must be repaired if the CO level exceeds 100 ppm in the oven, or 25 ppm on the range top burner. This is an "action level." If the action level is reached, action must be taken to retune or repair the burner. Under BPI standards, there is no action level for CO ppm for range top burners but the following does apply: 1) if the burners do not ignite properly or do not burn cleanly, cleaning and tuning of the appliance shall be recommended; and 2) if the appliance is located in a confined space and mechanical ventilation is not readily available, mechanical ventilation shall be recommended. Gas ovens need to be checked for CO level by placing the tip of the analyzer into the oven vent while the oven is on, see Figure 9-47 and 9-48. Typically, the vent is on the stovetop or back panel of the stove.

**Figure 9-47. This shows different methods of testing different parts of an oven/stove top with a combustion analyzer. A: Testing a burner with a CO "Hot Pot" B: Testing the combustion air from the oven from an oven vent on the back panel of the stove C: testing a burner without using a CO "Hot Pot" (test 6" above the burner).**

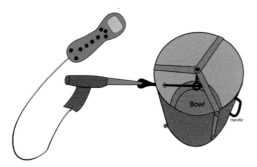

Figure 9-48. Close up of a CO "Hot Pot"—Notice the bowl at the bottom acts as a pan on the burner and the eyebolt to hold the combustion analyzer tip in position.

There are also action levels for other appliances, see Table 9-6. For instance, if a gas water heater, boiler, or furnace exceeds 100 ppm CO in its combustion byproducts, it needs to be retuned or repaired. Similarly, the action level for an unvented gas space heater is 200 ppm, air free. The action level for an oil-fired furnace, water heater, or boiler is 100 ppm. Thus, any time these action levels are exceeded, retuning or repairs must be made to bring the ppm of CO down below the action level. The following table summarizes various action levels.

It should be noted that boilers often show a higher CO level and in some cases will show a CO level of up to 105 to 110 even when operating efficiently although these levels are outside of the tolerance levels for some of the standards.

*Field*

When testing a gas cook stove oven, make sure all stored items, including pots and pans, aluminum foil, etc. are removed from the oven before beginning the combustion analyzer test. Also, make sure the self-cleaning features are not activated. Once the oven is turned on

Table 9-6. CO Action Levels for Combustion Appliances-Weatherization Appliance Type Action Level

| Appliance Type | Action Level |
|---|---|
| Gas Cook Stove – Range top burner | 25 ppm CO as measured |
| Gas Cook Stove – Oven | 100 ppm CO as measured |
| Gas furnace, boiler, or water heater | 100 ppm CO as measured |
| Oil-Fired furnace, boiler, or water heater | 100 ppm CO as measured |
| Unvented Gas or Oil Space Heater | 200 ppm CO air free |

and warmed for at least 5 minutes, place the probe of the combustion analyzer into the oven vent to read the undiluted CO content. The oven vent is typically represented by multiple openings or one larger opening somewhere in the back of the stove top or visible somewhere on the back panel of the oven. "Zero" or calibrate the combustion analyzer outside before testing (see operator's manual). Read your personal CO monitor to use 25 ppm CO in ambient air as a "stop test" cutoff or maximum, or in the case of BPI, use 35 ppm for a cutoff. With the weatherization program, remedial action is required if the CO level from the combustion analysis for the oven exceeds 100 ppm. Under BPI standards, if the oven CO measurement is in the 100 to 300 ppm range, the homeowner must install a CO detector and a recommendation for service must be made to the homeowner. If the oven CO measurement is over 300 ppm, under BPI standards, tell the homeowners they must have the appliance serviced before work can begin. If the level is still over 300 ppm after servicing, exhaust ventilation must be installed to the extent of 25 CFM continuous or 100 CFM intermittent.

### Testing for CO in Water Heaters

To test for CO in water heaters, first calibrate or "zero" the combustion analyzer outside.

Place the combustion analyzer probe into the throat of the internal vent inside the water heater and take two readings, one on each side of

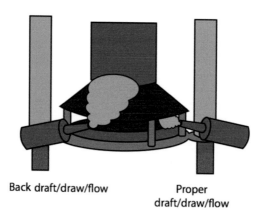

Back draft/draw/flow                    Proper
                                       draft/draw/flow

**Figure 9-49.**
**Differences in Proper and Improper Draft on a Water Heater Bell Housing**

the baffle that runs up through the center of the flue below the top of the water tank. You may have to drill a hole in the draft hood to get the probe down into the throat. It should be noted that the purpose of the baffle is to slow the flow of the water heater's hot flue gases to better transfer heat to the inside walls of the tank that surround the flue. Also, test for proper negative draft pressure with the probe placed through a hole positioned 1 to 2 feet above the draft diverter or shortly after the first elbow in the flue.

Check for spillage using smoke, mirror, etc., 1/2 inch below and 1/2 inch out from the lip of the draft diverter all the way (360) around it. If this spillage does not stop after one minute of water heater operation, the water heater fails the spillage test, see Figure 9-49.

It is common to find a home where the original installation included a gravity fed water heater tied into a common chimney or flue with a gravity fed furnace. Often the furnace has been upgraded and a high efficiency, direct vent furnace has replaced the gravity fed or medium efficiency furnace. The high-efficiency furnace requires its own PVC vent that cannot be shared with the water heater. As a result, the water heater no longer shares the chimney with another combustion appliance, potentially making the chimney an oversized vent for this "orphan" water heater. The best treatment here may be to bring the size of the chimney down to a smaller diameter by installing a stainless steel liner to exhaust the water heater combustion products alone ("stand alone" water heater) If a medium efficiency furnace has replaced the gravity-fed furnace, too often the new furnace causes backflow of the water heater flue. Watch for this in these situations.

### Worst-case Combustion Safety Testing

Worst-case combustion testing involves several types of CAZ (combustion appliance zone) tests that are done to verify that combustion appliances will still properly vent their flue gases even under the worst possible conditions and to check for CO levels in the appliance's combustion gases. If there is a draft reversal of the flue gases so they bleed into the CAZ, the occupants are in an unsafe environment. If the pressure in the flue of the appliance is not negative enough, it would backdraft under different circumstances. Acceptable levels for CO have been developed by the weatherization program and by BPI. These standards are discussed below. Not only is worst case testing done as part of the audit, it is also done at the end of each workday when retrofitting a home, and when the retrofit work is done.

Worst-case conditions involve creating the most negative pressure possible in the home by turning on exhaust fans, dryers, and other items that enhance negative pressure, etc. This type of CAZ testing is only done for appliances that use vented systems. When conducting any combustion safety test or CAZ test, regularly check your personal CO monitor to be sure the ambient CO level does not exceed 20 ppm under the weatherization standard or 35 ppm under the BPI standard. If the CO levels exceed 20 ppm, or 35 ppm under BPI, discontinue the tests, ventilate the CAZ, deactivate the combustion appliances, and arrange for or conduct repair of the CO problems.

## CONDUCTING WORST-CASE TESTING

### In General

Once the home has been placed in worst-case conditions, three different tests are conducted:

1.    a backflow or spillage test to make sure all the combustion byproducts are exiting out the vent or flue;

2.    a flue draft pressure test to see if there is enough negative pressure in the flue for it to properly draw;

3.    and a CO test to make sure there is an appropriate mixing of fuel and air in the appliance.

The general steps to completing worst-case combustion safety testing includes completing the following sequence in this order:

1.    Set up the house for worst-case conditions and get a "baseline" reading to determine any difference in pressure between the inside and outside of the house. (Some will use the baseline feature on their gauge instead of the procedure recommended here).

2.    Determine which combination of these conditions creates the most negative pressure in the CAZ,

3.    Evaluate whether the negative pressure created in the CAZ under these worst-case conditions could cause appliance failure by falling below the standards,

4.   Regardless of whether the CAZ passes or fails the initial standards, leave these worst-case conditions in place and, beginning with the lowest Btu appliance in the CAZ, test for spillage or backflow from the flue, adequate negative draft pressure in the flue, and the CO levels in the flues for all combustion appliances,

5.   If at any time any appliance fails the spillage test, take the house out of worst-case conditions and put it under natural conditions and start that appliance's tests over again,

6.   Follow the same procedure until you have finished testing all the vented combustion appliances in the CAZ. If the house has been placed under natural conditions to complete a test on one appliance, keep the house in natural conditions when checking all the remaining appliances. If there is more than one CAZ, use the same procedure to test each CAZ.

See Figures 9-50, 9-51, 9-52, and 9-53 for steps in worst case testing.

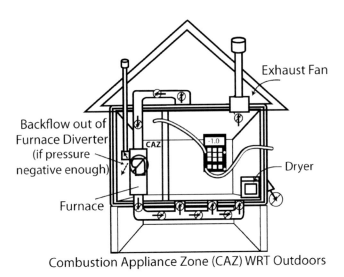

Combustion Appliance Zone (CAZ) WRT Outdoors

**Figure 9-50. First Step in Worst Case Testing—Baseline**
*Exhaust, Dryer, Bathroom, Kitchen, etc., Fans=Off*
*Furnace Fan= Off*
You might expect a slightly negative number like this in the winter due to stack effect. (If you were inside the CAZ, you would not need to have a tube on the upper left connector.)

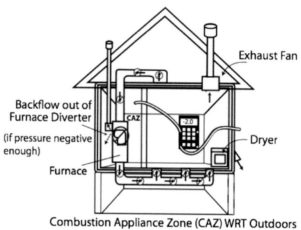

Combustion Appliance Zone (CAZ) WRT Outdoors

**Figure 9-51. Second Step in Worst Case Testing**
*Exhaust, Dryer, Bathroom, Kitchen, etc., Fans=On*
*Furnace Fan= Off*

(If you were inside the CAZ, you would not need to have a tube on the upper left connection.)

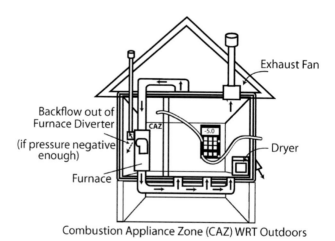

Combustion Appliance Zone (CAZ) WRT Outdoors

**Figure 9-52. Third Step in Worst Case Testing**
*Exhaust, Dryer, Bathroom, Kitchen, etc., Fans=On*
*Furnace Fan= On*

You would see a more negative number like this the more the supply ducts were leakier than the return ducts. (If you were inside the CAZ, you would not need to have a tube on the upper left connector)

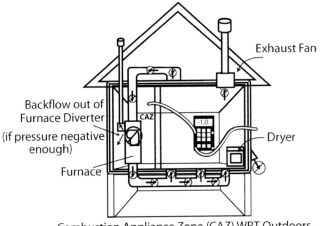

Combustion Appliance Zone (CAZ) WRT Outdoors

**Figure 9-53. Alternative in Worst Case Testing—
Natural Conditions (Like Baseline)**
*Used Only if Furnace Fails One of the CAZ Tests Such as Backflow
Exhaust, Dryer, Bathroom, Kitchen, etc., Fans=Off
Furnace Fan= Off*
(If you were inside the CAZ, you would not need to a have a tube on the upper left connector)

7.   Use the results from all three tests to decide whether to immediately shut down the appliance and *abandon* the audit for safety reasons, to *require* the homeowner to have the appliance serviced, to *recommend* the homeowner to have the appliance serviced, or to *allow* the retrofit work to proceed.

**Worst Case Testing In Detail**
*Setup*
          To preliminarily set up the house for the worst-case test, use the following procedures (Figures 9-54 and 9-55, Table 9-7):

**EC Combustion Appliance Zone (CAZ) And Worst Case Testing When Gauge is in CAZ - Such as Inside a Furnace Closet**

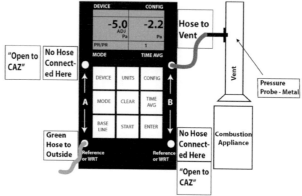

**EC Combustion Appliance Zone (CAZ) And Worst Case Testing When Gauge is NOT in CAZ - Such as Furnace Closet too Small**

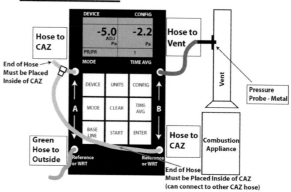

Figure 9-54. EC CAZ and Worst Case Testing

## Table 9-7. Setting Up the House for Worst-Case Conditions For EC DG-700 and Retrotec DM-2 Gauges

To be completed similarly for each CAZ in the home along with the applicable combustion appliance test in each CAZ (see below)

1.  Make sure your gauge batteries are fully charged or that you have adequate outlets and extension cords to connect the pressure gauge to its recharger/transformer.

**Retrotec Combustion Appliance Zone (CAZ) And Worst Case Testing
When Gauge is in CAZ - Such as Inside a Furnace Closet**

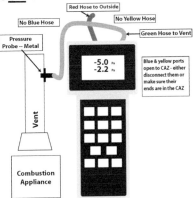

**Retrotec Combustion Appliance Zone (CAZ) And Worst Case Testing
When Gauge is NOT in CAZ - Such as Furnace Closet too Small**

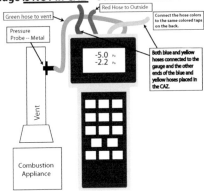

**Figure 9-55. Retrotec CAZ and Worst Case Testing**

Set water heater to "pilot" setting; turn off furnace, boiler, space heater, etc. electrical disconnects. If no electrical disconnect exists, then turn off the fuel disconnects.

2. If not already done, turn off all fans, regardless of whether they exhaust or supply air—bathroom and kitchen fans, clothes dryers, central vacuums, broiler hoods, whole house fans, swamp coolers, window air conditioner units, ceiling fans, solar air panel fans, attic/crawlspace powered ventilators, HRV/ERV or other fresh air ventilation systems, etc.

3.    If not already done, make sure house is in winter conditions: close exterior doors, latch or lock windows, open interior doors including closet doors.

4.    If not already done, close all operable vents (i.e.: fireplace damper).

5.    If have not already done so, record outdoor temperature.

      Temperature:_____

6.    If dirty: clean, remove or replace dryer lint, central vacuum, broiler hood, and furnace filters.

7.    Check your personal monitor for ambient CO level now.

      CO level: _____ppm

      If CO is greater than 35 ppm in BPI-related work or greater than 20 ppm in Weatherization (WX)-related work:

      Turn off the appliance, ventilate the space, and evacuate the building. The building may be reentered once ambient CO levels have gone below 35 ppm for BPI and 20 ppm for WX. The appliance must be cleaned, tuned, repaired and the problem corrected prior to completing the combustion safety diagnostics. If the ambient levels exceed 35 ppm for BPI and 20 ppm for WX during testing under natural conditions, disable the appliance and instruct the homeowner to have the appliance repaired prior to operating it again. This procedure should be used anytime your personal monitor reads greater than 35 for BPI and greater than 20 for WX.

8.    Set up a tube from inside the CAZ (combustion appliance zone) to the outside (use a small non-conducting rigid tube connected to the tube and place the tube under a closed exterior door). If you have to leave the hatch or door open to a significant degree that compromises your reading, consider using a smaller non-conducting rigid tube, adapted to fit the hose.

      *EC* Turn on the gauge—press "On/Off." If "PR/PR" is not showing in the lower left corner of the display, keep pressing the "Mode" button until it appears. Connect the other (inside) end of the tube to the gauge's Channel A reference tap (lower left Tap). Disconnect all other hoses to the gauge. Record the reading below.

      *Retrotec* Turn on the gauge by pressing and holding down the "On/Off" button. Once the gauge is on, press "Auto Zero" button until "On" shows in the upper right corner of the display. If both "PrA" and "PrB" do not show up on the left edge of the display, repeatedly press the "Mode" button until they do. Connect the other (inside) end of the tube to the gauge's red tap on the back). Disconnect all other hoses to the gauge. Record the reading. Record this number in Pascals as the "Baseline test" here and in the first line of the next table below.

_____Pa

*Determining the Level of Negative Pressure under Worst Case Conditions*

Once the house is set up for worst-case CAZ testing, establish just how much of a negative pressure or vacuum in the CAZ has been created under these worst-case conditions. To find out, use the following procedure (Table 9-8):

**Table 9-8. Establishing the Pressure Under Worst Case Conditions**

1. Baseline test. Read and record the CAZ to outdoor pressure difference (see previous section, step 8)
   _____Pa

2. Exhaust Fans On
   a. Turn on all exhausting appliances and exhaust fans (kitchen fans, bathroom fans, clothes dryers, central vacuums, broiler hoods, window air conditioner units, ceiling fans, solar air panel fans, attic/crawlspace powered ventilators, HRV/ERV or other fresh air ventilation systems, etc.) except for whole house fans or swamp coolers (swamp coolers are not exhaust fans as they pressurize the house). If you are not turning on the whole house fan, be sure to tell and report to the client that the whole house fan should be removed because they can prove hazardous if not enough windows are opened when they are on—if this is a weatherization audit, check with your weatherization standards.

      Do not turn on air handler at this point yet.

   b. If there is a fireplace(s) that is regularly used, set up and turn on the blower door to exhaust to the outside at 300CFM per fireplace (optional). Do this to simulate a fireplace(s) in use. Add 300CFM for each fireplace to give the rough equivalence of a fireplace(s) in use. Read and record CAZ to outdoor pressure difference again now that the exhausting fans have been turned on.
      _____Pa

3.    Air Handler On

Turn on furnace air handler without turning on the heat—set the furnace to the "fan" setting.

a.    **Rooms other than the CAZ:**

Beginning with the room farthest away from the CAZ (and with your back to the CAZ), check the pressure from the hall side of each and every door, one at a time (except for closet doors that should always be open and the door to the CAZ—see below). Check whether smoke goes under each closed door (negative pressure inside the room—leave the door open) or whether the smoke is blown back towards you from under the closed door (positive pressure in the room—leave the door closed). "If it blows on your nose, keep it closed." Use your gauge to do this by putting the tube inside the room under the door, but it is often not as accurate as the qualitative smoke technique. If you use the manometer, disconnect the tube to outside from the manometer and connect a short tube to the gauge (EC—upper left nipple of DG-700; blue nipple on the Retrotech DM-2). Typically you will find that almost all rooms that only have a heat supply source but not a return grille in them will create a positive pressure in the room and therefore, the room door is likely to be closed. Also, if a bathroom is connected to the room (for example, a master bath), keep the exhaust fan on in the bathroom. Identify the rooms in which the pressure is more positive inside the room by >2.5 Pa to identify where restrictions to return air exist (temporarily turn off any exhaust fans in the room or a bathroom connected to the room before taking this reading and, once the reading is taken turn the exhaust fan back on): Bedrooms 1 2 3 4 5 6 Baths: Master Main Basement Other:_____ Kitchen/Dining Rm/Living Rm/Family Rm /Storage room/Other: 1 2 3 4 _____

b.    **The CAZ room:**

Standing inside the CAZ

When you get to the last door—the door to the CAZ, furnace, or boiler closet—go inside the CAZ and check with smoke or measure the pressure under the door after closing off the door to the CAZ from the rest of house. If the smoke is drawn under the door or the pressure is negative, leave the door open. If the smoke pushes back towards you from under the door or the pressure is positive, close the CAZ door.

Standing outside the CAZ

If you are unable to physically fit in the CAZ or reach the bottom of the door from inside using smoke, etc., do the test from outside of the CAZ in the hall while remembering to reverse our test results in this rare case of testing the CAZ door itself from the hall. If the smoke is drawn under the door or the pressure shows negative for inside the CAZ using a gauge, then leave the CAZ door closed. If the smoke pushes back towards you into the hall from under the door, or the pressure inside is positive, leave the door open.

c.   Disconnect the short tube and then reconnect the longer tube to outside (EC—bottom left nipple of DG-700; Retrotech—red nipple of DM-2). Read and record CAZ to Outdoor pressure difference now.

_____Pa

4.   You have found the worst-case conditions for the CAZ—it is the lowest of Step 2 or 3 (remember that if it is a negative number, the number to put in here is the largest of the negative numbers). *Make a record of that lowest pressure here and return the house to the conditions you found had the lowest pressure. If it was Step 3, do not change anything. If it was Step 2, turn the air handler off and reopen all interior doors.*

_____Pa

5.   Now subtract the pressure you found at Step 1 from that of Step 4 to get the worst-case depressurization.

_____Pa

Remember, if you subtract a negative number, it is the same as adding that number as if it were positive.

Example #1: Pressure Step 4 =-10 Pa
          Pressure Step 1 =   -(-5) Pa  (subtract)
                              -5 Pa

Example #2: Pressure Step 4 = -5 Pa
          Pressure Step 1 =        -(+5) Pa  (subtract)
                              -10 Pa

6.   a. Now compare the worst-case depressurization in Step 6 with the BPI CAZ Depressurization Limits Table shown below or the Weatherization Standard (-5 Pa). Keep the house in worst-case conditions.

b. If the CAZ is more negative than the appropriate depressurization limit in the Table, it "fails."
Passed/Failed

7.   Leave the house in worst-case conditions and begin testing for spillage, draft pressure, and CO using the test procedures below, starting with the

lowest Btu appliance and going up. Be sure to do a separate CAZ test beginning with the first procedures above if there is more than one CAZ in the home.

After completing this part of the test, consider giving your opinion on the following items as part of your report (Table 9-9):

**Table 9-9. Possible Causes of and Remedies for Significant Depressurization**

| Are there some dominant force(s) that could be causing depressurization? | Inadequate return air path Kitchen Exhaust<br><br>Whole House fan Duct/Return leaks   Other: |
|---|---|
| Possible ways to alleviate a CAZ pressure that is too | Add supply air to and insulate CAZ |
| negative | Remove exhaust fan(s)<br><br>Seal duct/return leaks Cut more off the bottom of room doors<br><br>Install direct vent appliance (high efficiency)   Jumper duct Installation    Other: |

### Comparing the Home's Worst-Case Pressure Against the Standards

To compare against the standards, use the BPI CAZ Depressurization Limit Table for BPI (Table 9-10) oriented work or, if doing Weatherization oriented work, use the Weatherization standard of a negative pressure no lower than -5.0 Pascals as the cutoff—a negative pressure of -6.2 would fail the Weatherization test. Getting the baseline pressure difference allows for comparing the pressure in the CAZ to an independent reading outside that will not be affected by what is going on inside the house. For example, see if and by how much the CAZ pressure gets more negative as you proceed.

Table 9-10. BPI Appliance Depressurization Limits in Pascals

| Venting Conditions | Limits (Pascals) |
|---|---|
| "Orphan" natural draft water heater - used to share larger flue with furnace, etc. but no longer does (oversize flue, including outside chimneys) | -2 |
| "Stand alone" natural draft water heater (appropriately sized flue as per NFPA 31 (oil), 54 (gas), 58 (propane) and 211 (solid fuel) or the venting tables of the chimney liner manufacturer) | -5 |
| Natural draft boiler or furnace commonly vented with water heater | -3 |
| Natural draft boiler or furnace with vent damper commonly vented with water heater | -5 |
| Individual natural draft boiler or furnace | -5 |
| Mechanically assisted draft boiler or furnace commonly vented with water heater | -5 |
| Mechanically assisted draft boiler or furnace alone, or fan assisted DHW alone | -15 |
| Exhaust chimney-top draft inducer (fan at chimney top); high static pressure flame retention-head oil burner; and sealed combustion appliances | -50 |

## Testing for Spillage, Negative Draft Pressure in the Flue, and CO in Appliances

Once the house has been set up for worst case CAZ testing and you have established just how much of a negative pressure or vacuum has been created in the CAZ under these worst case conditions, do three tests—spillage, draft pressure, and CO—on each of the combustion appliances, starting with the lowest Btu appliance. This next section tells where to check and where to place the analyzer probe in gas water heaters, in low, medium and high efficiency forced air furnaces and boilers, and in oil-fired furnaces, boilers, and water heaters.

## Spillage Check

• Water Heaters: Check for spillage 1/2" below and 1/2" outside the lip of the bell housing and all 360 degrees around it with a smoke, mirror, etc.

• Up to 70%+/Atmospheric/Standard/Low Efficiency Furnace—Test at vent openings, such as dilution air/mixer/bell housing/draft diverter openings, barometric damper (oil-fired), with smoke, mirror (it fogs it), etc. 1/2" below and 1/2" all along the lip. Also, check for burner flame/diverter smoke disturbance both before and after fan turns on.

• 80%+/Medium Efficiency/Fan Assist Furnaces: Check for "spillage" at flue joints where it enters the base of the vertical chimney/vent, or, if the burners are not enclosed, smoke draw where burner flames enter the heat exchangers. If the 80%+ has a common/shared flue with hot water heater (DHW), check for spillage at hot water heater draft diverter. Check for burner flame/diverter smoke disturbance both before and after fan turns on.

• 90%+/Condensing/High Efficiency furnaces—Only check for burner flame disturbance both before and after fan turns on. Do not test for spillage; it cannot be checked because it is a closed system.

• Oil-fired—Check for spillage at the barometric draft damper.

## Draft Pressure Check

• Water Heaters: Drill a hole 1 to 2 ft above the bell housing/diverter or just beyond the first elbow in the flue and test, (Figure 9-56).

• Up to 70%+/Atmospheric/Standard/Low Efficiency Furnace (Figures 9-57 and 9-58): (If probe(s) will reach, check the draft by entering the probe up through the draft diverter on furnaces. If the probe does not reach, drill a hole into the flue just large enough for the analyzer probe at 1 to 2 ft above the draft diverter/vent connection or just beyond the first elbow.

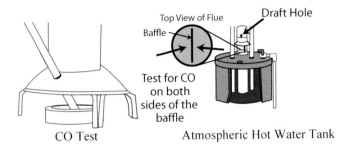

**Figure 9-56. Testing CO In a Water Heater and Draft**

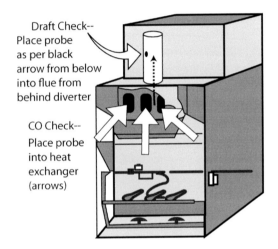

**Figure 9-57. 70%+ Atmospheric Furnace (Also Referred to as a Low Efficiency or Standard Efficiency Furnace)**

- 80%+/Medium Efficiency/Fan Assist Furnaces: Single Wall Flue (Figure 9-59)—Drill a hole into the flue just large enough for the analyzer probe at 1 to 2 ft above the vent connection or just beyond the first elbow. Do not drill into corrugated flues.
  — Two-wall flue. Do not drill into any two-wall flues. You can only check for spillage when these furnaces have two wall flues. You cannot check for draft or CO on these furnaces. Some weatherization programs drill two wall flues. Check with your supervisor first.

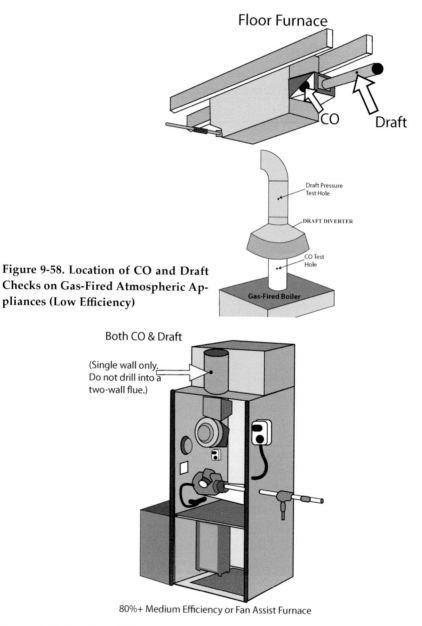

**Figure 9-58. Location of CO and Draft Checks on Gas-Fired Atmospheric Appliances (Low Efficiency)**

**Figure 9-59. Medium Efficiency Furnaces and Boilers-CO and Draft Checks are at the Same Location on Gas-Fired Medium Efficiency Furnaces and Boilers**

- 90%+/Condensing/High Efficiency furnaces (Figures 9-60 and 9-61)—No draft test is required. These have a positively pressurized PVC plastic flue and so you cannot check for a negative draft in the flue.

- Oil burners (Figure 9-62)–Place the draft hole before the barometric draft damper.

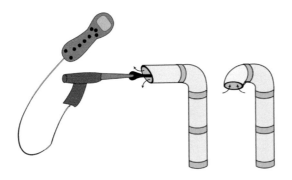

Figure 9-60. Check for CO only at output of flue gas at PVC outside for high efficiency or medium efficiency units with a two-walled flue.

Figure 9-61. Do not drill into PVC flue pipes of high efficiency gas-fired furnaces or boilers.

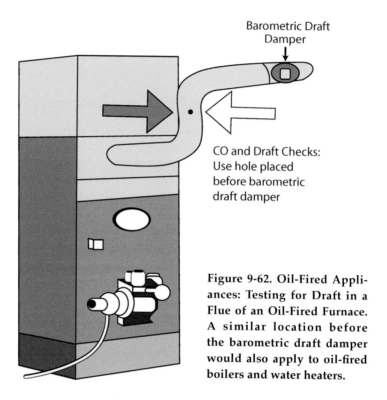

Barometric Draft
Damper

CO and Draft Checks:
Use hole placed
before barometric
draft damper

**Figure 9-62. Oil-Fired Appli-
ances: Testing for Draft in a
Flue of an Oil-Fired Furnace.
A similar location before
the barometric draft damper
would also apply to oil-fired
boilers and water heaters.**

**CO Check**

- Water Heaters: Drill hole in bell housing, if necessary, to send probe at least 1″ inside the throat of the internal vent—before air becomes diluted.

- Up to 70%+/atmospheric/standard/low efficiency furnace—Test inside of heat exchanger at the top of the heat exchanger (you do not have to drill into the flue, but in some smaller draft diverters you may have to drill through the draft diverter to get to the space at inside the top of the heat exchanger).

- 80%+/fan assist/medium efficiency furnace—Single Wall Flue (Figure 9-63)—Drill a hole into the flue just large enough for the analyzer probe at 1 to 2 ft above the draft diverter/vent connection or just beyond the first elbow. Do not drill into corrugated flues.

— Two-wall Flue

Do not drill into any two-wall flues. You can only check for spillage when these furnaces have two wall flues. You cannot check CO on these furnaces unless you can safely access the output port for the vents. Some weatherization programs drill two wall flues. Check with your supervisor first.

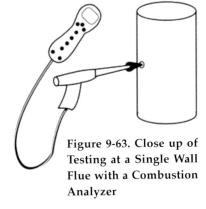

- 90%+/Condensing/High Efficiency furnaces—Go to outside pipe opening on the sidewall of house or on the roof (if safely accessible)—CO is checked inside this vent termination— Test the pipe where the air exhausts/flows out of the PVC pipe.

Figure 9-63. Close up of Testing at a Single Wall Flue with a Combustion Analyzer

- Oil-Fired: same location as for draft test.

- Seal all test holes afterwards with high temperature silicon sealant or a high temperature aluminum foil tape, (Figure 9-64).

### Detailed Field List For Spillage, Draft Vacuum Pressure, and CO Testing

Table 9-11 shows even more detail of the sequence for doing spillage, backdraft and 416-44 on water heaters and furnaces.

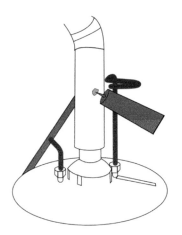

Figure 9-64. All Drilled Test Holes in Flues Must Be Sealed with a High Temperature Material. This diagram also shows approximately where to drill the hole to get a draft pressure and CO reading in a water heater

**Table 9-11. Water Heater Backflow/Spillage Test, Negative Draft Pressure Flue Test, and CO Test**

Make sure the house is still in worst-case condition.

Make sure your gauge batteries are fully charged or that you have adequate outlets and extension cords to connect the pressure gauge to its recharger/transformer.

1.   **Spillage Test:** Mark where the temperature setting is on the water heater dial so you can return the water heater setting to its original temperature. If the appliance is not already on, turn the appliance on by turning the temperature setting up until the water heater burners turn on. Observe the flames as it turns on, then immediately go to the top of the water heater and do a spillage test a full 360 degrees around and 1/2" below and 1/2" outside the lip of the draft hood with smoke, mirror (it will fog up where the combusted air is), etc. If it is oil-fired, check for backflow out of the barometric draft damper. If it spills, record how long it takes to stop spilling.

        _____seconds

    a.   If it takes less than 60 seconds, skip to the draft test, step 3 below.

    b.   If it takes more than 60 seconds before it stops spilling, it fails the spillage test. If it fails the spillage test, turn off the water heater, put the house into natural conditions (open all interior doors, and turn off all the exhaust fans), and, after allowing the vent to cool, retest the spillage, draft test, and carbon monoxide tests under these natural conditions. Do not test the water heater any further under worst-case conditions—do the spillage, draft test, and carbon monoxide test under natural conditions. Also, measure the net change in pressure when going back to natural from worst case, confirm the "baseline pressure." The new pressure should be close to the very first "Baseline" pressure you took when setting the house up for worst-case conditions (see "Establishing the Pressure Under Worst Case Conditions" table above). In addition, if you have to go back to natural conditions when you are testing any appliance, test all the remaining appliances only under natural conditions.

        Roll out of flames?     Y       N

|  | **Spillage Time in seconds** |
|---|---|
| Worst Case | Natural (only if fails Worst case) |
| Sec. _____ | Sec. _____ |
| >60 sec.: Fail | >60 sec.: Fail |

Check your personal monitor for ambient CO level after spillage test/retest:

CO level: _____ppm
If CO is greater than 35 ppm in BPI-related work or greater than 20 ppm in Weatherization (WX)-related work:
Turn off the appliance, ventilate the space, evacuate the building, and follow the remaining instructions included in Step 7 of the "Setting Up the House for Worst-Case Conditions."

If you had to revert to natural conditions:
Natural Conditions pressure close to Baseline pressure?   Y       N

2.   **Draft Vacuum Pressure Flue Test Setup**:

See the appropriate water heater drawing in previous section to understand this section better.

Do not drill holes in double walled flues (B-vents), accordion-style single wall B-vents, or sealed combustion units (high efficiency units with PVC pipe as a flue). If a unit has a double wall metal flue, an accordion-style single wall B vent, or a PVC flue, the only step you can take is to record CO at the exterior outlet of the flue, if accessible. You are not expected to get on unsafe or damageable roofs to check the output of exterior outlets of these furnaces (for example, roofs greater than 5 in 12 pitch, ice/rain/snow, rigid roofs like clay, asbestos, concrete, etc.). Skip to step 4 if the appliance has one of these flues.

b.   If it is a single wall flue that is not accordion-style, drill two holes: Drill a hole in the flue 1 to 2 feet downstream of (up above) the appliance draft diverter or just beyond the first elbow in its flue. Drill the other hole right into the draft diverter over an edge of the baffle/turbulator in the water heater vent coming up from inside the water heater so you can place the combustion analyzer on both sides of the baffle. The tip of the combustion analyzer should be able to go at least 1" below the top of the water heater down into the interior water heater vent. If it is oil-fired, be sure to place the hole before the barometric draft damper.

3.   **Draft Pressure Test**: Wait up to 10 minutes, at most, from initial firing to reach a stabilized efficiency and test the water heater for draft pressure, using either a digital manometer or a combustion analyzer on which you can also test draft pressures. Compare this number in Pascals to the number in the Minimum Acceptable Appliance Draft Test Ranges Table (see below).

a. If the appliance drafts at a pressure more negative than the minimum Acceptable Draft Test Range, it has passed the draft pressure test. Go on to test for CO.

b. If the draft comes in less negative (that is closer to zero) than the minimum Acceptable Draft Test Range, then the appliance fails the draft test. If it fails, turn off the water heater, put the house into natural conditions (open all interior doors, and turn off all the exhaust fans), and, after allowing the vent to cool, retest the spillage, draft test, and carbon monoxide under these natural conditions. Do not test the water heater any further under worst-case conditions if it fails—do the spillage, draft test, and carbon monoxide test under natural conditions. Also, measure the net change in pressure when going back to natural from worst case—the new pressure should be close to the very first "Baseline" pressure you took when setting the house up for Worst Case conditions (see the "Establishing the Pressure Under Worst Case Conditions" table above). In addition, if you have to go back to natural conditions when you are testing any appliance, test all the remaining appliances only under natural conditions.

### Draft pressure in Pascals (Pa)

| Worst Case | Natural (only if fails Worst Case) |
|---|---|
| _____ Pa | _____ Pa |
| Pass/Fail | Pass/Fail |

Check your personal monitor for ambient CO level after spillage test/retest:

CO level: _____ ppm

If CO is greater than 35 ppm in BPI-related work or greater than 20 ppm in Weatherization (WX)-related work:

Turn off the appliance, ventilate the space, evacuate the building, and follow the remaining instructions included in Step 7 of the "Setting Up the House for Worst-Case Conditions."

If had to revert to natural conditions: Natural Conditions pressure close to Baseline pressure?      Y      N

4.  **CO Test:** To get an undiluted CO sample of the flue gases place the tip of the combustion analyzer at least 1 inch down inside the throat of the water heater, on both sides of the baffle/turbulator (one sample on each side of the baffle).

*Fyrite® Insight Analyzer steps for draft pressure and CO analysis*. Connect the probe and tubes (the electrical connection plugs into the T-STK plug and the other tubes line up in order to the right of the electrical plug—they are connected side-by side to keep them in the right order).

Go outside: Press the PWR button and wait until calibration/warm up is complete.

Go inside: Press F2 (Menu), the down arrow (pressure should be selected), the middle green button, F2 (zero), Disconnect the hoses, press the middle green button, reconnect the hoses. Insert the tip so it is situated at about the middle of the diameter of the flue, press Esc, press Run twice, press the down arrow and wait until the T-STK equilibrates to about its hottest temperature, stop after about 5 minutes or the temperature is stable, and press Hold. Turn the printer on and line it up so the laser communicates. Press F1, allow the printer to complete printing and gently tear off the printed paper. Remove the probe tip from the flue and place it on flat non-flammable surface (such as a concrete floor), run it until it is cool and O2 reads about 20.9%. Disconnect all connections and hold down the PWR button until it shuts off. Empty the clear plastic water trap if there is enough dispensable water in it.

    c.    Evaluate the CO reading, draft and CO the spillage, and the draft results with the action levels table below and record the result. Also, remember, if you ever get a reading of over 400 ppm CO (as measured) and an appliance has failed either the spillage or the draft test, it is an emergency: Shut off the fuel to the appliance and have the homeowner call for service immediately. Do not finish the audit nor allow other retrofit work to continue.

    **Highest CO (Carbon Monoxide) reading in ppm**

| Left | ppm | Right | ppm |
|------|-----|-------|-----|
| Pass/Fail | | Pass/Fail | |

5.    After completing this test, turn the water heater to pilot. If however, you are done with all appliance testing, leave the water heater in normal operating mode.

6.    Seal all test holes afterwards with high temperature silicon sealant or a high temperature aluminum foil tape.

## Standards for Draft Testing

Tables 9-12, 9-13, and 9-14 show the Weatherization standards and BPI standards for the minimum acceptable draft negative pressure a vented appliance can have. If it has a more negative pressure than the one listed, it passes (for example, on a 50°F day, a gas fired furnace has a -5 Pascal draft, lower than the -3 Pascal standard).

**Table 9-12. Weatherization Minimum Acceptable Appliance Draft in Pascals and Inches of Water Column (Pa/IWC)**

|  | Outdoor Temperature | | | | |
|---|---|---|---|---|---|
| Vented Appliance | < 20°F | 21-40°F | 41-60°F | 61-80°F | >80°F |
| **Gas** fired furnace, boiler or water heater with atmospheric draft - Pa/IWC | -5/ -0.016 | -4/ -0.02 | -3/ -0.012 | -2/ -0.006 | -1/ -0.004 |
| **Oil**-fired furnace, boiler or water heater with atmospheric draft- Pa/IWC | -15/ -0.06 | -13/ -0.053 | -11/ -0.045 | -9/ -0.038 | -7/ -0.030 |

| Degrees, F | Pascals, Pa |
|---|---|
| <10 | -2.5 |
| 10-20 | -2.5 to -2.25 |
| 20-30 | -2.25 to -2.0 |
| 30-40 | -2.0 to -1.75 |
| 40-50 | -1.75 to -1.5 |
| 50-60 | -1.5 to -1.25 |
| 60-70 | -1.25 to -1.0 |
| 70-80 | -1.0 to -0.75 |
| 80-90 | -0.75 to -0.5 |
| >90 | -0.5 |

The BPI-equation used to calculate the values in the above table is actually: (Temp outside/40) – 2.75 Pa = Minimum Negative Draft Pressure Reading in Pascals needed to pass.

BPI standards are less strict than the Weatherization standards. A stronger draw or negative pressure draft in the flues of combustion appliances is needed with the Weatherization standards than with the BPI standards. For instance, for 10 °F, -2.5 Pa is the minimum with BPI while it is -5.0 Pa under the Weatherization standards. Notice that BPI does not have oil-fired appliance standards.

**BPI Specific Minimum Acceptable Appliance Draft (converted for specific outdoor temperature ranges)**

| Degrees, F | Pascals, Pa |
|------------|-------------|
| <10 | -2.5 |
| 10-20 | -2.5 to -2.25 |
| 20-30 | -2.25 to -2.0 |
| 30-40 | -2.0 to -1.75 |
| 40-50 | -1.75 to -1.5 |
| 50-60 | -1.5 to -1.25 |
| 60-70 | -1.25 to -1.0 |
| 70-80 | -1.0 to -0.75 |
| 80-90 | -0.75 to -0.5 |
| >90 | -0.5 |

The BPI-equation used to calculate the values in the above table is actually: (Temp outside/40)—2.75 Pa = Minimum Negative Draft Pressure Reading in Pascals needed to pass.

**Table 9-13**
**Weatherization Carbon Monoxide (CO) Action Levels: Vented Appliances**

| Vented Appliance Type | Suggested CO Action Levels | |
|------------------------|------------------------------|--------|
| | WX | BPI |
| Gas furnace, boiler, or water heater AND Oil-fired furnace, boiler, or water heater (Same standard for all) | 100 ppm, as measured | Depends on Spillage & Draft Test - see below |

**Table 9-14. Carbon Monoxide Action Levels: Vented—BPI**

Note that "Stop Work" means you can continue with the audit but the *retrofit* work may not proceed on the home until the problem is corrected. "Emergency" on the other hand means that the *audit and the retrofit* work cannot proceed until the appliance has been properly serviced.

| VENTED APPLIANCE | BPI - ppm & Spillage/Draft (S/D) result and action to take |
|---|---|
| Gas furnace, boiler, or water heater AND Oil-fired furnace, boiler, or water heater (Same standard for all) | 0-25 ppm & passes S/D - Proceed with work

26-100 ppm & passes S/D - Rec'd CO problem be fixed but audit and retrofit work may proceed.

26-100 ppm & fails only worst case S/D - Rec'd service call/repairs to home to correct the problem but audit and retrofit work may proceed.

100-400 ppm OR fails under both worst case & natural conditions - Stop Work – retrofit work may not proceed until system serviced and problem corrected. The audit may continue.

>400 ppm & Passes all Spillage/Draft Test - Stop Work – retrofit work may not proceed until system serviced and problem corrected. The audit may continue.

>400 ppm & Fails any Spillage/Draft Test- Emergency: Shut off appliance and owner calls for service . Both audit and retrofit work must be stopped |

Note that the draft pressure must be more negative than the Pascals indicated for each range of temperature. For instance, if you had a reading of 35°F outside, under BPI standards you would need a negative pressure of about -1.82 Pa or more negative, such as -3.2 Pa, for the draft pressure reading taken from the flue to pass the BPI standard. The more negative the draft, the better the draw of combusted products out of the house. Here are some listings of the different action levels for CO in vented appliances for the weatherization programs and BPI:

Table 9-15 shows the sequence for doing spillage, draft, and CO tests on furnaces and boilers.

## Table 9-15. Testing Furnace/Boiler Backflow/Spillage Test, Negative Draft Pressure Flue Test, and CO Test. Make sure the house is still in worse case condition.

Make sure your gauge batteries are fully charged or that you have adequate outlets and extension cords to connect the pressure gauge to its recharger/transformer.

1.  **Spillage Test**: You must inspect forced air furnaces for flame interference. Turn the furnace on by first recording its current setting (_____°F) and by turning its thermostat 10 or 20 degrees higher and check the burners as the fan turns on. Follow this step and the Spillage step (Step 2) at the same time Roll out of flame?          Y      N

2.  With the appliance on, do a spillage check with smoke, mirror (it will fog up where the combusted air is), etc. Test 1/2" below and 1/2" outside the lip of the draft hood all along the draft hood where the flue gases would backdraft into ambient air. For medium efficiency units, check for spillage at the base of the chimney (where the flue enters the chimney, at the water heater draft hood/bell housing if a water heater shares the flue, or where air goes into the burner compartment if that area is open to air, etc.). You cannot check for spillage on high efficiency units because they are direct vented—no openings to air inside the house. If it is oil-fired, check for backflow out of the barometric draft damper. Record the time it takes for the spilling to stop.

    a.  If it takes less than 60 seconds, skip to the draft test, step 4 below.

    b.  If it takes more than 60 seconds before it stops spilling, it fails the spillage test. If it fails, turn off the appliance, put the house into natural conditions (open all interior doors, and turn off all the exhaust fans), and, after allowing the vent to cool, retest the spillage, draft test, and carbon monoxide under these natural conditions. Do not test the appliance any further under worst-case conditions—do the spillage, draft test, and carbon monoxide test under natural conditions. Also, measure the net change in pressure when going back to natural from worst case—the new pressure should be close to the very first "Baseline" pressure you took when setting the house up for Worst Case conditions (see the "Establishing the Pressure Under Worst Case Conditions" table above). In addition, if you have to go back to natural conditions when you are testing any appliance, test all the remaining appliances only under natural conditions.

**Spillage Time in Seconds**

| Worst Case | Natural (only if fails Worst case) |
|---|---|
| _____Sec. | _____Sec. |
| >60 Sec.:Fail | >60 sec.:Fail |

Check your personal monitor for ambient CO level after spillage test/
retest:

CO level: _____ppm

If CO is greater than 35 ppm in BPI-related work or greater than 20 ppm
in Weatherization (WX)-related work:

Turn off the appliance, ventilate the space, evacuate the building, and follow the remaining instructions included in Step 7 of the "Setting Up the
House for Worst-Case Conditions."

If the flames burn differently and continue to burn differently after the
blower fan turns on, this could be a sign of a cracked heat exchanger. A
cracked heat exchanger is not repairable. Typically the furnace must be replace. Request a more extensive heat exchanger test to be done, or conduct
additional heat exchanger tests yourself if qualified.

Evid. Heat Exchanger Cracked?          Y          N

Any CO detected at Supply registers?   Y          N
N/A (no reason to test)

If had to revert to natural conditions: Natural Conditions pressure close to
Baseline pressure?          Y          N

3.    **Draft Vacuum Flue Pressure Test Setup**
      a.   See the appropriate furnace drawing in the earlier section to under-
           stand these tests better. Do not drill holes in double walled or cor-
           rugated flues or accordion style (B-vents), or sealed combustion units
           (high efficiency units with PVC pipe as a flue)—no draft test can be
           done. Skip to step 5 if the appliance has one of these flues.

           If a unit has a double wall metal flue, an accordion-style single wall
           B-vent, or a PVC flue, the only step you can take is to record CO at the
           exterior outlet of the flue, if accessible. You are not expected to get on
           unsafe or damageable roofs to check the output of exterior outlets of

these furnaces (roofs greater than 5 in 12 pitch, ice/rain/snow, rigid roofs like clay, asbestos, concrete, etc.). Skip to step 4 if the appliance has one of these flues.

b. If the appliance uses a single wall metal flue that is not accordion-style, drill a hole into the single wall flue 1 to 2 feet downstream of (up above) the appliance draft diverter or just beyond the first elbow in its flue. If it is oil-fired, be sure to place the hole before the baro-metric draft damper. If it is gas-fired, you may also have to drill an-other hole right into the draft diverter so that you can later place the combustion analyzer into the top of each of the heat exchangers to get undiluted combusted air for the CO test. The tip of the combustion analyzer should be able to reach inside of each heat exchanger to the test on low/standard efficiency units (atmospheric/gravity-fed)—see below.

4. **Draft Pressure Test**: Wait up to 10 minutes to reach a stabilized tempera-ture and test the appliance for draft, using either a digital manometer or a combustion analyzer that you can also test draft pressures on. Compare this number in Pascals to the number in the *Minimum Acceptable Appliance Draft* Test Ranges Table (see table above).

a. If the appliance drafts at a pressure more negative than the minimum Acceptable Draft Test Range. It has passed the Draft Pressure Test. Go on to test for CO.

b. If the draft is less negative (that is, closer to zero) than the minimum acceptable draft test pressure from the table, then the appliance fails the draft test. If it fails, turn off the appliance, put the house into natu-ral conditions (open all interior doors, and turn off all the exhaust fans), and, after allowing the vent to cool, retest the spillage, draft test, and carbon monoxide under these natural conditions. Do not test the appliance any further under worst-case conditions—do the spillage, draft test, and carbon monoxide test under natural condi-tions. Also, measure the net change in pressure when going back to natural from worst case—the new pressure should be close to the very first "Baseline" pressure you took when setting the house up for Worst Case conditions (see the "Establishing the Pressure Under Worst Case Conditions" table above). In addition, if you have to go back to natural conditions when you are testing any appliance, test all the remaining appliances only under natural conditions.

**Draft pressure in Pascals (Pa)**

Worst Case                          Natural (only if fails Worst case)

_____Pa                         _____Pa

Pass/Fail                           Pass/Fail

Check your personal monitor for ambient CO level after spillage test/retest

CO level: _____ppm

If CO is greater than 35 ppm in BPI-related work or greater than 20 ppm in Weatherization (WX)-related work:

Turn off the appliance, ventilate the space, evacuate the building, and follow the remaining instructions included in Step 7 of the "Setting Up the House for Worst-Case Conditions."

If had to revert to natural conditions: Natural Conditions pressure close to Baseline pressure?     Y     N

5.   **CO Test**: To get an undiluted CO sample of the flue gases from a standard or low efficiency appliance, place the tip of the combustion analyzer inside each heat exchanger and record the reading. If drilling a hole in the draft diverter is the only way for you to access the heat exchangers on a low efficiency furnace before the combusted air mixes in the diverter, then drill a hole in the draft diverter for the combustion analyzer. If it is a medium efficiency furnace and you have been able to drill a hole in the flue for a draft test (such as a single wall flue only), do a CO test in the draft test hole. If it is a double walled or corrugated flue (B-vents), or high efficiency units with PVC pipe as a flue, the only step you can do here is to record CO at the exterior outlet of the flue if safely accessible. It is the only exterior pipe where the air exhausts/flows out of the PVC pipe. You are not expected to get up on unsafe or damageable roofs to check the output of exterior outlets of these furnaces (roofs of greater than 5 in 12 pitch, ice/rain/snow, rigid roofs like clay, asbestos, concrete, etc.).

Connect the probe and tubes (the electrical connection plugs into the T-STK plug and the other tubes line up in order to the right of the electrical plug—they are connected side-by side to keep them in the right order).

Go outside: Press the PWR button and wait until calibration/warm up is complete.

Go inside: Press F2 (Menu), the down arrow (pressure should be selected), the middle green button, F2 (zero), Disconnect the hoses, press the middle green button, reconnect the hoses.

a. Evaluate the CO reading, draft and CO and the draft results with the Combustion Safety Test action levels table above. Also, remember, if you ever get a reading of over 400 ppm CO and an appliance has failed either the spillage or the draft test, it is an emergency: Shut off the fuel to the appliance and have the homeowner call for service immediately. Do not finish the audit nor allow other retrofit work to continue.

*Fyrite® Insight Steps for Draft Pressure and CO Analysis*: Place the tip so it is situated at the top of the heat exchanger, press Esc, press Run twice, press the down arrow and wait until the T-STK reaches a steady temperature, stop after about 5 minutes or the efficiency is stable, and press Hold. Turn the printer on and line it up so the laser communicates. Press F1, allow the printer to complete printing and gently tear off the printed paper. Remove the probe tip from the flue and place it on flat non-flammable surface (such as concrete floor), run it until it is cool and O2 reads about 20.9%. Disconnect all connections and hold down the PWR button until it shuts off. Empty the clear plastic water trap if there is enough dispensable water in it.

### Highest CO (Carbon Monoxide) reading in ppm in each heat exchanger (P=Pass, F=Fail)

| #1 | #2 | #3 | #4 | #5 | #6 |
|-----|-----|-----|-----|-----|-----|
| P/F | P/F | P/F | P/F | P/F | P/F |

6. CO level: _____ppm

   If CO is greater than 35 ppm in BPI-related work or greater than 20 ppm in Weatherization (WX)-related work:

   Turn off the appliance, ventilate the space, evacuate the building, and follow the remaining instructions included in Step 7 of the "Setting Up the House for Worst-Case Conditions."

7. After completing this test, keep the furnace on if the furnace is tied into a common flue with a water heater and then go to the next section. If the furnace is not tied into a common flue with a water heater, turn down the thermostat to its original temperature.

8. Seal all test holes afterwards with high temperature silicon sealant or a high temperature aluminum foil tape

Table 9-16 shows the procedure for checking a water heater that is tied into a common flue with the furnace or boiler:

**Table 9-16. Supplemental Water Heater Testing When on a Combined Flue with a Furnace or Boiler**

Make sure the house is still in worst case conditions.

1.  After the furnace has been turned on, turn on the water heater as well.

2.  Spillage Test: Retest only the water heater for spillage and record the time it takes to stop spilling in seconds. If it takes more than 60 seconds before it stops spilling, it fails the spillage test. If it fails, turn off the appliance, put the house into natural conditions (open all interior doors, and turn off all the exhaust fans), and, after allowing the vent to cool, retest the spillage, draft test, and carbon monoxide under these natural conditions. Also, measure the net change in pressure when going back to natural from worst case—the new baseline pressure should be close to the very first baseline pressure you took when setting the house up for worst case conditions (see the "Establishing the Pressure Under Worst Case Conditions" table above).

**Spillage Time in Seconds**

| Worst Case | Natural (only if fails Worst case) |
|---|---|
| _____Sec. | _____Sec. |
| >60Sec.:Fail | >60 sec.:Fail |

Check your personal monitor for ambient CO level after spillage test/retest:

CO level: _____ppm

If CO is greater than 35 ppm in BPI-related work or greater than 20 ppm in Weatherization (WX)-related work:

Turn off the appliance, ventilate the space, evacuate the building, and follow the remaining instructions included in Step 7 of the "Setting Up the House for Worst-Case Conditions."

If had to revert to natural conditions: Natural Conditions pressure close to Baseline pressure?    Y    N

3.  Draft Flue Pressure Test Setup: Retest the draft pressure on the water heater, record the reading, and indicate whether it passed or failed (see

previous water heater test section). If the draft is less negative (that is, closer to zero) than the minimum Acceptable Draft Test Range (see table above), then the appliance fails the draft test. If it fails, turn off the appliance, put the house into natural conditions (open all interior doors, and turn off all the exhaust fans), and, after allowing the vent to cool, retest the spillage, draft test, and carbon monoxide under these natural conditions. Also, measure the net change in pressure when going back to natural from worst case—the new pressure should be close to the very first "Baseline" pressure you took when setting the house up for worst case conditions (see the "Establishing the Pressure Under Worst Case Conditions" table above).

**Draft pressure in Pascals (Pa)**

| Worst Case | Natural (only if fails Worst case) |
|---|---|
| _____Pa | _____Pa |
| Pass/Fail | Pass/Fail |

Check your personal monitor for ambient CO level after spillage test/ retest

CO level: _____ppm

If CO is greater than 35 ppm in BPI-related work or greater than 20 ppm in Weatherization (WX)-related work:

Turn off the appliance, ventilate the space, evacuate the building, and follow the remaining instructions included in Step 7 of the "Setting Up the House for Worst-Case Conditions."

If had to revert to natural conditions: Natural Conditions pressure close to Baseline pressure?     Y     N

4.  Seal all test holes afterwards with high temperature silicon sealant or a high temperature aluminum foil tape.

In addition, under BPI standards, the homeowner must be: 1) notified of the results of all combustion tests—draft negative pressure, spillage/backflow, "worst case," and both appliance and ambient CO; and 2) *recommended* to get a CO detector and *recommended* to place a CO detector on all floors, and 3) *required* to obtain a CO detector in cases where there is an attached garage, or in the highly common situation, when the house has at least one combustion appliance.

## UNVENTED APPLIANCES

In the case of unvented appliances such as stovetop burners, ovens and unvented space heaters, there is no need to place the house under worst-case conditions because there is no vent on the appliance that could backflow. Nevertheless, check for the CO levels that are created by each of these appliances. Testing is not done on kitchen appliances and ovens until you make sure the self-cleaning features have been turned off. Some recommend that you turn on an exhaust fan and open a window when testing ovens, but you may then be testing the appliance in a different way than the homeowner uses it. Whether you ventilate the kitchen or not, be sure to check your personal CO monitor or set it for alarm at 20 ppm (WX) or 35 ppm (BPI) when you are checking any appliance. All combustion appliances must reach steady state (5-10 minutes) before testing for CO or draft negative pressure, so wait at most 10 minutes before testing.

When it comes to unvented appliances, the standards are different as you can see from Table 9-17.

## TESTING EFFICIENCY OF COMBUSTION APPLIANCES

The efficiency of combustion appliances is important. Some combustion appliances draw too little or too much gas relative to the volumes of air they take in during the combustion process. Often, modern combustion analyzers will also tell you the efficiency of a combustion appliance by simply reading the screen when doing combustion testing. If you do not have a combustion analyzer, here is a method to check whether the manufacturer's designated Btu input matches the actual gas used by the appliance. First, make sure the unit being checked is the only one turned on. Next, go to the gas meter and time at least two revolutions on the smallest dial on the gas meter. To determine the time for one revolution, divide by the number of revolutions that were timed—if two revolutions were timed, divide the time by 2. Use a chart, such as shown in Appendix K, that requires the time for one revolution along with the volume in cubic feet represented by one revolution on the dial to compute the Btus per hour utilized by the appliance, see Figure 9-65. Look for the most rapidly moving dial on the meter and check what its volume per rotation is (usually ½, 1, or 2 cubic feet per minute). The

**Table 9-17. Carbon Monoxide Action Levels—Unvented**

| Unvented Appliance | Weatherization | BPI |
|---|---|---|
| Unvented Gas or Oil <u>Space Heater</u> | Rec'd removal, cannot serve as <u>primary heat</u>, 200 ppm, air free | <u>Not allowed</u> at any ppm |
| Gas cook stove: <u>Range-top burner</u> | 25 ppm, as measured | <u>35 ppm</u> ambient exposure, as measured unless confined space, then <u>rec'd</u> 25 cfm continuous or 100 cfm intermittent |
| Gas cook stove: <u>Oven</u> | 100 ppm, as measured | <u>Always</u> rec'd ventilation<br><br><u>100-300 ppm</u>, as measured: Install <u>CO detector</u> & <u>recommend</u> service before start<br><br><u>Over 300 ppm</u>: <u>Must</u> be serviced before work starts, 25 cfm continuous or 100 cfm intermittent if still 300 ppm after service |

table in Appendix K gives the resulting predicted Btu use of the furnace in Btus/hour. Thus, if you timed a dial designated as showing 1 cubic foot per minute to take 40 seconds for two revolutions (20 seconds for one revolution), you would estimate the appliance runs at "180" on the chart, which is actually read to mean 180,000 Btus per hour.

With this information, you can determine the true Btu input on the combustion appliance. If it is ±10% or more beyond the manufacturer's input rating, you need to refer the unit to an HVAC technician. The technician can inspect it and adjust the gas pressure so that it falls back within acceptable levels. If the 180,000 Btu appliance in this example is a medium efficiency furnace and its rating on the manufacturer's data plate output Btus is 190,000 Btu/hour, the real meter input reading

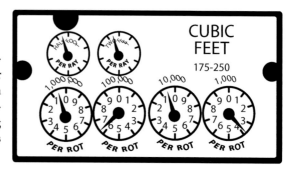

**Figure 9-65. Calculating the Btus used over a given time frame can tell whether a combustion appliance is using more or less than it is rated.**

would be within ±10% of the rating. More specifically—

$$(\text{Manf. Output Btus—meter Btus/Manf. Output Btus.})$$
$$(190{,}000\text{-}180{,}000)/190{,}000 = 0.053 \text{ or } 5.3\%,$$

which is well within the ±10% cutoff. For high-efficiency condensing units, the actual fuel input should be within ±5%. If the combustion appliance under-fired or over-fired beyond its input rating in Btu per hour, it could produce higher than allowable levels of CO. This would be the result of the ratio of fuel to oxygen being too high or too low to permit full and complete combustion.


POSSIBLE REPAIRS

**Fixes for Combustion Safety Problems in Vented Appliances**
    Caution: Local, state, etc. laws and/or rules may require that you hold a license in HVAC, etc. before conducting any repairs on combustion appliances. Some possible recommendations for resolving excessive depressurization or inadequate draft pressure include properly sizing connectors, vents, or liners, and repairing any chimney or vent disconnections, obstructions, or leaks. Other possible recommendations include sealing any return ducts or return leaks that are located in the CAZ, reducing the capacity of large exhaust fans providing makeup air for exhaust fans and dryers etc., balancing the supply and return air on the furnace by adding return air openings to the hall of the house or adding new returns, or installing a metal chimney liner in oversized chimneys and a wind rated cap at the top of the chimney.

Here is an inexpensive method that may resolve excessive CAZ depressurization or inadequate draft pressure in the flue, and at the same time better provide for a more even temperature throughout the house. Cut the bottom edges of doors in the house up higher so return air has a less restrictive route to get back to the furnace. Another effective measure for resolving excessive depressurization in the CAZ is to provide a permanent combustion air inlet from the outside to the CAZ (see Combustion Appliances, Background, Inspection and Testing).

While this method can help reduce the safety issues with a depressurized CAZ, it does not help provide a more even temperature throughout the house. Creating an air inlet from the outside to the CAZ also requires insulating the walls and ceiling of the CAZ (for example, the water heater or furnace closet) from the rest of the house to avoid having the cold air entering the CAZ from outside during the winter defeat the purpose of heating the home. This typically means insulating the interior walls of the furnace closet. Having a return air vent or an unintended opening in the return ductwork within the CAZ can create an even more severe problem than having leaky return air ducts in other areas. The negative pressures created by the return air system on the CAZ cause flue gases to backflow or spill into the CAZ, which can then be drawn into the conditioned house air through return air openings and expose the occupants to CO.

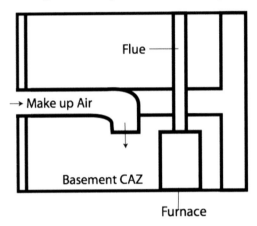

**Figure 9-66. Combustion or Make Up Air System in a CAZ**

COOLING MEASURES

**Central Air-conditioning Systems**

Central air-conditioning systems are defined as electrically operated refrigerant type systems used for cooling and dehumidification.

They typically operate with a 220 V capacity and require a 30 to 70-amp overcurrent protection. Air conditioners are typically rated in "tons" or tonnage. Heat pumps are similar to central air conditioners, but use a reversing valve and, as a result, can also be used as heating devices. Air-conditioning systems should be tested only when the outside air temperature is above 60°F. Below that temperature, the systems can be damaged when operated.

There are two types of central air-conditioning systems: integral (Figure 9-67) and split (Figure 9-68). In the integral system, all mechanized components—compressor, condenser, evaporator, and fans—are contained in a single unit. The unit may be located outside the building (usually on the roof or side of the house) with its cold air ductwork extending into the interior, or it may be located somewhere inside the building with its exhaust air ducted to the outside.

In the split system, the compressor and condenser are located outside the building and are connected by refrigerant lines to an evaporator inside the building's air distribution ductwork. Split systems in buildings that are heated by forced warm air usually share the warm air system's circulating fan and ductwork. In such cases, the evaporator is placed either directly above or below the furnace, depending on the furnace design. It is always placed downstream of the furnace.

Sometimes you will find air conditioning and heat pump systems that have ductwork and blowers that are completely independent of the primary or auxiliary heating units. If the system is not running, it is usually caused by power not getting to the unit.

If the system is running, but it is not cooling, the unit could be hav-

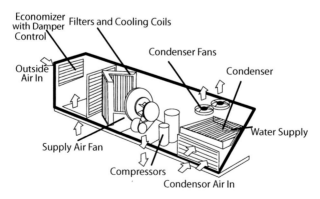

**Figure 9-67. Package Air Conditioning System**

ing airflow problems, or it may need refrigerant. If the system cycles on and off too often, the thermostat may be poorly adjusted or defective or there could be airflow problems. If some rooms are too cool and others too warm, the registers or duct system dampers may need adjustment or the ductwork may need resizing.

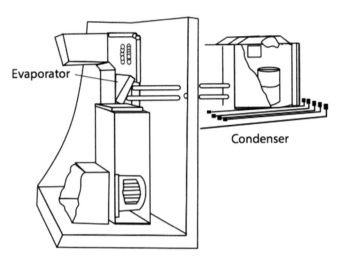

**Figure 9-68. Split System Air Conditioning**

### Compressor and Condenser

The compressor "pumps" refrigerant gas under great pressure through a condenser coil, where it gives up heat and becomes a liquid. The heat is exhausted to the outside air by the condenser fan blowing outside air past the condenser coil. Compressors have a service life of 5 to 10 years, depending on the maintenance they receive, and are the most critical component in the air-conditioning system.

If the compressor, condenser, and condenser fan are part of a split system and are located in a separate unit outside, check the airflow around the outside unit to make sure it is unobstructed (not surrounded by shrubs, etc.). Look for dirt and debris inside the unit, particularly on the condenser coils and fins, and check all electrical wiring and connections. The unit should be level and well-supported with its housing intact and childproof. An electrical disconnect switch for use during maintenance and repairs should be located just adjacent to the units within 6 feet and within eyesight of the unit. A good integral system located somewhere on or in the building should have its compressors placed on vibration mountings to minimize sound transmission to inhabited building space.

**Refrigerant Lines**

Refrigerant lines form the link between the interior and exterior components of a split system. The larger of the two lines carries low-pressure (cold) refrigerant gas from the evaporator inside to the compressor outside. It is approximately the diameter of a broom handle and should be insulated along its entire length to help avoid condensation. The smaller line is uninsulated and carries high-pressure (warm) liquid refrigerant back to the evaporator inside. Check both lines for signs of damage and make sure the insulation is intact on the larger line. On the exterior, the insulation can be protected from ultraviolet damage by a covering or by white paint. Sometimes a sight glass is provided on the smaller line. If so, the flow of refrigerant should look smooth through the glass. Bubbles in the flow indicate a deficiency of refrigerant in the system. Frost on any exposed parts of the larger line also may indicate a refrigerant deficiency.

**Evaporator or Plenum Coil**

The evaporator is enclosed in the air distribution ductwork and usually can only be observed by removing a panel or part of the furnace plenum. High-pressure liquid refrigerant enters the evaporator and expands into a gas, absorbing heat from the surrounding air. Air is pushed past the evaporator coil (Figure 9-69) by the system's circulation blower (for example, the heater fan in ductwork systems that is shared by both the furnace and air conditioner). In the process, water vapor or humidity from the air condenses on the evaporator coil and drips into a drain pan. This reduces the humidity of the air in the house and, in the summer when the air conditioning is running, can increase the comfort level in the house as a result.

From the drain pan, it is directed to a condensate drain line that empties in a house

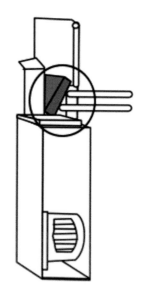

**Figure 9-69. Location of Evaporator Coil (grey).**

drain or somewhere on the building's exterior. Water stains or corrosion on top of the furnace or in the fan box may indicate a plugged condensate drain line or evaporator coil fins, or a cracked condensate tray under the evaporator coil. If this is occurring, it is possible that the water flowing out of the condensate drain or directly off the coil fins is rusting the heat exchanger that may lie below the evaporator coil. Sometimes this results in stains or permanent rusting in the base of the fan box if the furnace is an up-draft type. If, while you are checking for proper filter placement in such a furnace, you notice these signs, consider the possibility of plugged evaporator coil fins, especially if there is no evidence the moisture came from water in the basement.

Sometimes, especially on houses with no lower level floor drain, a condensate pump is used to divert air-conditioning condensate water to a sewer line. Examine the ductwork around the evaporator for signs of air leakage, and check below the evaporator for signs of water leakage due to a blocked condensate drain line. Such leakage can be a serious problem if the evaporator is located above a warm air furnace, where dripping condensate water can rust the heat exchanger, or above a ceiling, where it can damage building components below. You can follow the condensate line and make sure that it terminates in a proper location.

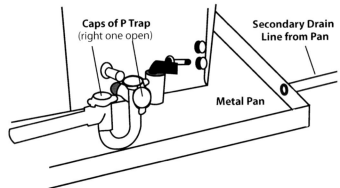

Figure 9-70. Sediment and dust in a "P" trap on a condensate drain can be cleared with a flexible bottle brush (the caps at the tops of each end of the P trap can be popped open). Often the P trap is clear plastic so it can be easily checked for buildup. Notice that the secondary drain above the pipe connected to the P trap does not have a pipe connected to it. This is often the case when the unit is protected from damaging the house below by having a metal pan under it. The metal pan has a pipe running from it to serve as a secondary drain.

In split systems where the evaporator is located in an attic or closet, the condensate tray should have an auxiliary condensate drain line located above the regular drain line, or an auxiliary drain pan beneath the condensate tray) that is separately drained. The drain lines should be sloped to drain properly. The best configuration sends the auxiliary drain line overflow water out over a window so that the dripping will be recognized and the drain system cleared so that it will drain properly using the primary drainage system. The connection of a condensate drain line to a plumbing vent in the attic may violate local codes, especially if it is not U-trapped to prevent sewer gases from potentially backing up into the conditioned air. Check for such considerations.

**Distribution Duct Work and Controls**

Cool air distribution ductwork and controls, including zone controls, can be inspected similarly to those for forced warm air heating systems. Because hot air rises (the warmest rooms in the house are on the upper floor) and the cold air sinks (the coolest rooms in the house are usually on the lower floor), air flow distribution may require adjustment when the shared distribution systems are switched from heat to cooling and vice versa. Register grille (Figures 9-71 and 9-72) or ductwork damper adjustment can help with this correction. In some homes, you may find independent ductwork for the air-conditioning system in the ceilings so that more effective cooling is provided as the cool air descends. Insulation should exist around air conditioning ductwork in unconditioned spaces, especially attics.

An *estimate* of flow from each register can be obtained using an airflow hood (Figure 9-73), but these should generally be treated only as estimates. These are also known as flow hoods, flow capture hoods, capture hoods, or balometers. A general comparison can then be made between flows in different rooms and corrected (to at least some degree) for room volume and cubic footage of exposure of the room to outside walls, ceilings, etc. For instance, a room with twice the cubic footage might require twice the flow to keep the area warm. An equivalent sized room, but with one extra wall exposed to the outside and a ceiling exposed to an unconditioned attic, should require more flow, but you may not be sure how much more.

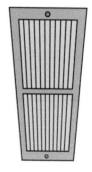

**Figure 9-71. Ceiling Register/Vent**

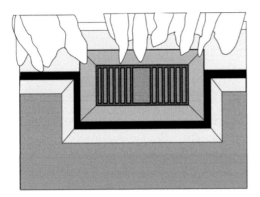

**Figure 9-72.**
**Floor Register/Vent**

**Figure 9-73. Flow Hood for**
**Estimating Register Flow**

Other, somewhat less expensive techniques you might use to estimate flow on registers include:

1.  Canadian bag method—This method uses a bag of a given volume where you time how long it takes to fill with air, and then back calculate what the flow is. Large traditional garbage bags are not big enough to check the airflow in heat registers, but they can be used for smaller exhaust fans. To check heat registers, a larger bag, such as a wet weather pallet cover bag, can be used. These bags are almost exactly 100 cubic feet in volume, which makes for an easy calculation. No matter what type of bag you use, it works best when you have a frame of some kind attached around the bag opening that allows you to quickly move it over the opening and start timing. For smaller bags this could be a reconfigured hanger wire. For larger bags such as the pallet cover bag, you could attach the door frame from your blower door set-up to the bag opening to make it an easier application. More information on this method can be found at: http://www.cmhcschl.gc.ca/en/co/maho/yohoyohe/inaiqu/inaiqu_003.cfm

2.  Anemometer—This device is used to measure air velocity. However, if you use it to compare the flows in rooms that are comfortable in the winter or summer with those that are not as comfortable in the winter or summer you may be able to eliminate undersized/

disconnected ductwork as a cause. If you find that the air velocity from the different registers seems to be very close, then the colder/hotter room situation more likely precludes damaged/undersized/disconnected ductwork.

## Central Gas Absorption Cooling Systems

Gas-absorption cooling systems may occasionally be found in older residential buildings. Such systems use the evaporation of a liquid such as ammonia or liquid bromide as the cooling agent. Like a gas refrigerator, these systems are powered by a natural gas or propane flame instead of a condenser. Because a gas-absorption system operates under several hundred pounds of pressure, it should be tested only by a specialist. The local gas or fuel supplier may maintain the unit. However, you can operate the system if it is in the summer. It should start smoothly and run quietly. Examine the physical condition of the system's exterior and interior components and inspect ductwork.

HEAT PUMPS

Electric heat pumps are electrically operated, refrigerant-type air-conditioning systems that can be reversed to extract heat from outside air and transfer it indoors. A reversing valve reverses the flow of the refrigerant when heat is demanded. Heat pumps are normally sized for their air-conditioning load, which in most parts of the country is smaller than the heating load. Below certain temperatures (such as 15°F), it is not cost-effective to have heat pumps be the source of heating in the home, at least for units that rely solely on outside air as a source for heat in the heat pump (Figure 9-74 shows various sources for heat exchange). Auxiliary electric heaters are used to provide the extra heating capacity the system requires in the heating season when the weather is too cold outside. In some heat pumps, a burner outside under the coils helps provide the heat needed to keep the heat pump operating in very cold weather. Some heat pumps use the ground (the evaporator coils are imbedded in the ground below the frost line) or water (the evaporator coils are exposed to water in a lake, river or well under the ice, etc.) to obtain heat in cold periods of the winter.

Heat pumps are typically evaluated in the mode you find them being used. However, the cooling mode should not be used if the outdoor temperature is below 65°F, while the heating mode should not be used if the outdoor temperature is above 65°F.

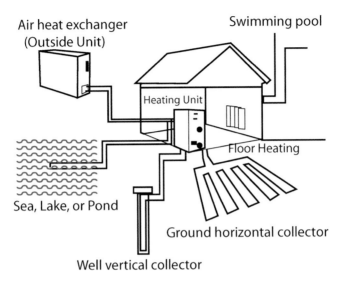

**Figure 9-74. Various Sources for Heat Exchange on Heat Pumps**

Test heat pumps as discussed earlier for central air-conditioning systems. Testing in one mode is usually sufficient. The sequence of operation of heat pumps to warm the house is generally as follows: When the thermostat calls for heat, the compressor turns on, and at the same time, the exterior evaporation fan and interior furnace fans turn on. Warmth passes from the gases in the coils in the interior furnace compartment to the air forced by these coils in the furnace circulation fan. This warm air is typically circulated to the rooms via ductwork.

Like air-conditioning systems, heat pumps can be either split or integral. Integral systems located outside the building should have well-insulated air ducts between the unit and the building. If they are located on or within the building, they should be mounted on vibration dampers, thermally protected, and have an adequate condensate drainage system.

**Auxiliary Heater Failure**

Electric resistance auxiliary heaters (Figure 9-75) are designed to activate (usually in stages or in sequence) below about 30°F (outside

**Figure 9-75. Electric resistance back up heat on heat pumps (circled) only provides expensive electric heat if the outdoor temperature is too cold for the heat pump to work effectively.**

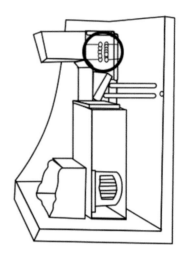

temperature) when the heat pump may not be able to produce enough heat to satisfy the thermostat. Switching to "emergency heat" at the thermostat usually turns on electric auxiliary heaters on heat pumps. Examine the physical condition of all auxiliary heaters. If possible, activate the auxiliary heaters to observe their operation. Operating failures may be caused by a faulty heater element, faulty relays, or a faulty thermostat.

### Improper Defrosting

During cold, damp weather, frost or ice may form on the metal fins of the coil in an outdoor unit. Heat pumps are designed to defrost this build-up by reversing modes either at preset intervals or upon activation by a pressure-sensing device.

### Faulty Reversing Valve

In most heat pumps, a reversing valve changes modes from heating to cooling (some heat pumps use a series of dampers instead). Generally speaking, you should never switch between the heat mode and cool mode on a heat pump just to check the system.

### Unit Heat Pump Systems

Unit, ductless, or mini-split heat pump systems are simply heat pumps that are used to cool and heat only one room. These are not central systems, nor are they window units. Rather, they are permanently installed miniature heat pumps whose refrigerant lines go directly to a coil placed up on a wall in the room they serve. With the use of a fan, air blows directly from one side of this evaporator coil to the other without the use of any ducts, thus the phrase ductless. They are often equipped with oscillating fins to spread the air evenly around the room.

## EVAPORATIVE COOLING OR "SWAMP" COOLER SYSTEMS

Evaporative cooling systems are simple and economical devices that use wetted pads or screens to cool, through evaporation, air passed through them. Such systems are used to a great degree only in dry climates where evaporation readily takes place and where dehumidification is not required (such as Inland Southern California, Arizona, Nevada, Utah, and New Mexico). Occasionally they will be installed in less arid climates, but typically prove useful for only a handful of summer days in those areas. Thus, you may find them in some of the most unexpected areas.

Evaporative coolers (Figure 9-76) consist of evaporator pads or screens (usually mounted on all four sides of the unit by attachment to removable panels), a means to wet them (a small metal water pump and distribution line that provides water at the top of the pads), an air blower (much like a fan blower), and a water reservoir in the base of the unit that has an overflow drain and float-operated water supply valve. These components are contained in a single, typically box-like housing, usually located on the roof, and connected to an interior air distribution system, usually one large plenum-like vent that carries

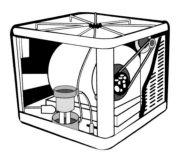

**Figure 9-76. Evaporative Cooler**

the cooled air directly to a large register in the ceiling of the uppermost floor of the house. A fan may distribute as many as 5,000 or more cubic feet per minute from the unit.

This is an especially good arrangement as the cold air descends naturally ("cold air sinks") to the rest of the house from the upper floor and the operating costs of these units are very low. In wetted-pad coolers, evaporator pads are wetted by a circulating pump attached to tubing that continually trickles water over them. In some coolers, a spray wets evaporator pads. In rotary coolers, evaporator screens are wetted by passing through a reservoir on a rotating drum.

The water in evaporative coolers often contains algae and bacteria that emit a characteristic "swampy" odor and as such, these coolers are sometimes referred to as "swamp coolers." These can easily be removed with bleach. Some systems help counteract this by treating the water or adjusting the float valve setting (much as a toilet float is adjusted) so that a small flow of water continually exits the overflow port, thus requiring a refreshing of the water. Another disadvantage of these systems is that they do not cool well when the humidity gets high (such as during a rain storm).

You can check evaporative cooling systems by examining the condition of each component that can be inspected with or without removal of any covers. From the roof, you can look for signs of leakage and corrosion, especially at the reservoir in the base of the unit, and check for a drain line (usually no more than a garden hose) to carry overflow water from the drain underneath to the rain gutters or another location away from the house. Without such a tube to run overflow water off the roof, the roof can be severely damaged in only a few years, depending on the nature of the water coming from the cooler. Drainage of the water from the drain line to a plumbing vent on the roof may cause sewer gases to back up into the conditioned air, although this is rarely seen. Check to be sure this has not happened. If the unit has been winterized (covered with a tarp during the winter to avoid allowing cold air into the house and the water reservoir and lines drained to avoid freezing), note this on your report so you are not held responsible for evaluating the cooler. It is not recommended that even the fan be tested under these circumstances. However, you can still note some problems that are noticeable without turning on the unit. Such items would include emission of flue gases too close or below the top of the evaporative cooler, no drain line from the overflow, or no step flashing around the cooler penetration into the attic. Far too often, cooler installers take advantage of the proximity of one of the surrounding cavities in a multiple wall flue to run the copper water supply tubing to the cooler. As a result, they place the cooler too close to a nearby flue and do not raise the flue opening above the top of the cooler in the process—leaving the home owner with the risk of exposure to combusted air as a result of the air drawn into the cooler.

Activate the system from the controls and listen for unusual sounds or vibrations. The controls are typically located in the hall around the area below the cooler ceiling register on the uppermost floor. Check to make sure the appropriate settings seem to operate properly. First, turn the settings to "pump" to determine if it appears that the pump is operating (you should hear a low level noise without the fan operating at this point). Then turn the knob to all the other settings to see if they seem to be appropriately triggering the cooler (for example, the "high vent" and "high cool" settings should provide higher flow rates than the "low vent" and "low cool" settings). The high vent and low vent settings simply operate the fan without the pump pumping water over the pads (for example, no cooling effect, just drawing in outside air), but do so with a high and low flow, respectively. If the cooler was not on when you began inspecting it, you may have to let it run (Figure 9-77) for a while for the pads to become fully wetted and thereby produce a cooling differential with the outside air. You can inspect all distribution ductwork and evaluate the system's overall ability to cool the building.

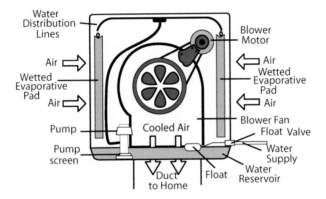

**Figure 9-77. Flow of Air in a "Swamp" Cooler**

The bottom pan of the cooler is waterproof with two to four inches of water in it. A pump lies in an open area of this pan and pumps water up tubes (typically routed to the center on the inside top of the cabinet and then out to the trough on the top of each panel in a cluster of tubes) and into perforated troughs at the top of each removable panel. Placed flat on the inside face of each removable panel and below the perforated trough is a rectangular pad (made of any one of several types of filter-like materials) that becomes wet as the water seeps through the trough perforations and into the pad material. The fan is hidden inside the boxlike housing of the cooler and draws air into the cooler through the pads and down into ductwork that directs it into the uppermost ceiling. Once the pads become saturated with water and the fan is activated, the air passing through them is cooled. A copper tubing supply line replenishes water from the reservoir that evaporates with the inflow controlled by a float-activated valve. The bleed-off kit or overflow drain is typically nothing more than a tube that allows the water to flow out of the reservoir when the level is too high. The outside end of the overflow drain typically has garden tube threads on it for easy attachment of a garden tube overflow drain line.

## UNIT (WINDOW) AIR CONDITIONERS

Unit air conditioners are portable, integral air-conditioning systems without ductwork that are typically set in windows or walls. You can check their overall condition and the seal around each unit and its attachment to the window or wall. Insure that it is adequately supported and look for obstructions to air flow on the exterior and for proper condensate drainage. Make sure all electrical service is properly sized and that each unit is properly grounded. Bent fins on the condenser coils may be "combed." Operate each window unit long enough to determine its cooling capacity; after several minutes the air from the unit should feel quite cool. It should start smoothly and run quietly. Do not test if it is deactivated due to the season.

## HEAT GAIN IN THE SUMMER

Heat gain is the heat created inside the house. Different sources of heat gain include heat transmission, air leakage, internal heat and solar heat. To cool the home, heat gain has to be extracted. Perhaps surprisingly, heat transmission through the envelope (ceilings, walls, etc.) of the home is probably the largest contributor to heat gain. Air leakage that allows hot outdoor air into the house is probably the next largest contributor to heat gain in the summer. Internal gains are represented by body heat from the occupants of the home, as well as waste heat from appliances and fixtures in the home, including water heaters, refrigerators, and lighting (as shown in Figure 9-78).

Internal gains probably represent about the same amount of energy loss as air leakage. Solar gains can represent as much as half of the energy losses when sunlight hits the roofs and walls or comes in through windows. Walls tend to be less important primarily because the angle of the sun on them is not as great as it is on the roof in the summer.

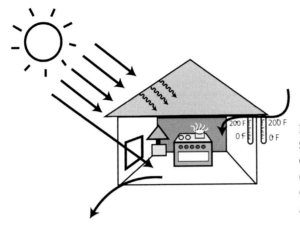

**Figure 9-78. Some Sources of Internal Gains in Homes: Solar (Window/Wall/Roof), Cooking, Lighting, and Air Leakage**

## SIZING AIR-CONDITIONERS

Central air-conditioner sizing is typically done by licensed air-conditioning contractors using Manual J (published by the Air-Conditioning Contractors of America). Sizing of air-conditioners is measured in "tons" of cooling capacity with 1 ton equaling 12,000 Btus per hour. A good air-conditioner sizing rule of thumb is about 1 ton for every 400 square feet in older, less efficient homes and 1 ton for every 700 to 1200 square feet in newer more efficient homes. Cooling load is more difficult to calculate than heating load because of the variability of solar heat, internal gains, and air leakage. Thermal mass of the building has an effect on calculating both the cooling and heating load—the greater the thermal mass the more it reduces daily temperature variation in the building and the energy demands. Air leakage has a double negative effect on cooling load in that it not only brings in warm air from outside but it also takes energy to remove the moisture in the that air.

Table 9-18 gives Energy Star guidelines for sizing window air conditioning units.

## AIR-CONDITIONER EFFICIENCY RATINGS

The efficiency rating for central air conditioners is referred to as the Seasonal Energy Efficiency Ratio or SEER. In the case of unit air conditioners placed in windows or walls, the efficiency rating is called the

**Table 9-18. Sizing of Window Air Conditioning Units**

| Area To Be Cooled, square feet | Capacity Needed, Btu's per hour |
|---|---|
| 100 up to 150 | 5,000 |
| 150 up to 250 | 6,000 |
| 250 up to 300 | 7,000 |
| 300 up to 350 | 8,000 |
| 350 up to 400 | 9,000 |
| 400 up to 450 | 10,000 |
| 450 up to 550 | 12,000 |
| 550 up to 700 | 14,000 |
| 700 up to 1,000 | 18,000 |
| 1,000 up to 1,200 | 21,000 |
| 1,200 up to 1,400 | 23,000 |
| 1,400 up to 1,500 | 24,000 |
| 1,500 up to 2,000 | 30,000 |
| 2,000 up to 2,500 | 34,000 |

Energy Efficiency Ratio or EER. SEERs should be between 10.0 and 17.0, while a good EER should be between 8.5 and 10.0. The SEERs and EERs are usually found on the AC data plate but may be hidden in a model or serial number rather than specifically identified on the plate.

Whether talking about the SEER or the EER, the definition is the same: it is the Btus of heat per hour that are removed divided by the watts of electrical power used by the air conditioning system.

$$\frac{\text{Btus per hour (Btuh) of heat removed in a season}}{\text{Watt-hours of electrical power used during the season}}$$

Since this equation does not compare watts of heat removed to watts of electrical power used, it does not indicate a percent efficiency. Instead, because units are being mixed in the equation, it is best used to compare the efficiency of one air-conditioner to another. Thus an old air-conditioning unit from the 1970s with a SEER of 7 will cost twice as much to operate as a brand new air-conditioner with double the efficiency, a SEER of 14. Air conditioners have improved dramatically over the years because of improvements in motor design for better efficiency, closer spacing of aluminum fins on coils to improve heat transfer, increased surface area on the inside of the copper tubing coils by putting grooves in them, use of two-speed compressors, use of two-speed for variable speed fans, and use of special relays that time delay the operation of the evaporator fans.

Faulty installation can reduce the benefits of a high SEER air-conditioning system as leaky ducts, inappropriate refrigerant charge, and low airflow keep an air-conditioning system from reaching its highest efficiency. Low airflow can be caused by a dirty evaporator coil, a low fan speed, or damaged, blocked fiberglass duct boards, pinched flexible duct, or inadequately-sized ducts, including return air ducts. In fact, as with heating, if there is not an adequate opening under the doors of rooms (obtained by cutting off the bottoms), then you may have to install grilled openings in the doors or ducts through the walls to the hall.

## MOISTURE REMOVAL CAPACITY

An air-conditioning system is also rated based on its ability to remove moisture—the moisture removal capacity, also referred to as the Sensible Heat Factor or SHF. This number falls between 0.5 and 1.0 in an inverse relationship. The higher the SHF, the lower the ability of the air-conditioner to remove moisture. Since a higher humidity in the air can make it feel warmer, it could be important to have a low SHF air-conditioning system in the home if it is in a high humidity area. Fortunately, special thermostats can now be provided that take into account both temperature and humidity in controlling the air-conditioner's operation.

## HEAT PUMP EFFICIENCY RATINGS

The older rating system for heat pumps is the Coefficient of Performance (COP) that tells the relative efficiency of a heat pump compared with electrical resistance heating. For instance, if a heat pump delivers 3.0 kWh of heat for every kilowatt-hour of electricity used, it has a COP of 3.0. Since the air nearest the equator contains more heat, the COP is lower the farther north one goes. The newer heating efficiency rating for heat pumps is called the Heating Seasonal Performance Factor or HSPF. This efficiency rating, much like the SEER rating for air-conditioners, tells the relative Btus per hour the heat pump moves for every one hour

**Table 9-19. Energy Efficiency In Air Conditioners and Heat Pumps**

| Rating Acronym | Name | Purpose | Application |
|---|---|---|---|
| SEER | Seasonal Energy Efficiency Ratio | Efficiency rating based on a season | Central AC's and AC sides of HP's |
| EER | Energy Efficiency Ratio | Efficiency rating based on a season | Room/Unit AC's or AC side of HP's |
| HSPF | Heating Seasonal Performance Factor | Efficiency rating based on a season | Heat Pumps (Heat Side Only) |
| COP | Coefficient of Performance | Old Efficiency rating at 47° F | Heat Pumps (Heat Side Only) |
| SHF | Sensible Heat Factor | Ability to remove moisture (water vapor) from air | Any AC or AC side of an HP |

of electricity used. Typically, the HSPF maximum is about 10. You can evaluate the efficiency of a heat pump when it is in the air-conditioning mode as well. This involves establishing its SEER as a central air conditioner, which typically does not reach much higher than 12. EER is used for rating heat pumps that serve as individual room units.

## WEATHERIZATION RECOMMENDATIONS

In the weatherization program, an energy auditor recommends cooling measures when a family uses 1000 kWh or more per year on air conditioning costs, or the utility usage profile suggests there is a spike in summer electricity usage. Exceptions for this guideline can be made if special circumstances exist such as an elderly person cooling only a single room or the home is small. If these exceptions arise, it may make more sense to repair or replace existing room air conditioner units or install window shading, even though less than 1000 kWh per year are being used for air conditioning, as long as the measures taken are cost-effective (they have a SIR of 1 or greater).

Some of the measures allowed for cooling under the weatherization program include

1)  insulating and air sealing the building shell,
2)  insulating ceiling ductwork in unconditioned spaces,
3)  applying a white or reflective roof coating, Figure 9-79,
4)  applying shading and other applications on South, West, and East facing windows,
5)  applying general waste heat reduction measures,
6)  cleaning the cooling equipment, and
7)  tuning or replacing the cooling equipment.

**Cooler Roof**          **Warmer Roof**

**Figure 9-79. The color of the roof affects the temperature of the roof—darker will be hotter.**

In the private arena, these and other measures can be used depending on the client's desires. Numbers 1, 2, and 3 have been covered elsewhere in this book. The other measures are discussed below.

APPLYING SHADING

Window or solar films and shades can dramatically reduce home heating during the cooling season. Because of the changing path of the sun from summer to winter, window overhangs can be built on the south side that will protect these windows from the sun in the summer. These shade line factors can reduce the effect of the glass load factors, in particular the solar heat gain that occurs with these windows. These will still allow the windows to have exposure to the sun in the winter when the sun travels lower in the sky and when solar gain may actually be a benefit. Before placing solar film or shading on south, west, or east facing windows, evaluate the solar exposure to determine if these measures will be cost-effective. For instance, if a nearby tree already shades a window, the benefit of installing shading or a solar film may be very limited. However, if you are confident that solar gain is overheating a room and causing an increase in air conditioner use, recommend those measures.

Special tools such as the Solar Pathfinder (Figure 9-80) can determine the actual window exposure year-round. While standards will vary by the climate region of the home, careful research can tell what the shading coefficient, the U-value, and the resulting total solar energy rejected will be for a window in your area. The shading coefficient describes how much solar energy will be transmitted through window openings when the film is installed compared to a clear single glass (the highest shading coefficient of 1.0). If a shading coefficient is 0.9, that would indicate that the film allows quite a bit of solar energy through. A more reflective glass would have a shading coefficient of around 0.2 to 0.45. The shading coefficient is always less than 1.0 but higher than the solar heat gain coefficient (SHGC).

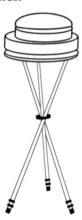

**Figure 9-80.
Solar Pathfinder**

Make sure the film is manufacturer approved to be used on insulated panels or insulated glass. Some clients may be concerned about rooms becoming too dark with the installation of window films, louvers, or shading fabric. Simply providing the client with a sample of materials so they can see what it will look like and how it will be applied may calm these fears. In applying film, first clean the interior surface of an eligible window and then apply and smooth the film. There are some other tools for keeping homes cooler although they may not be commonly used in the weatherization program. White or other highly reflecting shades or blinds can also help to some degree to shade windows and block solar heat. Furthermore, small portable fans used to circulate air throughout the house can create a "wind chill" effect to keep the occupants of the home feeling cooler at a very low cost. These small fans can help a person feel 3 to 7°F cooler. Ceiling fans have a similar effect.

## Planting Trees

Careful placement of trees can provide protection from the sun in the summer and exposure to the sun in the winter if the trees shed their leaves in winter. Plant trees that are tall enough to block the higher angle of the sun on the south side of homes, but short enough to block the lower angle of the sun on the east and west faces of homes. This can reduce the solar gain from sunlight entering the home through windows and absorption of the heat from solar rays on the walls. If trees that shed leaves are chosen, then you can also obtain some cost free heating from the sun during the winter when the trees have no leaves, Figure 9-81.

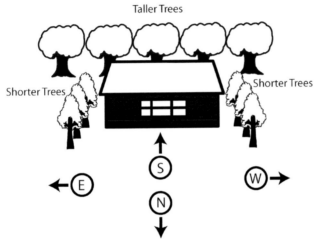

**Figure 9-81. Trees that shed their leaves in the winter should placed on 3 sides of a home with taller trees on the south side.**

In addition, evaporation of water from these trees and the shading of the soil beneath them significantly reduces the air temperature around the home anywhere from very little up to about 20°F cooler—there is always some cooling effect. Since this is more of a long-term benefit because it takes some time for the trees to grow, these are not common recommendations in weatherization programs even though they should be recommended to homeowners in the private arena.

## Waste Heat Reduction

Generally, reducing baseload can reduce waste heat. Methods to reduce baseload include insulating the water heater and hot water

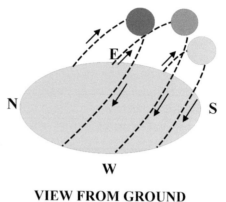

**Figure 9-82. Changes in Sun Trajectory Depending on Season**
The trajectory for winter is on the far right, the one for the equinoxes is in the middle and the one for summer is on the left.

**VIEW FROM GROUND**

pipes, installing low flow showerheads and faucet aerators, replacing old inefficient refrigerators, and converting to fluorescent bulbs throughout the house, etc. (see Baseload Section). Educating your client on the following items can also help save electricity: turn off lights when not being used, use the dishwasher or dryer only when they are full, clean dust from around the refrigerator coil, and avoid standing in front of the refrigerator with it open.

## COOLING SYSTEM CLEANING

### Cleaning Guidelines On Central Air Conditioner Systems

If airflow is hindered through the evaporator or condenser coils of a central air conditioner system, there will not be efficient heat transfer when the fan is blowing air across the coils. Therefore, make sure the filters are clean. If the evaporator coil is dirty, it should be specified in the report and the retrofitter should conduct the cleaning.

To clean an indoor evaporator coil, the following steps should be taken. Make sure the main switch or breaker to the air handler is shut off, open the door to the blower compartment and, using a flashlight, look into the blades of the blower. Run a brush along one of the fan blades to see how much dust is present. If the blower fan is dirty, the best way to clean it is to remove it. However, only a qualified technician can remove the blower. If the motor can be removed from the blower fan, the blower can be cleaned with hot water or a household cleanser.

If the blower fan is dirty, then the indoor evaporator coil is also probably dirty. If possible, inspect the evaporator coil visually to evaluate it for dirt. If there is no access hatch to do this, you may have to create one. Use a mirror to view the entire underside areas of the evaporator coil. If the coil is dirty, clean it very carefully using water and indoor coil cleaner and a brush. Remember the aluminum fins on a coil are very delicate and can easily be damaged to the point where they do not allow as much air through the coil. If you find some damaged fins or damage some fins yourself, use a special fin comb to straighten bent fins and thus prevent them from reducing airflow. Clean the drain pan and the drain line that sit at the bottom of the evaporator coil. Attempt to clean clogged weep holes with a properly sized wire, and pour a tablespoon of bleach into the weep holes to help prevent fungus growth.

Protect the indoor evaporator coils by making sure there is a good, well fitting, clean filter. This helps keep the blower as well as the evaporator coils clean. Obviously, it is easier to clean or replace filters than it is to clean or replace a blower or an evaporator coil. If the client is available during the audit, this is a good chance to educate them on what it takes to make sure the coils, etc. remain clean and free from obstruction. If allowed, consider a clear plastic condensate P trap, which typically comes with a special bottle brush to clear the trap, to allow the homeowner to more easily clear plugs in the condensate drain.

If the condenser coil is dirty, it should be specified in the report and the retrofitter should conduct the cleaning. When cleaning a central air-conditioner condenser, cut vines, weeds, or grass that obstruct airflow (including shrubs that have been placed too close to the unit), clean the condenser coils with an outdoor coil cleaner, clean the fins with a soft brush, level the unit on the concrete pad where it sits, straighten any bent fins very carefully with a fin comb (Figure 9-83), and educate the client to put a cover over the condenser unit in winter to help prevent leaves from accumulating or damage to occur to the unit from the elements.

Never use a hose to clean the fins as dirt can easily turn to mud and the mud could splash into the unit and cake on the fins. If there is no insulation on the cooling line to the outdoor unit or if there is significant damage to the insulation, make sure to specify that new insulation be placed on the line if you are an auditor, or place new insulation on the line if you are a retrofitter. The frequency for cleaning outdoor condenser coils depends primarily on the levels of pollen and dust in the area. If the home is located in an area with high levels

Figure 9-83. A fin comb can help straighten coil fins to open for airflow if they have been damaged.

of dust and pollen, the coils should be cleaned at least once a year. If the area is low in dust and pollen, then cleaning the outdoor coils once every three years should be sufficient. Changing the coolant charge or replacing an air conditioner unit needs to be done by a qualified refrigeration technician who has an EPA section 608 certification. If you are doing this work, see Appendix L.

*Cleaning Room Air Conditioners*
It has been found that cleaning room air conditioners can improve their performance by about 5% on the older models. This cleaning is typically a very low cost measure. Parts of a room air conditioner include the housing that must be removed to clean the outdoor coil, the outdoor condenser coil that helps dissipate the heat to the outdoor air, a compressor that compresses the refrigerant after it has passed through the evaporator, an indoor evaporator coil that allows the refrigerant to collect heat from the indoor air as it evaporates, controls that are used to set the temperature or time the length of time that the air-conditioning unit is on, a filter to help remove dirt and dust from the indoor air before it passes through and potentially plugs the indoor evaporator coils, a removable indoor grille that typically needs to be removed to clean or replace the filter, and a condensate drain that collects the moisture that condenses on the evaporator coil and directs it to run off away from the unit in the building to the outside.

*Field*
There are several steps to follow in cleaning a room air conditioner. First, unplug and remove the air conditioner from the wall or window. Remove the grille on the interior side of the unit to access and remove the filter. Follow the manufacturer's instructions to clean the filter. After taking the unit to an outdoor area that drains well, such as a concrete

pad, cover the electrical components, the fan motor, and the compressor with plastic bags that can be held in place with rubber bands, etc. Dampen the coils with a water spray and then, with an old hairbrush, rake as much of the dirt off the coils as possible. Spray a special indoor coil cleaner into the indoor coil and a special outdoor coil cleaner into the outdoor coil. Let these cleaners sit for a few minutes and then rinse the dirt out of the coils with a very gentle spray from a water hose. Keep repeating this process until the water draining from the coils is clean. Then use a fin comb to straighten bent fins so that they do not reduce airflow. Now reverse the dismantling process and put the unit back together again and into the wall or window.

If moisture drips from the bottom of the unit when it is operating, that could indicate a clogged condensate drain. If the clog is not cleared, it can lead to moisture damage. Once a window air conditioner has been reinstalled, make sure there is a good air sealing between it and the window and trim in the case of a window unit, and between it and the wall in the case of a wall unit.

For the weatherization program, be sure to include the cost of incidental repairs associated with a room air conditioner, such as damaged window frames, sills, or trim, in the SIR calculations. If applicable, be sure to follow the weatherization program standards when you specify incidental repairs related to a window air conditioner. Since a home is only weatherized once, it is very important to leave the home as efficient as possible and still follow the program guidelines.

## Replacing Room Air Conditioning Units

If it is cost-effective, replace an existing room air conditioning unit with a high-efficiency properly sized unit. See the Energy Star sizing table given previously in this chapter. In addition, when replacing a whole-home unit, do not oversize the system beyond what Manual J recommends. If installing the unit in a corner, choose a unit that can direct the airflow out into the open area of the room. Use some sizing adjustments to correct for the Energy Star numbers. If the area outside around the unit is heavily shaded, reduce its capacity by 10%. If it looks to be very sunny on the outside around the unit, increase its capacity by 10%. In addition, add an additional 600 Btus for each occupant over two that normally occupies the room serviced by the unit. If the room air conditioner is going to be used to cool the kitchen, add 4000 Btus.

**Table 9-20. Correction Factors for Room Air Conditioners**

| Room AC Situation | Change in Sizing |
|---|---|
| Heavily shaded area | Reduce capacity by 10% |
| Very sunny area | Increase capacity by 10% |
| Each occupant over 2 | Add 600 Btu's |
| AC unit used to cool kitchen | Add 4000 Btu's |

RENEWABLE ENERGY

Renewable energy is naturally occurring and theoretically inexhaustible. Fossil fuels do not represent renewable energy because they are limited in their availability and they use energy from a different timeframe. This creates an unbalanced carbon dioxide burden in our day that may contribute to global warming. There are a number of examples where renewable energy methods can be applied in homes, although many of these are seldom used in weatherization retrofits—typically because of their long payback.

**Wind Power**

Wind powered energy created by generators with propellers (Figure 9-84) attached to them can work well, but they must be installed in a location that has strong enough winds for extensive enough periods during a day to generate electricity.

**Solar**

Solar photovoltaic (PV) panels (see Figure 9-85) that create direct current (DC) electricity (converted to alternating current (AC) electricity for household use by an inverter) represent another example of renewable energy. However, there must be enough sunny days per year, no shading where they are installed, and the panels must be tilted to the proper angle to the sun during the season of highest demand. PV panels only

**Figure 9-84. Wind Power Energy**

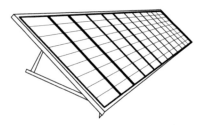

**Figure 9-85. Solar PV Panel**

work when the sun is out unless there is a "stand-alone" system that stores the energy in batteries. However, if the PV is oversized for the needs during the day and the panels are properly "tied" into the utility electrical grid, you can obtain "zero net metering" if they can give more electricity back to the grid than they take from it.

Solar PV and solar thermal techniques represent "active" solar technologies. If they are used for heating the conditioned areas of a home they can be referred to as active space heating. "Passive" or "climatic" techniques do not involve the use of fans, pumps, or electrical devices. If they are used for heating the conditioned areas of a home they can be referred to as passive space heating. An example of a passive technique is designing a home with large windows on the south side to take advantage of the low-lying sun in the winter to help heat the home. Likewise a solar thermal collector system, as shown in Figures 9-86 and 9-87, typically heats water or air inside tubes or ducts, respectively, that are painted black (for better solar gain). This hot air or water is used for heating the home or the domestic hot water. These solar thermal systems can typically only be used when the sun is out, unless there

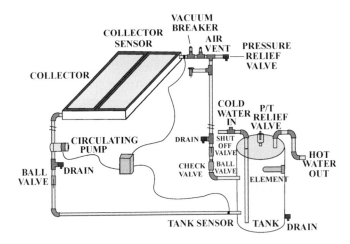

**Figure 9-86. Solar Thermal System**

is a large enough storage tank for water (or gravel, etc. for solar air), then the water can be heated throughout the night. Solar air systems are widely recognized as some of the most cost-effective systems to install.

**Geothermal Sourcing**

Another application of renewable energy involves geothermal sourcing for heating and cooling through a geothermal or ground source heat pump. Unlike the traditional air source heat pump or air conditioner system whose outside refrigerant is cooled by outside air, pipes are run deeply enough into the ground so the antifreeze-laden water circulating through them will ultimately match the temperature of the soil around them: 55°F. Since 55°F is warmer than the air often is outside in a cold weather area in the winter, this heat can be used to warm the house and its hot water with little supplemental energy. Likewise, since 55°F is cooler than the air often is outside in the summer, the water can be used to cool the house. This is an expensive proposition since there is a great deal of cost to even just excavate to install piping far enough into the ground that remains at 55°F year-round. Other options for geothermal include drawing water from deep water in a lake or a water well. A heat pump can also be set up to use a combination of sources for heating or cooling the refrigerant, such as one that uses the cooler lake water in the summer to cool while using deep soil to get heat energy during the winter. A heat pump that uses different sources for heat and cold could be referred to as a "combo" as could any other heating system that has the option of using different types of heating sources in the same house.

**Reducing Heating and Cooling Load First**

An often overlooked aspect to adding renewable energy systems such as solar, wind power, etc., to a home is reducing the heating and cooling load on the home prior to calculating how much of a renewable energy application to put on the home. Thus, it is critical to retrofit the home for energy purposes prior to deciding how much of a renewable energy application to install. Otherwise, you may over-install the amount you need and pay much more for the application than was necessary for the home. This can become quite a juggling act when you consider that many are installing PV, etc. based on the expected tax credits, rebates, and other forms of subsidies that have deadlines or "fading" reimbursements for installation. Some have found it best to try

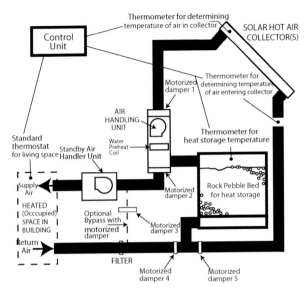

**Figure 9-87. Solar air systems are by far the least expensive systems to put in and tend to be the most cost-effective system with the faster payback.**

to estimate their savings from retrofit and base the amount of PV, etc. to install on those estimates rather than forego the sometimes highly incentivized, but fading subsidies. Furthermore, for installations that involve potentially having to change the structural capacity of parts of an existing home (roof structures supporting solar panels, parts of the home supporting holding tanks, etc.) the installation of renewable energy applications becomes even more complicated.

# Chapter 10

# Baseload and How to Improve It

## MODIFICATIONS TO IMPROVE BASELOAD MEASURES

Baseload energy is energy used in a home that is not used for conditioning the air and therefore is not a seasonal load, (it does not peak in the winter or summer). As seen in Table 10-1, space cooling and heating make up only 39% of the energy use in homes that are eligible for the weatherization program. The remaining 61% of the total energy use is considered nonseasonal baseload that can be reduced by either replacing or modifying existing appliances to help increase efficiency.

Even homes that use electric heat for space heating still have baseload energy use that is about 50% of all the electricity used in the home. In addition, there are many houses heated with oil or gas whose annual electric bills are higher than their fossil fuel bills. One component of electrical bills you cannot help your client with is "peak demand." This is a "demand charge" or service charge on the electric utility bill that is based on the largest amount of electricity used during a set period of time, (15, 30 or 60 minutes, depending on the utility's rule) during the entire month of the billing. Thus, the utility attaches a supplemental charge to the bill based on the highest 15 minutes use of electricity over the entire billing month.

**Table 10-1. Types of Energy Costs**

|  | % Of Total Energy Costs by End Use in Weatherization Households |
|---|---|
| Space Heating | 30% |
| Space Cooling | 9% |
| Appliances | 27% |
| Water Heating | 15% |
| Refrigerator | 9% |
| Lighting | 6% |
| Cooking | 4% |

Typically, the five biggest users of baseload energy in a home are refrigerators, clothes dryers, hot water heaters, lighting, and other various electronics, particularly flat screen televisions.

Some of the measures under the weatherization program and in the private arena that are typically used for reducing baseload energy use include retrofitting the lighting with compact fluorescent bulbs, replacing or repairing the refrigerator, modifying the water heater, and installing low flow fixtures such as low flow faucet aerators and low flow showerheads. Some auditors in the private arena also suggest that any appliance should be considered for replacement with an Energy Star rated appliance if it is over 15 years old.

Energy Star products are high efficiency general appliances, electronics, and heating/cooling appliances. To be Energy Star labeled (Figure 10-1), a product must meet the minimum energy star criteria for that particular product as set by the Energy Star program. In addition, the DOE-based EnergyGuide label on an appliance allows you to compare both the estimated yearly operating costs and an estimate of how much electricity an appliance will use in a year.

These numbers are determined from tests established by the DOE. However, a number of appliances are not eligible for the EnergyGuide label, including clothes dryers, ranges, ovens, humidifiers, and dehumidifiers.

FINDING
THE BASELOAD

Baseload energy use is found by analyzing an entire year of utility bills. This can help separate the baseload from the seasonal loads tied to cooling and heating. To determine base-

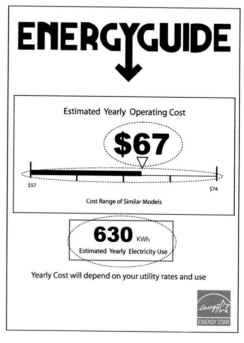

**Figure 10-1. EnergyGuide Label.**

load, average the three lowest months of energy use in the home, thus assuring that little or no heating or cooling occurred during that period. Take this average and multiply it by 12 to estimate the annual baseload usage. Below are three examples of finding the baseload.

### Example 1. Cold Climate With All Electric Heat

Table 10-2 shows the actual meter reading in kilowatt-hours from roughly one month to the next. For purposes of simplifying this example, assume that the meter was read at the first of each month, but use the kilowatt-hours for the actual time frame on the bills from one month to the next (see listing of monthly readings below).

First, find the three lowest months usages, which are 442, 522, and 538 kWh. Average these three numbers for a total of 1,502/3 or about 501 kWh per month. In this case, the 501 kWh baseload per month represents 31% of this home's total energy bill over a year. (Multiply it by the 12 months in the year to compare it with the yearly total).

$$\frac{501 \times 12}{19631} = .31 \text{ or } 31\%$$

That means that the remaining 69% is suspected to be cooling and heating loads. The fact that baseload represents a relatively small percent of the total energy bill suggests that this house probably uses electric heating in a cold climate and uses very little electric-based cooling in the summer since the lowest energy use months are in the middle of summer.

### Example 2. Moderate Climate in Alternative Fuel Heated Home

Once again, assume for purposes of simplification that the meter was read on the first day of the month but nearest

**Table 10-2. Example 1 Meter Readings**

| Date | # Of Days | Reading | kWh |
|---|---|---|---|
| 5/7/99 May | 29 | 06248 | |
| 4/8/99 April | 30 | 05217 | 1031 |
| 3/9/99 March | 29 | 03249 | 1968 |
| 2/8/99 Feb. | 31 | 00106 | 3143 |
| 1/8/99 Jan. | 32 | 96750 | 3356 |
| 12/7/98 Dec. | 33 | 93102 | 3648 |
| 11/4/98 Nov. | 29 | 90896 | 2206 |
| 10/6/98 Oct. | 32 | 86569 | 1327 |
| 9/4/98 Sept. | 29 | 88993 | 576 |
| 8/6/98 Aug. | 29 | 88551 | 442 |
| 7/8/98 July | 30 | 87667 | 874 |
| 6/8/98 June | 32 | 87155 | 522 |
| 5/7/98 May | 30 | 86617 | 538 |
| Total | | | 19631 |

when the billing periods end. In this example, it appears that there is much less seasonal fluctuation in the electric bill, see Table 10-3. While it is difficult to say whether this home is in a cooling or heating dominated climate, it is possible that it is either in a very moderate climate or that the home uses an alternative fuel for heating the house. It apparently does not use a natural gas or oil furnace with an air handler or the electric bill would be noticeable in the winter. It could be

**Table 10-3. Example 2 Meter Readings**

| Date | # of Days | Reading | kWh |
|------|-----------|---------|-----|
| 1/29/99 Jan. | 30 | 65389 | |
| 12/30/98 Dec. | 35 | 65045 | 344 |
| 11/25/98 Nov. | 29 | 64627 | 418 |
| 10/27/98 Oct. | 29 | 64280 | 347 |
| 9/28/98 Sept. | 32 | 63950 | 330 |
| 8/27/98 Aug. | 28 | 63528 | 422 |
| 7/30/98 July | 30 | 63127 | 401 |
| 6/30/98 June | 32 | 62711 | 416 |
| 5/29/98 May | 30 | 62256 | 455 |
| 4/29/98 April | 30 | 61866 | 390 |
| 3/30/98 March | 28 | 61554 | 312 |
| 3/2/98 Feb. | 32 | 61292 | 262 |
| 1/29/98 Jan. | 30 | 61037 | 255 |
| Total | | | 4352 |

heated by a wood or coal burning stove or some other non-electric appliance.

It appears that February, March, and April are when the least electricity is used in this home. Total these three months of 255, 262, and 312 kWh to obtain 829 kWh and divide that by three months to obtain the average baseload per month 829/3 = 276. To estimate the total annual baseload usage, multiply this average baseload per month by 12 months to get 276 x 12 = 3316 kWh per year of baseload energy use. To find what percent this baseload energy represents of the total electrical energy, add the total kilowatt-hours used over the twelve months in a year, which in this case is 4352. Thus, the 3316 kWh per year of baseload energy use divided by the 4352 total kilowatt-hours used for the year gives 3316/4352 = .76 or 76%. Thus, 24% of the annual energy usage in this home is used for cooling and heating loads.

### Example 3. Cold Climate Heating with Air Conditioner and Non-electric Heat

Look at yet a third bill analysis from another home using Table 10-4.

In this example, it appears there is much more seasonal fluctuation in the electric bill than either of the two prior examples. This home

**Table 10-4. Example 3 Meter Readings**

| Date | # of Days | Reading | kWh |
|------|-----------|---------|-----|
| 2/24/00 Feb. | 29 | 10036 | 1904 |
| 1/26/00 Jan. | 30 | 08132 | 2097 |
| 12/27/99 Dec. | 40 | 06035 | 2341 |
| 11/17/99 Nov. | 30 | 03694 | 1382 |
| 10/18/99 Oct. | 31 | 02312 | 695 |
| 9/17/99 Sept. | 30 | 01617 | 1334 |
| 8/18/99 Aug. | 28 | 00283 | 1175 |
| 7/21/99 July | 34 | 99108 | 1314 |
| 6/17/99 June | 29 | 97794 | 888 |
| 5/19/99 May | 28 | 96906 | 752 |
| 4/21/99 April | 30 | 96154 | 1182 |
| 3/22/99 March | 29 | 94972 | 1826 |
| 2/24/99 Feb. | 26 | 93146 | |
| Total | | | 16870 |

is in a heating dominated climate because the highest use months are in December, January, February, and March. It appears May, June and October are the months where the least electricity is used. Total these three months of 695, 752, and 888 kWh to obtain 2335 kWh and divide by three months to obtain the average baseload per month which is 2335/3 = 778. To estimate the total annual baseload usage, multiply this average baseload per month times 12 months to get 778 x 12 = 9340 kWh per year of baseload energy use. The total kilowatt-hours used over a year's time in this home are 16,870. Thus, the 9340 kWh per year of baseload energy use divided by the 16,870 total kilowatt hours used for the year gives 9340/16,870 = 0.55 or 55%. Therefore, 55% of the electricity over a year is used for baseload. Also, heating and cooling loads represent 45% of the annual energy use. Estimate the total summer air conditioning load by subtracting the average baseload for these months (778 kWh x 3 months = 2334 kWh) from the total for the same three summer months (1314 + 1175 + 1334 = 3823 kWh) to get 3823 – 2334 = 1489 kWh. This represents 1489 kWh out of 16870 kWh for the year, 0.088 (8.8%) or about 9% of the annual electric load could be attributed to the summer air conditioning load.

## ENERGY USAGE LEVEL OF APPLIANCES

Table 10-4 showed that appliances and lighting account for over 50% of energy use. Thus, improving their efficiency can reduce energy use. Table 10-5 shows some generally accepted appliance energy use on an annual basis.

Table 10-5. Appliance Energy Use, per year

| Appliance | Low kWh | Mid kWh | High kWh |
|---|---|---|---|
| Refrigerator | 400 | 900 | 1,500 |
| Clothes Dryer | 500 | 900 | 1,200 |
| Clothes Washer | 500 | 900 | 1,200 |
| Indoor Lighting | 350 | 900 | 1,200 |
| Air Conditioning | 300 | 500 | 750 |
| Cooking | 300 | 500 | 750 |
| Television | 200 | 350 | 600 |
| Outdoor Lighting | 50 | 250 | 350 |
| Space Heating Motors | 150 | 250 | 350 |
| Stereo | 100 | 200 | 300 |
| Hair Dryer | 25 | 100 | 200 |
| Vacuum Cleaner | 25 | 50 | 75 |
| Miscellaneous | 100 | 200 | 400 |

Many factors can affect appliance energy efficiency including the following:

1)   the age of the appliance (appliances lose their efficiency every year they age),

2)   the standards in place at the time of manufacture (the older standards did not require as energy efficient appliances as today),

3)   the size of the appliance,

4)   special features (water service and "ice through the door" increase energy use while moisture sensors on clothes dryers reduce energy use), and

5)   user habits, like standing in front of an open refrigerator, running air conditioners with the windows open, or leaving lights on in unoccupied rooms.

As a rule of thumb, there can be a good potential for energy savings if a household of four individuals uses more than 1000 kW per month if they have an electric water heater, or more than 600 kW per month if they do not have an electric water heater. Obviously, baseload is affected by the number of occupants in the home, the habits of those occupants, and the efficiency and size of the appliances. There will be many cases where appliances will not qualify for replacement in the weatherization program and yet, the clients may have great potential to reduce their energy use by changing their habits.

## REFRIGERATORS

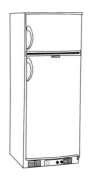

**Figure 10-2. Newer Style Refrigerator**

Generally, refrigerators can be a major source of baseload electrical usage. The older the refrigerator, such as in Figure 10-3, the more likely it is an "energy hog." Whether to replace a refrigerator should be on the agenda of any auditor in the weatherization program. When a refrigerator is replaced, it must be removed and decommissioned to make sure it will not be plugged in again. To decommission a refrigerator, remove and properly store the refrigerants, and then if possible, give the refrigerator to a metal recycler. Thus, the expenses for decommissioning must be included in the cost evaluation to calculate SIR in the weatherization program. The SIR has to be ≥ 1. If it is not, the weatherization program cannot fund that retrofit measure. To calculate the energy usage of an existing refrigerator, meter the refrigerator with an electrical metering device or find data from a recognized database such as the Home Energy, the Weatherization or WAP Refrigerator Guide, or The Association of Home Appliance Manufacturers (AHAM). Whether checking the energy usage on a refrigerator with a meter or a database, be sure to correct for temperature and other factors to establish a final number. Even if all the refrigerator units are in a database, it is still required under the weatherization program to meter at least 10% of the refrigerators that are evaluated in the program overall or 1 in 10 refrigerators replaced.

The older a refrigerator is, the more likely that it will be replaced. If the gaskets are worn on the door, it will make the refrigerator run more often and use more energy. Check to see how efficient the insulation is by pushing on the door when it is closed. If the front face of the door can be easily pushed in, it is very likely to be fiberglass and therefore less efficient. However, if pushing on the front face of the door is not easy and it

**Figure 10-3.
Older Style Refrigerator**

seems more firm, is very likely to be a solid foam and therefore more efficient. Another strong indicator for efficiency is the color of the refrigerator. If it is an avocado green, harvest gold, or a Coppertone brown, then it is very likely that replacement would be cost-effective. Those colors were popular before the current efficiency guidelines were established. Round-edged refrigerators are even older.

**Using a Refrigerator Database Rating**

If you tend to use the database ratings, you may be interested in a special refrigerator study conducted from June 2003 to May 2004 wherein electricity usage and room temperature were monitored on refrigerators for two weeks. For the northeastern United States, it was found that adding 11% to a database rating was a fairly reliable predictor of actual energy use if the unit was located in the living space and not in an unconditioned basement, garage, etc. It was recommended that 5% be added to the database rating for every occupant in the home for primary, not secondary, refrigerators. If the refrigerator's anti-sweat switch was on, it was recommended that 20% be added to the database energy use rating. Furthermore, if the refrigerator had a through-the-door ice system, then it was recommended to add 15% to the energy use rating. If there are visible gaps in the refrigerator's door seal, it was recommended to add 15% to the rated use. If the refrigerator was bought used, it was recommended to add 20% to the rated use. Because room temperature can affect energy efficiency, it was recommended to add 5% to the rated use if the winter room temperature was kept in the low 70s, and subtract 5% from the rated use if the winter room temperature was kept in the low to mid 60s. If none of the conditions mentioned above existed, it was recommended that the actual energy use on the refrigerator be estimated at 85% of that refrigerator's rating on the rating table. It should be mentioned that these correction factors are not considered official in the weatherization program, but may prove useful. This information comes from a study by Michael Blaznik entitled "Measurement and Verification of Residential Refrigerator Energy Use" published in 2004.

If the refrigerator cannot be metered for any of the following reasons, then use a database to estimate the energy use: if the floor would be damaged by moving it to access the plug, if it is difficult to access the refrigerator because of a stove, furniture, cabinets, etc., if the refrigerator cord cannot be plugged into the metering device or if the cord appears unsafe, then use a database to estimate the energy use. If you cannot find the brand, model and serial numbers of the refrigerator in the database, use the chart from the appendix of the "Refrigerator Info Toolkit" to convert the model and serial numbers to a date of manufacture. The "Refrigerator Info Toolkit" is available through the weatherization program. As a general rule, if the date of manufacture is 2003 or later, the required SIR to replace the refrigerator cannot be reached.

Because efficiency declines over time, remember to adjust for the age of the refrigerator when using the database. Refrigerator efficiency can be expected to degrade 1% to 2% per year. In the weatherization program, the approved computerized audit tools will automatically account for this. Only use this correction when relying on the database for the estimated energy use, as in the private arena.

### Metering Refrigerators

While there is some debate on the minimum time a refrigerator must be metered to get an accurate estimation of its annual energy use, it is generally thought that conducting a three-hour test will give accuracy 90 times out of 100. The weatherization program recommends metering for at least two hours. Adjust for temperature variation as well and disable the defrost cycle while the test is being conducted. Use a correction factor that makes up for turning off the defrost function.

**Wattmeter**

**Figure 10-4. A wattmeter evaluates the energy use of electrical appliances, including refrigerators**

If the defrost cycle is left on during metering, it will significantly affect the accuracy of the results. A demand level of about 380 W or more is usually an indicator that the defrost heaters are on. It is recommended to turn off the defrost cycle and add 8% to the results to take into account the additional annual energy used by the defrost cycle. To avoid the defrost cycle turning on during the test, advance the defrost timer through the defrost cycle. This way metering will begin at the start of a new cycle with no risk that the defrost will kick in during the test. The defrost timer is most commonly found behind the kick plate below the door. It is also sometimes found behind the lighting panel, inside the freezer on some bottom freezer models, or on the rear side of the refrigerator in a detachable mounting box. Use a broad tipped screwdriver to advance the defrost timers. Be sure to check the frequency of the defrost cycle on the timer. Sometimes the defrost cycle occurs every 180° on the timer, while on other models it may be only every 360°.

After a refrigerator has been metered, the temperature variations can be accounted for by applying a correction factor. For instance, if the metering is done on an uncommonly hot day, the refrigerator will be calculated to use much more energy per year than it actually does. The difference in temperature between the ambient air outside the refrigerator and the temperature inside the refrigerator will affect efficiency by 2.25% to 2.5% per degree Fahrenheit. If the room the refrigerator is in is not at its normal temperature, multiply the difference in temperature between the normal kitchen temperature and the temperature the day the refrigerator was metered by 2.5% as a correction factor. A kitchen that happens to be cooler than normal while metering also requires a correction factor. As an example, consider a kitchen that normally is at 72°F but happens to be at 66°F during metering on a particularly cold day.

$$1 + (2.5\% \times (72°F - 66°F)) = 1 + (0.025 \times 6) = 1.15$$

In this case, multiply the estimation of the annual energy usage based on the metering by 1.15 to obtain a more accurate idea of the true energy consumption as a result of the house normally being warmer than when metered.

To calculate the annual usage from a meter reading, use the following formula:

$$\frac{\dfrac{\text{metered usage (kWh)}}{\text{metering duration (min.)}} \times\ 60\text{ min./hr} \times 8760\text{ hrs/yr.}}{0.882}$$

With this formula, divide the metered usage by the minutes of test time. Multiply by 60 to convert the results to hourly usage and then multiply by 8760 hours per year to obtain the total annual energy usage. Divide this final number by 0.882 to take into account that the metered energy use will be lower because it will not be opened and closed during the test, as it would be during normal usage.

As an example, say the meter showed 0.18 kWh had been used over a three-hour period of metering. Use the following numbers and formula to obtain the annual energy use number for the refrigerator:

$$\frac{\dfrac{0.18 \text{ kWh}}{180 \text{ min.}} \quad \times \quad 60 \text{ min./hr} \times 8760 \text{ hrs/yr.}}{0.882}$$

$$\frac{0.001 \times 60 \times 8760}{0.882} = 525.6/0.882 = 596 \text{ kWh}$$

Thus, the refrigerator would be expected to use about 600 kWh of energy per year.

An indoor/outdoor thermometer that displays relative humidity readout can help simplify refrigerator data gathering. If the remote sensor is placed in the freezer and the unit is placed in the fresh food compartment while monitoring electrical use with a wattmeter, you can obtain the temperature data useful for being able to evaluate for refrigerator replacement.

Even if refrigerator replacement cannot be justified, noticeable energy savings can be achieved by changing the settings on the refrigerator. For example, in a hot climate with an ambient temperature of 90°F in the kitchen, changing the setting on the fresh food compartment from 36.4°F to 43.9°F and the freezer setting from -9.1°F to 12.4°F could save the client over 200 kW per year (Larry Kinney, E Source, Boulder, CO, 2001). Cleaning the refrigerator coil can also reduce the annual energy use. Before cleaning the coil, be sure to turn the refrigerator off by unplugging it or by turning off the circuit breaker in the case of built-in refrigerators. Clean the coils with a long flat vacuum cleaner attachment and/or a long narrow brush, much like a bottlebrush design. Cleaning a refrigerator once a year can generate a 3% savings in energy use. If a refrigerator that has never been cleaned is cleaned, 20% or more can be saved in every use. This is about 200 kWh, or roughly the equivalent of $22 per year in savings.

### Replacement of Refrigerators

If refrigerators are replaced, make sure the replacement complies with the federal regulations of UL 250. You should be able to find a sticker or plate on the refrigerator that shows this is the case.

Dispose of all old refrigerator units in compliance with the Clean Air Act of 1990, section 608, including its amendment 40 CFR 82, dated 5/14/93. This means that the old unit cannot remain at the home nor can it be moved to another home. Inefficient refrigerators are replaced to get them off the electrical grid and you do not want to be responsible for the creation of a black market in used refrigerators. In addition, in the weatherization program the cost of refrigerator disposal needs to be reflected in the SIR calculations.

DRYER VENTS

Dryers need to be vented all the way to the outside of the home. It is common to find them vented into a crawlspace or attic, and in some cases, even into the living space. Some of the considerations for dryer vents, Figure 10-5, include keeping the vent as short and straight as possible while still getting the vent to the outdoors. Damp warm air must be vented to the outside so it cannot damage the home or expose the occupants to gas or propane combustion gases if these fuels are used. Prevent leaves or animals from getting back up into the dryer hose by installing a pressure dependent valve that opens only when the dryer is being used. Make sure there is enough airflow through the vent to help avoid the build up of lint in the ducts.

**Figure 10-5. Dryer Vent**

For the least resistance to airflow in vent pipe materials use a smooth metal pipe that has an opening of at least 16 in$^2$ or 5-inch diameter. Much more lint, dust, and moisture are moved through dryer vents than any typical cooling or heating ducts, so it is important that there are no restrictions. The smoother the interior surface of the duct, the least resistance and the less risk of lint buildup. Flexible metal duct can be used, but it is not as good as smooth metal duct. In some areas, you may be allowed to use a flexible foil duct, which is even less sturdy. Plastic flex ducts should never be used as a vent pipe inside of walls. Plastic is considered inappropriate not only because of possible temperature issues but also because the pipe can generate static electricity that attaches the lint to the pipe walls and causes lint buildup. Flexible plastic can be, and is typically used as a connector between the dryer itself and the short length metal connector at the wall.

If dryer ducts are not properly installed, the result could be fires, poor indoor air quality, moisture problems, poorly performing dryers, and drafts. The Consumer Product Safety Commission reports that over 13,000 home fires per year are a result of dryer vents that have a buildup of debris or lint. Another concern is the use of screws to fasten ductwork sections together. If screws are placed from the outside into the ductwork, they can leave a protrusion that will catch and collect lint and ultimately block the vent. This not only reduces the effectiveness of the dryer, it is also a major fire hazard.

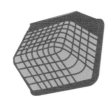

**Figure 10-6.
Dryer vent screens
should not be used**

The proper way to connect vent duct is with aluminum foil tape, not duct tape. Duct tape cannot tolerate the higher temperatures and always fails.

The terminations where the dryer air exhausts to the outside can be at a sidewall or out the roof. With roof terminations, be sure to have properly placed, caulked, and sealed the roof vent and flanges into the shingles (see Roof Section).

WATER HEATING (SEE FIGURE 10-7)

Inspect the condition of the water heater by evaluating it for damage, rust, corrosion and leaks. If any of these problems are found, require replacement based on health and safety. Also, test it as a combustion appliance (see combustion appliance testing sections).

Adjust the temperature setting to 120°F. Many water heaters have been set to as high as 140°F. Nowadays, dishwashers have small electric booster heaters to heat the 120°F water to the 140°F water required. Thus, there is no longer any need for water heaters to be set at 140°F. Having the water heater set to the lower temperature helps prevent scalding and saves energy.

To determine the proper temperature setting for the water heater, mark the original dial position so there is a reference point. Even if the water heater has temperature settings on the dial, these can be inaccurate.

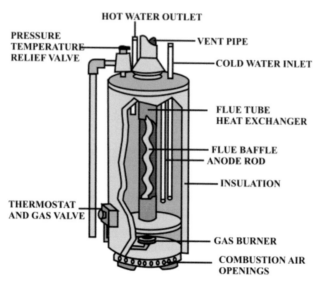

**Figure 10-7. Domestic Hot Water System**

Next, check the actual temperature of the water coming from the hot side of the faucet that is closest to the water heater. If the temperature is greater than 120°F and it is a gas water heater, adjust the temperature dial down. Let the water heater water cool down for a while and then recheck it. If it is an electric water heater, first shut off the electrical supply to the water heater, remove the two element covers (confirm that it is off by using an electrical point tester), and then, using the appropriate tool (screwdriver or nut driver), adjust the thermostat for each element. Whether adjusting a gas or electric water heater, it sometimes takes several attempts to get the proper temperature setting. Do not go below 120°F. In terms of energy use, if the hot water heater is using 4500 kWh per year, 343 kWh per year can be saved (at $.11 per kWh this represents an almost $40 savings) by turning the temperature setting down from 140°F to 120°F.

"Standby losses" or "storage losses" through the water heater tank walls can represent 20 to 50% of total water heater energy usage. Reduce this loss by insulating the water heater body, as shown in Figure 10-8, and the first 6 feet of pipe coming from the water heater. This can give a payback in savings within the first year or two.

Guidelines for insulating water heaters are as follows: For gas water heaters, keep the water heater blanket at least 3 inches above the floor so that the unit can draw in oxygen for the burners from underneath

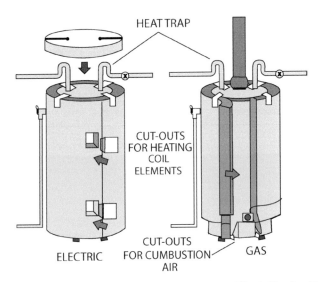

**Figure 10-8. Insulation on water heaters can cut utility bills significantly.**

and so that if the tank leaks it will be noticeable. In some areas, a 6-inch clearance may be required. Keeping the blanket insulation off the floor and above the bottom edge of the water heater allows the combustion air intake openings underneath the water heater to be unobstructed. Make a mark where the access panel is on the water heater blanket for future reference. Do not cover the top of a gas water heater tank with insulation and make sure the pilot light is always accessible. Do not place flammable pipe insulation closer than 6 inches to the single wall flue pipe on the hot water heater. Insulate the under side of elbows correctly by cutting the insulation sleeves at a 45° angle to cover the joint. Often the pipes coming off the top of a gas water heater may be too close to the flue to install pipe insulation in that location. In this situation, do not install flammable pipe insulation until it is at least 6 inches away from a single wall flue pipe.

In the weatherization program, replacement of water heaters can be justified because of health and safety concerns, software-based audit results, or criteria in the Water Heater Info Toolkit (available from your local Weatherization program). If the water heater needs replacing, replace it with a standard tank type water heater, a whole home tankless water heater, a heat pump water heater, or a solar water heater. With electric water heaters, the insulation needs to stay 3 inches, and in some cases 6 inches, above the floor depending upon the standard in the area. However, unlike a gas water heater, the top of an electric hot water heater tank can be covered with insulation.

Table 10-6 shows where hot water is used in a typical home.

**Table 10-6. Hot Water End Use**

| Appliance | Percent of Total |
|---|---|
| Showers | 37% |
| Bath Filling | 20% |
| Clothes Washer | 14% |
| Sink Filling | 14% |
| Dishwasher | 10% |
| Faucet Flow | 6% |

As seen, showers represent more than a third of the total hot water use. Furthermore, installing low flow fixtures, such as those shown in Figure 10-9, at showers and faucets is a potential area for cutting down on hot water use.

Measure the showerhead flow using a 1 gallon bucket or milk container. Turn on the water and time how long it takes to fill the bucket to a predetermined level and then calculate how many gallons per minute the shower is putting out. For instance, if it takes the shower 20 seconds to fill a 1 gallon bucket half-full, multiply the 1/2 gallon by the fraction 60 sec/20 sec to get:

1/2 gal x 60 sec per min/20 sec = 1-1/2 gal/min

**Figure 10-9. Water Saver Shower Heads**

It is often difficult to distinguish a water saver (left) from a regular (right) showerhead. A measurement of flow is typically required.

This showerhead is performing in the proper range--less than 3 gallons per minute.

When replacing showerheads be careful not to turn the shower arm or pipe connected to the showerhead. Too often, when the shower arm is turned even slightly, the pipe seal at the joint behind the wall is loosened. This can result in serious leakage of water behind the wall. If you move the shower arm even slightly, when you get around to installing the new showerhead, take the shower arm out and be sure to put new pipe caulk or Teflon tape on the end of the shower arm before reinstalling to better assure a tight seal.

In the private arena, if your client needs both a new water heater and a new furnace, consider installing a high-efficiency (Figure 10-10) condensing water heater that, although expensive, can be combined with a warm air or hot water to air heating system. This can result in significant savings, both in replacement and in utility bills.

### Efficiency of Water Heaters

A number of different ratings are used to establish the efficiency of water heaters. The "energy factor" is an efficiency rating for water heaters used in single-family dwellings. The "energy factor" is the fraction of the energy entering the water heater that stays in the water heater storage tank, assuming there are 64 gallons per day of hot water usage. Typically, this number is between 0.5 and 1.0 with 1.0 representing a 100% efficient water heater, which only electric water heaters can approach. The most

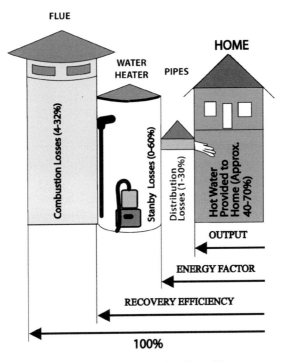

**Figure 10-10. Efficiencies in Water Heaters**

efficient oil or gas water heaters typically have an energy factor in the 0.60 to 0.68 range.

The "recovery efficiency" rating only takes into account the losses occurring while the water is being heated. Thus, the recovery efficiency is always higher than the energy factor except with tankless, pilotless water heaters (Figure 10-11). In that case, the two efficiencies are the same because there are no storage losses.

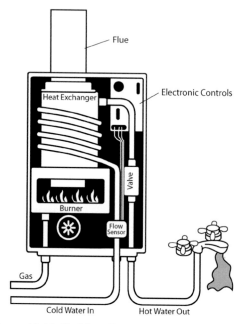

**Figure 10-11. Tankless Domestic Hot Water System**

The "overall system efficiency" rating (or "output" on the diagram) includes all losses including those through the distribution system down to the fixture level.

### *Recovery Efficiency ≥ Energy Factor*

The "storage capacity" of a water heater is the number of gallons held in the storage tank. The "energy input" for a water heater is represented by the number of Btu's per hour that are burned by the water heater when it is on. In addition, the "recovery capacity" of a water heater is the number of gallons per hour of hot water it can generate.

## LIGHTING

Lighting can represent up to 10% of total home energy use. Converting incandescent lights to compact fluorescent lights can cut lighting energy use by 75%, which results in an overall energy reduction of about 7.5% for the home. Table 10-7 shows the dramatic differences in wattage use between incandescent and Energy Star CFL bulbs with the same light or lumen output.

**Table 10-7. Lumen Table**

| Incandescent (watts) | Minimum Light Output (lumens) | Common ENERGY STAR Qualified Bulbs (watts) |
|---|---|---|
| 40 | 450 | 9 to 13 |
| 60 | 800 | 13 to 15 |
| 75 | 1,100 | 18 to 25 |
| 100 | 1,600 | 23 to 30 |
| 150 | 2,600 | 30 to 52 |

Table 10-8 shows what the annual savings would be per bulb when an incandescent bulb is replaced with a CFL. This table only shows the savings using a bulb on the average of four hours a day. Obviously, if lights are used more than this, the savings will be even greater. This table also assumes a cost of $0.11 per kilowatt-hour. If more expensive electricity is used, the savings will be even greater. This analysis also does not take into account the cost of replacing incandescent bulbs, which last only about 1/10 the time of CFLs because they burn out much faster.

**Table 10-8. Lighting Savings Table**

| Incandescent (watts) | Average Usage (Hrs/Day) | CFL Replacement (watts) | Annual Savings* ($/ bulb replaced) |
|---|---|---|---|
| 40 | 4 | 9 - 13 | 4.98 |
| 60 | 4 | 13 - 15 | 7.55 |
| 75 | 4 | 18 - 25 | 9.15 |
| 100 | 4 | 23 - 30 | 12.37 |
| 150 | 4 | 30 - 52 | 19.27 |

Some clients may show reluctance to have their incandescent bulbs replaced with CFLs because of sensationalized news programs involving the mercury content of CFLs or even just because of past experience. You can specify special replacements to make sure that the clients will enjoy the results of the lighting retrofit. For instance, regular CFLs do not work with dimmer switches, and so make sure that a light replacement is "dimmer compatible" if it needs to work on a fixture that uses a dimmer switch. Another example would be bulbs that operate with photocells such as outdoor lighting.

Only dimmer compatible bulbs will work in these fixtures.

Furthermore, specialty bulbs like reflectors, candelabras, etc. may need to be specified to fit the fixture properly, for example, because of a lampshade or other fixture hardware. In the past, there were also complaints that CFL light seemed less flattering than light from incandescent bulbs. Avoid this problem by specifying CFLs that have a CRI (color rendering index) of between 80 and 90 and a Kelvin temperature that is pleasing to the client. (Table 10-9 shows CRI values for various lights). For instance, if the client wants warm white/soft white light, use a bulb with a Kelvin temperature of between 2700-3000 K. If they want a cooler bright white light, use a bulb with a Kelvin temperature of between 3500-4100 K. If the client wants natural or daylight lighting from the bulb, use a bulb with a Kelvin temperature of 5000-6500 K.

**Table 10-9. Color Rendering Index By Type of Lighting**

| Lighting Type | Color Rendering Index |
|---|---|
| Fluorescent, Standard | 52-62 |
| Fluorescent, T-8 and CFL | 81 90 |
| High-Pressure Sodium | 25-65 |
| Incandescent | 97-100 |
| Mercury Vapor | 22-52 |
| Metal Halide | 60-90 |

If handled properly, the danger of mercury exposure can be greatly reduced by following EPA's cleanup guidelines when a bulb breaks. To follow this procedure, get animals and people out of the room (especially pregnant women and children), open a window, use cardboard to sweep up the fragments and tape to pick up smaller pieces, wipe the area with a damp paper towel, and dispose of all the materials in a closable jar. Dispose of CFL bulbs properly to avoid contaminating water supplies. It should be noted that electrical power plants actually release mercury as a byproduct of producing electricity. Using CFLs to save energy more than offsets the release of mercury into the atmosphere by utility plants. Most CFLs have about 4 mg of mercury in them. A milligram of mercury is about the size of the tip of a pencil lead.

In the private arena, there may be more opportunity to use motion sensors for light switching. Timers and photoelectric cells can also be useful to make sure security lights, etc. automatically turn off when the sun comes up and do not turn on until the sun goes down. Make sure you are installing a CFL that can be used in conjunction with the timers and photoelectric cells. Avoid magnetic ballasts, and instead use electronic ballasts when installing fluorescent lamps. Many other types of lighting bulbs and fixtures can be used. Table 10-10 shows their characteristics.

OTHER
BASELOAD ISSUES

A hot-water tap leaking two drops per minute can cost almost $16 a year assuming a cost of $0.11 per kilowatt-hour. Stopping larger leaks can have an even more substantial effect. For instance, a hot water leak that drips at the rate of one drop per second can cost over $300 per year assuming the same energy costs. In addition to repairing leaks, consider replacing any showerhead that has a flow greater than 3 gallons per minute, and both kitchen and lavatory faucets with low flow fixtures or aerators (if they do not already have these) regardless of the flow.

Removing any restrictions in the dryer vent, including the build up of lint, can save up to $50 a year because of the improved dryer performance that results from a smooth, open pathway for the exhaust.

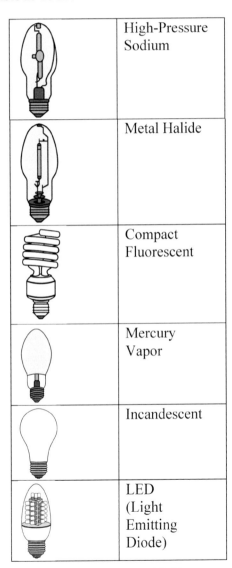

| | |
|---|---|
| | High-Pressure Sodium |
| | Metal Halide |
| | Compact Fluorescent |
| | Mercury Vapor |
| | Incandescent |
| | LED (Light Emitting Diode) |

**Table 10-10. Types of Bulbs**

Also in the private arena, pool pumps are known to run 24 hours a day even though about a third of that time is usually enough for adequate circulation and filtration. In many cases, having a timer installed and putting an insulated cover over the water can represent some of the most cost-effective savings in this situation.

WATER SAVINGS

Energy loss and waste is only aspect of water usage. The loss of wasted water is another aspect. Almost 350 gallons can be wasted each year from a dripping faucet that drips at a rate of only 10 drops per minute. The EPA WaterSense® program provides a labeling program for toilets, faucets, showerhead, etc. that use water at lower rates than are typical. It is much like the Energy Star program for electrical appliances, but instead provides a label based on water conservation.

# Chapter 11

# New Construction Energy Evaluations

## CODE

While many codes have an effect on the energy rating of a new home being built, perhaps the most important series of codes relating to energy is the International Energy Conservation Code (IECC). Every 3 to 5 years this code is updated, with the 2009 and 2012 IECC being the most recent. Unfortunately, each state adopts its own version of the code, including the IECC, and, as a result, each state's code can vary dramatically from other states.

There are some programs that rely on the IECC to set standards nationally. One such system is the HERS rating program established by RESNET (see below for details). The HERS rating is recognized in many states as an evaluation of energy code compliance. Whether the state has approved a HERS rating as one method of showing energy code compliance or not, there are three generally accepted ways to satisfy the requirements of the energy code, see Table 11-1: software analysis or the performance method, the component path, and the prescriptive path. The software analysis is very similar to the HERS software modeling. In fact, if you can show through software analysis that the home being evaluated would have lower utility bills than the energy code's "Reference Home," then you have documented compliance with the energy code using software analysis.

The component path to the satisfaction of the energy code requirements involves making sure the components being used have a lower energy usage than those required by code. If so, you have shown compliance with the energy code. The prescriptive path to satisfying the energy code involves following the exact requirements of the energy code for the climate zone where the home is located. If you can show that you have satisfied every requirement for that zone in the home as built, then you have shown that the home satisfies the energy code requirements.

**Table 11-1. Satisfying Requirements of Energy Code**

| | |
|---|---|
| Performance Method (Software Analysis) | Uses approved software to show satisfaction of energy code requirements |
| Component Path | Use components that all have a lower energy usage than required by code |
| Prescriptive Path | Follow exact requirements of the energy code for the climate zone |

RESNET RATINGS

  This section covers topics that are important to conducting RESNET HERS ratings on new homes or new additions to existing homes. RESNET is an acronym for The Residential Energy Services Network while HERS is short for the Home Energy Rater System that has been established by RESNET. RESNET is the recognized national accrediting entity for "Providers" that provide oversight for the "Raters," the individuals that actually conduct energy ratings on homes. The HERS Index gives a home a score of 0 to 100 (or more than 100 if it is more efficient). It is a reverse rating system. The lower the score, the more energy efficient the home. A home with a score of 100, for the most part, satisfies the requirements of the 2009 International Energy Conservation Code (IECC). While this may seem backwards, it is actually a very good way to try to estimate the percent savings or losses from a home that satisfies the 2009 IECC. A house that goes beyond the energy efficiency of the 2009 IECC can score below the 100 mark on the HERS rating index. If the house is so energy efficient that it uses no net energy year round (a "zero energy" home), then it scores a zero on the HERS Index. In addition, for every number that the index goes down from 100, a savings of 1% is expected to be realized by the homeowner over and above what would be achieved if the home only satisfied the 2009 IECC. Thus, a home that receives a HERS Index of 75 is 25% more efficient than a 2009 IECC home with a score of 100. On the other hand, a home that has a HERS Index of 125 is 25% less efficient than the home that satisfies the code.

The HERS Index, a sample chart is shown in Figure 11-1, developed by RESNET is used in a number of arenas:

1) It can be used to determine utility rebates or tax credits or deductions, although these programs can vary greatly by state and locality,

2) It is used to confirm that a home qualifies for special energy mortgages,

3) It is used to determine if the home can use the EPA's "Energy Star" label when it is sold by the contractor, and

4) The index is used in qualifying homes for the Department of Energy's "Challenge Home" program (formerly "National Builder's Challenge" program). Typically builders must build the home so it scores 70 or lower on the HERS Index to satisfy the requirements of this program.

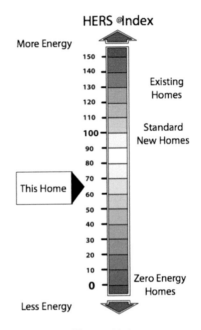

**Figure 11-1.**
**Sample HERS Index Chart**

There are other benefits to having a HERS Index evaluation of a home. For instance, a builder can discover from a HERS rating what changes could be made in the design or building of the home to provide a home whose utilities will cost less, even if there is not an energy-efficient mortgage involved in the sale, etc. Builders can also benefit from an evaluation of the quality of the work that subcontractors in the various trades have provided in building the home. Finally, some builders have found the independence of the HERS rater, or the idea of a third party conducting the evaluation, helps sell the home, even if the HERS rating is not used to qualify the home for programs that rely on a HERS rating such as Energy Star, EarthCraft (a sustainable/energy program used primarily in the southern states), or the LEED-H program (a LEED certification, provided by the U.S. Green Building Council, specifically designed for homes).

There are benefits to the owners of homes as well. While there continues to be a debate about how much higher the appraisal will be for a home that provides significant energy savings, in some cases the multiplier in appraised or added appraisal value can be as high as 10 to 20 times the annual savings in dollars. Furthermore, a homeowner's concerns over the utility bills, comfort, and indoor air quality, and the cost and inconvenience of maintenance, can be alleviated by the knowledge that an independent third-party evaluated the home for issues related to these types of concerns.

While your independence from others is critical, it can be important that others who are in a position to provide you with referrals or jobs are aware of the services you offer in the energy arena. Thus, you should provide a prompt and conscientious service. This includes being polite, respectful, dressing appropriately, and being responsive. However, you need to remain fully aware that the relationships with those that are not going to become the homeowner must remain an arms length transaction that does not involve either receiving or giving anything of value in exchange for referrals or jobs. Of course this also means that there should be no formal or informal agreement, spoken or unspoken, that you will make sure even that one home is given special consideration as a result of referrals or jobs being provided by any builder. You must be able to confirm that your relationship to those that provide you with referrals or jobs is independent of those referrals or jobs and that you do not, in any way, compromise the independence of the rating process because of the receipt of those referrals or jobs.

You must also do your best to be a conscientious businessperson. That typically means that you are fiscally responsible by making sure that bills are paid, and for the most part, you are not spending more money than you are bringing in, and that you are planning in advance for business activities, etc. Make sure that you have adequate cash to pay the bills whether from income from the company or outside capital, including your personal savings account, etc. A good business plan must include who you expect your market to be, along with how you expect to market to them. Never overlook the opportunities that a good website and other inexpensive marketing techniques can provide for marketing directly to homebuyers. Careful consideration should also be given to assure you are appropriately educating your clients as to what they are receiving. It should also include a plan for exactly what services you expect to provide, including any services outside of a Home Energy Rating Service (HERS). Also, consider accessing the materials available on operating an energy business available on the BPI website at: http://www.bpi.org/qualitymanagement/qualitymanagement.html

As for pricing, you can base pricing on a flat rate for each rating completed, an hourly rate, or any number of methods. However, you need to make sure that you have an adequate income, or other source of funding, to get your business going and keep it running.

Regardless of the purpose for getting a HERS Index rating, the Home Energy Rater conducts the rating process the same way. Home Energy Raters are qualified by RESNET to conduct the actual rating of the home and give it a HERS Index number. Raters use software to calculate, first as an estimate based on the plans and then upon completion, what the HERS Index is for a home. After a plan review, the Rater list changes and improvements for the builder on what can be done to improve the rating on the home before it is built. Once the house is being built, the Rater then visits the site one or more times and, in the process, does an inspection, a blower door test and a duct leakage test. In the end, the Rater issues a HERS Index rating on the home. Home Energy Raters must take training, including field training, and pass a nationally recognized (ANSI certified) written exam in order to qualify to conduct Home Energy Ratings on new homes.

RESNET also qualifies individuals to conduct energy evaluations on existing homes much as BPI does. Their system for evaluating existing homes follows the BPI system very closely.

Many of the topics that are important to learn to become a Home Energy Rater have already been covered in this book. These include building science and diagnostics, duct leakage, calculation of area, en-

| |
|---|
| Home Energy Survey Professional - HESP |
| Diagnostic Home Energy Survey Professional - DHESP |
| Home Energy Rater - Rater - RESNET - New Homes only |
| Comprehensive Home Energy Rater - CHERS |

**Table 10-2. RESNET Certifications (In Order of Increasing Difficulty of Achieving).** RESNET Certifications (see Appendix M for more detail)

ergy units and definitions, health and safety concerns, envelope insulation, evaluating equipment efficiencies, comfort problems, household appliances, evaluating window and door efficiency, building durability, and US climate zones. Please review the coverage of these topics in other chapters or the appendices of this book. Also, for further details, many of the RESNET related documents are available on the RESNET web site—www.resnet.us. Additional topics needed to be a Rater are given below.

## MEASURING AND SKETCHING BUILDING DIMENSIONS

You should measure at least a few building dimensions on site. Drawing a rough sketch of the building can help keep track of dimensions. It can also help you reference specific problems when communicating with the builder or others. A scaled sketch will look proportionately more accurate but is probably not critical since you are not required to provide a sketch as part of a rating of a new home. Make sure that other individuals who will see the sketch will be able to identify, and check if necessary, the items you have identified in the drawing and their dimensions. While the sketch does not have to be to scale, you should not allow major distortions to exist in the drawing to the extent that others can no longer identify various sections of the home.

Perhaps the most important sketches to make are the side views of each side of the home, also called the elevation drawings, and a view looking down from above the home, called a plan view or flooring plan. The elevation drawings or side views should show the rough location of windows and doors on that side of the home, as well as all their dimensions. Identify the length of overhangs as shown in Figure 11-2, including eaves or covers on each elevation drawing of the house. Indicate horizontally how far out from the wall the overhang extends over a window as well as what the distance is from both the top and bottom of each window to the shading or overhang above.

A consistent order or convention for these dimensions will allow you to simply record these numbers in the sequence you prefer next to each window on your sketch as shown in Figure 11-3. If you are going to count a room as a bedroom, it must have a closet, a window or exterior door for egress, and be at least 70 ft² in area, the common definition of a bedroom under most local and national codes, including the 1012 IRC.

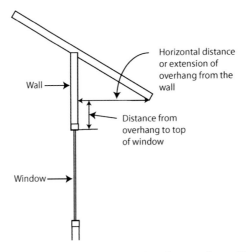

**Figure 11-2. Method of Measuring Overhangs above Windows**

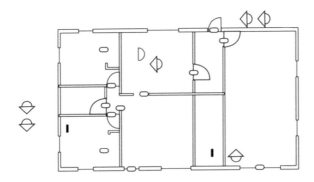

**Figure 11-3. Floor Plan**
You can indicate dimensions of windows, doors, walls and grades
next to their top view on a floor plan like this.

On sloped or vaulted ceilings, at least 50% of the floor area must have
a ceiling at least 7 feet high or you cannot count any of the floor area as
conditioned area or space. Count all the floor area as conditioned that
exists underneath a vaulted or sloped ceiling that is 7 feet high or higher
in at least half of the room area. A standard horizontal ceiling must be a
minimum of 7 feet tall, or 6' 4" with minor obstructions like ductwork,
false beams, etc.

As you go around the building, record the window, door, over-hang, wall, and roof dimensions (including height of each floor and the total height of all floors from the ground level) onto the sketch and appropriately identify the corresponding dimensions. Based on the ori-entation of the building to true north, identify each face of the building as north, south, east, or west. If none of the walls corresponds well to a standard orientation, simply identify the north elevation as the most northerly wall, etc., in your notes so that both you and others seeing your sketch will know how you designated the orientation of each el-evation drawing.

As for the floor plan drawing, it is not important to provide a lay-out of the interior of the home. However, since you are not required to provide a sketch anyway, you can simply use the floor plan, without any elevation drawings, to sketch the building. As long as you provide the same dimensions as you would in elevation drawings for each side of the home, such as door, window, etc. dimensions, you are serving virtu-ally the same purpose. Indicate a door or window in a wall by simply thinning the wall line to show their presence and record their dimen-sions next to the thickened wall line. An exception would be if there is uneven grading around the house. Here it may be more important to see how the grading changes using elevation drawings of each side of the home. As for dimensions, record dimensions to the nearest inch, and square footage to the nearest square foot. At the same time, this is a great opportunity to confirm the actual building with the plans, if plans are made available.


## READING BUILDING PLANS

Building plans have a variety of different views of the building as drawings. In reviewing the plans for evaluation before the building is built, at least look at the elevation plans and floor plans. The elevation plans should show any variations in grading around the home if they were designed for a specific site. If they have not been designed for a specific site, then they will not likely have a "site" or plot plan. The site plan would show variations in the topography or "lay" of the land from a "birds-eye" view so that you could identify any variations in the grading around the home. If the roof must be figured into any of the calculations, the site plan should also show the roof slope. If the plans do not include a site plan for obtaining this kind of information, ask the builder what the plans are for grading around the house, etc. If the elevation plans do not show dimensions for the height of the building, then you need to look at the sectional plans that typically give height dimensions. A sectional plan is a cross-section of the home at a specific "slice" through the home, just as a loaf of bread is cut through at different cross-sections.

While dimensions or measurements provided on the plans are obviously the most important reason for reviewing plans, symbols used in the plan, although not critical, may also be useful to understand the drawings better. Many plans have a legend or key to explain the symbols. In addition, you may be interested in the "schedule" or "notes," found either among the drawings or in a separate bound or stapled document called "specifications." Specifically, look for the window and door schedule that describes the different sizes and types of windows in the building, and may include the energy rating for them as well. This may include a requirement that the windows be NFRC rated. It may give the required maximum U-value and Solar Heat Gain Coefficient (SHGC) and the minimum visible light transmittance of the windows. Some other specifications that may be useful include roof shingle color and reflectivity requirements, the schedules for thermal barrier application (gives the minimum R value for walls, ceiling, and floors), residential commissioning (gives the maximum air leakage in cubic feet per minute per square foot of building envelope and maximum duct system air leakage to the outside), special house-to-garage air barrier and sealing requirements, the minimum efficiency factor for a water heater, the required sealing characteristics for ducts and limits on using panned joists as return ducts, special energy truss roof structure requirements, requirements that bath and kitchen ducts exhaust to the exterior, the required SEER rating for the air conditioning system, whether furnace sizing will be within certain limits of the most recently accepted version of the ACCA Manual J, required heating efficiencies of heat pumps or furnaces, some level sound and automation requirements, and whether unvented combustion appliances will be allowed. You typically can also find any important material properties and vapor barrier requirements and specifications. Some of the dimensions may have to be calculated indirectly from dimensions provided in the drawings, as it is unusual to have every possible dimension laid out in the plans.

It should be mentioned that all construction drawings are done to a specific "scale." Most home construction drawings are drawn to a 1/4-inch per foot scale. This means that every 1/4-inch on the drawing represents 1 foot of the actual size of the home. Since it takes four multiples of 1/4 inch to make 1 inch and there are 12 inches in a foot, this means that the drawing is done on a 1/48th ratio. The house in the drawings is 1/48th the size of the real home it depicts. Likewise, it also means that the drawing is 48 times smaller than the house. Figure 11-4 to 11-11 show examples of different site and elevation plans.

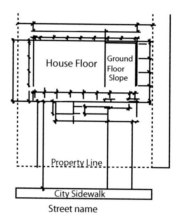

**Figure 11-4. Site Plan**

**Figure 11-5. Front Elevation**

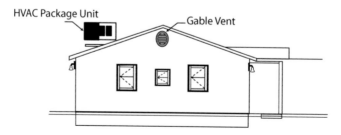

**Figure 11-6. Left Elevation**

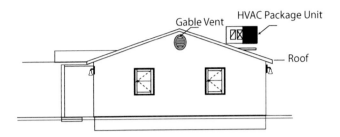

Figure 11-7. Right Elevation

Figure 11-8. Rear Elevation

Figure 11-9. Lengthwise Building Section

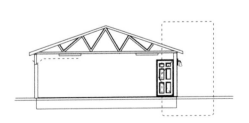

Figure 11-10. Widthwise-Building Section

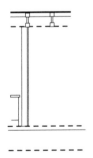

Figure 11-11. Wall Detail Section

ETHICS AND DISCLOSURE

See "Rating and Home Energy Survey Code of Ethics" on available on the RESNET web site—www.resnet.us. Ethics and disclosure cover conflicts of interest such as ownership in the home being evaluated, or receiving fees from third parties as part of a survey, or disclosing and obtaining consent from all parties involved regarding such conflicts. It also requires objectivity and honesty in disclosure and reporting as well as confidentiality of reporting. Accurate advance disclosure to clients of fees and work to be done is also required, as well as informing clients that they have a right to obtain competitive bids for any work recommended by the rater.

Ethics complaints against the Rater are filed with the entity that provides oversight of the Rater, the Rater's Provider (see below), and then, if the Provider does not investigate the complaint, a complaint against the Provider can be filed with RESNET.

REAL ESTATE FINANCING

Several types of real estate financing use RESNET HERS as a premise for the financing, as outlined in Table 11-3. An Energy-Efficient Mortgage or EEM can help *homebuyers* who are purchasing a new home obtain a mortgage. Thus, EEMs are only for buyers of new homes. An EEM takes into account the energy-saving features of the home and the resulting utility cost savings to the homeowner. These saving can be credited either to the purchaser's income or can reduce the monthly payment amount by the energy savings when calculating the ratio of monthly payment to income (otherwise known as the qualifying ratio). Thus, if the expected monthly payment on the loan comes to 35% (0.35 ratio) of the borrower's income, but the maximum allowed ratio is 32%, the buyer/borrower could still qualify if the monthly utility savings from the more energy-efficient home were large enough, when subtracted from the monthly payment, to bring the ratio down to 32%. Another way to benefit from the rating is for the appraisal to take into account the utility savings by increasing the value of the home by a proportionate larger amount. This would automatically increase the maximum monthly mortgage payment allowed as well.

Another financing tool that can rely on HERS is the Energy Improvement Mortgage or EIM. The EIM can be used by either *homebuyers or homeowners* to make improvements to the energy characteristics of the home. Thus, EIMs can be applied to both new and existing homes regardless of whether they are on the market for sale or not. Lenders recognize that the monthly utility savings that can be realized once the home is upgraded would free up money that can be used to pay the loan back on a monthly basis.

Yet another financing tool that can rely on HERS is the Energy Improvement Loan or EIL. This loan is intended to help current *homeowners* make energy improvements on their homes that will save monthly utility costs. Thus EILs are only for those who already own a home, not homebuyers. The savings can be used to make the monthly payments on the energy improvement loan. These financing tools are summarized in Table 11-3.

**Table 11-3. Types of Finance Tools**

| Finance Tool | Characteristics |
|---|---|
| EEM: Energy Efficient Mortgage | Takes into account energy saving features to reduce monthly payment (homebuyers of new homes only) |
| EIM: Energy Improvement Mortgage | Used by homebuyers or homeowners to improve energy efficiency of home |
| EIL: Energy Improvement Loan | Used by homeowners to make energy improvements to save utility costs. |

## IDENTIFICATION AND DOCUMENTATION OF FEATURES OF THE RATED HOME

The rater needs to identify the energy features of the home being rated to the minimum required by the RESNET standards, input this information into the appropriate software program for analysis, and generate a report for the client. Detailed specific on-site inspection procedures are found in Section B.5 and in Appendix A of the National Home Energy Rating Technical Guidelines, Section 303.4, and Appendix A of MINHERS. The MINHERS Technical Standards found in Chapter

3 are available from the RESNET web site—www.resnet.us. To obtain default values for combustion appliance efficiency, see Tables 303.7.1 (3) (age-based) and 303.7.1 (4) (non age-based).

It is important for a rater to be able to look up data on specific heating and air conditioning appliances in the home so this information can be integrated into the energy evaluation. To check efficiencies of air conditioning, heating, water heating, etc. equipment, collect as much information as is available from the equipment data plate or nameplate. Typically information about brand, model, serial number, type of system, etc. are often given on the appliance data plate. Another source for this information is the Air Conditioning Heating, Refrigeration Institute or AHRI (formerly the ARI). To find information, access the web site at http://www.ahridirectory.org and conduct a search of the "AHRI Directory of Certified Product Performance." Input the appropriate data from the web site into the computer program that you are using to evaluate the home.

As part of your reporting, you will be expected to provide the likely cost of utilities to be used in the home. Just be sure to get this information directly from the utility company if you have any doubts. Having a reliable source for this information is important. You may find in your area that since utility companies are guaranteed a profit, the utility bills may be based on a return on investment, a percent guaranteed profit, or on expected revenue (revenue based pricing).

Once all the data are properly input into the software, you are in a position to help guide the builder as to what changes could provide an improvement in the HERS score (improvement analysis).

Specifications (see "Reading Building Plans" above in this chapter).

QUALITY ASSURANCE

All HERS Raters must be part of a quality assurance process. This begins with Raters providing consistent, accurate ratings. The next step is for the Provider to evaluate the quality of the Rater's ratings by reinspecting homes that the rater has already rated. The Certified Quality Assurance Designee or QAD, who is responsible for all the assurance process for a HERS Provider, must have passed the national QAD exam. While the QAD must review 10% of the Rater's reports, they need only review 1% of a Rater's homes with a field visit and review. RESNET in turn reviews the provider's quality assurance program to better ensure quality ratings.

## COOLING AND HEATING DESIGN TRADE-OFFS

With new homes, an often-overlooked design and planning issue relates to the orientation of the home and placement of its windows. If the short walls of the home face east and west, and the long walls face north and south, then there could be some energy savings. Try to limit windows on the north, east, and west faces as much as possible. Have the largest window area on the south face with an appropriately sized overhang that blocks the summer sun yet allows in the winter sun. Doing this simple passive energy design helps in the optimal placement of the home and windows. Advanced framing techniques can also provide some savings in construction cost that can be redistributed to add more insulation or better windows (see section on thermal bridging).

### Testing Differences From BPI Standards

There are numerous differences between the RESNET and BPI standards for Combustion Appliance Zone/Closet (CAZ) testing (see CAZ section). The following are distinctions that apply to the RESNET standards only:

1. When developing the worst-case conditions, close the door to the CAZ.

2. The standard does not suggest to test appliances simultaneously that have connected flues.

3. The standard does not mention using mirrors, a lit match, or other methods other than smoke pencil to check for backflow on combustion appliances.

4. The standard does not mention any 60-second grace period before checking for backflow and it does not suggest checking for backflow until after the appliance has been on for at least 5 minutes.

5. Check for CO and draft backflow after waiting 5 minutes for the appliance to reach steady state.

6. There is no flue draft pressure testing within the flue, only CAZ Worst Case pressure testing.

7. There are different standards for negative pressures in the CAZ allowed under worst-case conditions. The CAZ pressure cannot go lower than:
   -15 Pa for pellet stoves with exhaust fans and sealed vents.
   -5 Pa for all other appliances

8. If the home fails worst-case conditions but passes under natural conditions, the work order must specify "building pressure remediation."

9. Personal CO monitors must be recalibrated annually by the manufacturer or use the manufacturer's instructions for calibration.

10. If the ambient CO is greater than 35 ppm, you need only turn the appliance off, not disable it.

11. When testing an oven, you must open a window or door to the outside.

12. The ambient CO level on an oven is conducted 5 feet away from the oven at countertop level.

13. If you measure greater than 200 ppm of CO in an oven vent, then the work order must specify replacement or repair of the oven.

14. If the appliance CO levels are above 100 ppm, you must stop the combustion appliance tests unless the appliance is required to be replaced.

15. Each report must include a statement that no warranty or guarantee of energy savings estimates is being made.

16. The work order by the auditor must specify that CAZ depressurization testing must be completed at the end of each workday or after any work that has been specified that might affect building tightness or pressures.

17. The work order must include a statement that work order items must be viewed as independent of each other rather than as a whole.

18. If there are any combustion appliances within the building envelope, including any ovens, then the work order must specify that a CO detector be placed in a main area of each floor.

19. If a combustion appliance fails the spillage or the house fails the CAZ pressure limit test, then
    —the work order must specify replacement of all atmospheric combustion appliances
    —recommend replacement of unvented appliances, and, in the event not removed, then the work order should specify that no tightening of envelope, etc. be conducted.

## REFERENCE HOME

The Reference Home is a home that uses the minimum energy code characteristics to satisfy the 2009 IECC. It is defined as "the geometric twin of the rated home, configured to a standard set of thermal performance characteristics, from which the energy budget, that is the basis for comparison, is derived." Using the Reference Home values, you can compare, using software, the energy budget with the expected energy budget of the rating on the home being evaluated. To better understand the nature of these comparisons, materials and construction code characteristics, see Section 303.4 and Appendix A of MINHERS and Sections B.4 and B.5 and Appendix A of the National Home Energy Rating Technical Guidelines for more detailed information. Technically, the HERS Score Computation is made using the Normalized Modified Loads Rating Method (see the Florida Solar Energy Center for an article by Philip Fairey on this topic).

## PROJECTED AND CONFIRMED RATINGS

Projected ratings are referred to as "plan reviews" since they involve calculating a rating, albeit an uncertified rating, based on the blueprints and specifications without the building being built, or if it is built without visiting the site. You can use the same software to conduct a projected rating as you do to conduct a final certified "confirmed" rating. The same information is needed to conduct a projected rating as to conduct a confirmed rating; however, with a confirmed rating, you include a site visit of the home.

## ENERGY CODE COMPLIANCE
## (SEE "CODE" SECTION ABOVE IN THIS CHAPTER)

## THE ENERGY STAR PROGRAM OF THE EPA

### Program Qualifications

The Energy Star Program is perhaps the most significant program that relies on HERS. Some of the requirements for a home to qualify as an Energy Star home include:

1) satisfying the requirements of a Thermal Enclosure System Rater checklist before drywall is installed

2)  having a RESNET rater evaluate both the envelope and duct leakage;

3)  having the home's characteristics analyzed by RESNET-approved software;

4)  having the quality of the insulation installation inspected and graded;

5)  having an appropriate level of Energy Star qualified appliances, fans, light fixtures, windows, HVAC equipment;

6)  satisfying an HVAC System Quality Installation Rater and Builder Checklist; and

7)  satisfying the Water Management System Checklist. The checklists, guides, and other technical publications for Energy Star can be found on the Energy Star web site—www.energystar.gov. Each Energy Star home must achieve a certain target HERS index score specific to the home. For instance, for one home, a score of 81 may be inadequate, while for another home, a score of 85 would be adequate.

**Sampling Protocols Under Energy Star**

If builders are major production builders, a sampling protocol allows them to less expensively evaluate their homes for Energy Star qualification through a "sampling" technique. Once any builder has completed seven homes and they have all been inspected, tested, and successfully satisfy the Energy Star protocol, the builder can opt for sampling conducted by a RESNET-Accredited Sampling Provider. Three of the builder's homes must pass the Energy Star requirements in a subdivision before any homes in that new subdivision are to be included under the sampling protocol. All homes included in the sampling protocol must have the same envelope arrangement and be of the same construction variety to be included as part of the sample. Each group of homes, known as a "batch," has an established benchmark. This typically involves evaluating each plan for a worst-case situation and orientation, and using that as a HERS index for each plan in the batch. One in every seven homes of the batch must be evaluated under the sampling protocol. If any of the sample homes fail, the builder can be asked to start over again with the next seven homes that are built, unless the builder can establish a specific cause for the failures and a well defined process for correcting the specific cause of previous failures.

## DETERMINING BUILDING ORIENTATION

As part of a rating, the building orientation must be identified relative to true north. Magnetic north is not the same as true north. The variation between the two is often referred to as the magnetic declination or variation. If you know the magnetic declination for the area where you conduct ratings, you can calculate the true north direction from the magnetic north direction. For instance, if you are on Long Island east of New York City, the magnetic declination is -14°. So you would have to pretend the magnetic arrow was about 14° more clockwise from where it is actually reading on the compass to find the true north direction. Below is a government map showing magnetic declination in North America (see Appendix J).

## PASSIVE HOUSE PROGRAM

The "Passivhaus" (German for "Passive House") or Passive House program is a high-performance building program that originated in Germany. It is similar to the RESNET program in that it is used to help design and build new buildings. But, Passive House can also be used to conduct retrofits on existing buildings. A specialized software, the Passive House Planning Package (PHPP), is used to evaluate building plans for energy efficiency. The program further utilizes the Lawrence Berkeley National Laboratory's Therm software for an even more detailed analysis of R-values for specific details of the construction, see example shown in Figure 11-12. Since design of the building and quality in construction on-site is critical in providing a high performance building, it is generally assumed that the architect or con-

Figure 11-12. An example of a thermal roof/wall juncture diagram created and evaluated by the free LBL Therm software program with temperature gradient lines embedded in the diagram. Notice with a highly insulated situation the gradient lines vary little despite the very challenging wall/roof juncture.

sultant would need to pay numerous visits to the building site while special quality training is available for the trades under the Passive House Program. (See Appendix Q for more information)

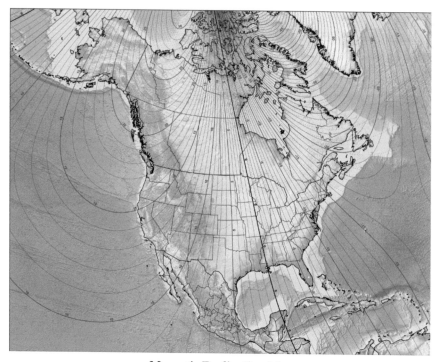

**Magnetic Declination Map**

HERS Score Computation (see "RESNET Ratings" section above in this chapter)

# Chapter 12

# Building Professional Training and Certification

## INTRODUCTION

There is a great need for up-and-coming qualified energy professionals within the building arena. Fully 31% of about 2000 respondents to a 2013 survey of Association of Energy Engineers members plan on retiring within the next 10 years. Sixty-seven percent of these respondents suggested that, within the next five years, they fully expect a shortage of qualified renewable energy and energy management professionals. Other evidence also indicates the need for professionals in the renewable energy and energy efficiency industries will be increasing. A few of those factors include:

- Executive orders issued by President Obama to help reduce climate change impacts.

- Programs established by numerous states and cities to reduce the production of greenhouse gases while promoting both renewable energy as well as energy efficiency strategies.

- The increase in electricity rates by as much as 40% when coal-fired power plants are converted over to a natural gas fuel supply.

The best way to keep up with the latest energy efficiency technology in all the arenas—residential, commercial, and institutional—is to get the training and continuing education needed for success in the field.

## THE ASSOCIATION OF ENERGY ENGINEERS

The Association of Energy Engineers (AEE) is a non-profit professional association founded in 1977 whose mission is to "promote the scientific and educational aspects of those engaged in the energy industry and foster action for sustainable development." AEE has a membership of over 16,000 in 89 countries today.

In 1981, AEE offered the first training-based certification in energy management called the "Certified Energy Manager" or CEM. Since then, it has added many other training programs that lead to additional energy-related certifications.

AEE designations hold the mark of distinction in the industry. AEE certifications single out qualified individuals who can help increase the energy performance of buildings.

## AEE CERTIFICATION

AEE is a certifying body under ANSI/ISO Standard 17024. Certification programs offered through AEE in the building arena are detailed at http://www.aeecenter.org/certification. Specific experience and education requirements are needed for an individual to become AEE certified in a specific area. In addition, the candidate must complete an AEE approved seminar and successfully pass a four-hour examination. At this time, well over 26,000 professionals have achieved one AEE certification credential or another.

AEE certification programs remain strong due to the fact that they:

- Create a professional competence standard

- Represent total quality management because the clients of an employer are assured of a high level of competence, experience, and specialized knowledge through the certified employees of the company.

- Advance quality by way of providing continuing education in ever-changing fields.

- Provide encouragement for meeting long-term career goals.

The certified energy manager (CEM) program is AEE's flagship certification. It is used widely in the building arena. The Residential Energy Auditor (REA) certification is a new certification that can be helpful for those performing energy audits.

The REA certification is an important tool for being recognized by the public for expertise in the residential energy-auditing arena. To achieve this certification, the candidate must do the following:

- Complete an AEE training seminar that covers residential energy auditing. The REA course is a 12-hour course offered online, and on some occasions, as a live classroom course.

- Pass a four-hour examination specifically designed for residential energy auditors with a score of at least 70%. (This exam must be proctored by an AEE approved Exam administrator. It is an open book exam and can contain both multiple choice and true false questions.) A free, downloadable study guide is available from AEE to help prepare for the examination. The study guide also includes some sample questions. The test can be taken at any one of the AEE's remote testing centers.

- Meet some eligibility requirements.

To meet the eligibility requirements for the certification, you must only satisfy one of these options:

- Be a Certified Energy Manager (CEM) or Certified Energy Auditor (CEA) in good standing.

- Have three years of confirmed experience relevant to residential energy auditing.

- Be a Military veteran with at least three years of confirmed technical experience.

- Have a two-year or higher degree in energy auditing, construction management, building management, engineering, science, or another specialty that is related.

## CONTINUING EDUCATION PROGRAMS

There are numerous training options offered by AEE for continuing education units (CEUs). Each CEU is the equivalent of 10 Professional Development Hours (PDHs). AEE provides CEU documentation for continuing education courses that have been completed successfully, as well as a certificate of participation for each course completed.

In 2012, over 39 states required CEUs or PDHs as prerequisites for renewing the Professional Engineer license. North Carolina, New York, and Florida approve AEE courses for continuing education. Furthermore, the American Institute of Architects also recognizes AEE as an approved education provider for continuing education.

The various types of AEE programs offered include:

- Expositions and conferences
- 24/7 asynchronous online training
- Live seminars
- In-house seminars
- Asynchronous self-paced online training
- Synchronous online real-time training

# Appendix A
# Weatherization Standards

Please see the "Appendix A" that is found in the DOE WAP Rule of the CFR Part 440.

# Appendix B
# Math Basics

An energy auditor and even a retrofitter perform many calculations as part of the weatherization process. Therefore, it is good to understand some math basics before attempting these calculations.

## Fractions and Conversions

Most people have a good understanding of addition and multiplication, but some people are thrown off by fractions. Here is a simple rule to remember about fractions that you may find useful. Dividing by a fraction is the same as multiplying by the reciprocal. For instance, the reciprocal of ½ is 2/1 or 2. Thus, if you were to divide 4 by ½, it would be the same as multiplying 4 by 2, which is equal to 8. Here are some examples:

$$10 \div \tfrac{1}{2} = 10 \times 2/1 = 10 \times 2 = 20$$

$$10/\tfrac{1}{2} = 10 \times 2/1 = 10 \times 2 = 20$$

$$\tfrac{3}{4}/\tfrac{1}{2} = \tfrac{3}{4} \times 2/1 = 6/4 = 3/2 = 1\tfrac{1}{2} = 1.5$$

Here is an example of a simple mistake to avoid. Just because there are 12 inches in a foot, 7'5" does not equal 7.5 inches. Instead, 7'5" is actually 7 and 5/12th feet. Divide 5 by 12 to get 0.416, which is roughly equal to 0.42. Thus, 7'5" is actually 7.42 feet as a decimal number.

Here is another tip on converting one number to another. Calculating numbers using inches and feet can be very tricky. Remember that a foot is 12 inches and a square foot contains 1 ft x 1 ft or 12 x 12 in² = 144 in². Therefore, to find the area in a rectangle that is 300 in x 200 in, the total area in square inches is 60,000 in². To find how many square feet there are in 60,000 in², divide by 144-in²/1 ft²:

$$60,000 \text{ in}^2 / 144 \text{ in}^2 / 1 \text{ ft}^2 = 417 \text{ ft}^2$$

### Measuring a House

Area and Volume: Typical units that you want to become familiar with when dealing with math in the energy arena include surface area and volume. Surface area is length multiplied by width, represented as L x W, or feet x feet, or ft². The units are in square feet (sq. ft. or ft²), see Figure B-1.

Volume (Figure B-2) adds one more dimension to surface area so it is length multiplied by width multiplied by height. The units are in cubic feet (cu. ft., ft³).

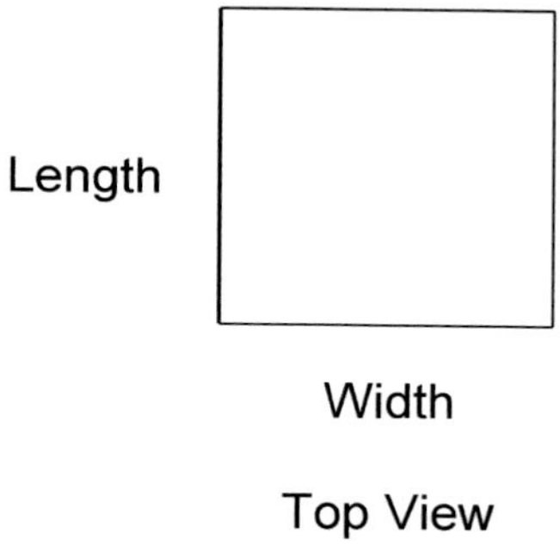

### Simple Shapes

To evaluate floor area, use a top view sketch of the floor plan, (sometimes called a "footprint") and note the areas that are conditioned and unconditioned. Calculations are usually

**Length**

**Width**

**Top View**

**Figure B-1. Area Dimensions**

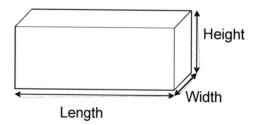

Height

Width

Length

**Figure B-2. Volume dimensions**

only made on conditioned areas. To calculate the conditioned area, multiply the length by the width. The best way to do this is to break into simple small shapes that can more easily be calculated in pieces and then added together.

As an example, look at a home that is one and one-half stories on the main house with a rear addition and an "L" section off the rear corner next to the addition, Figure B-3. The diagram below shows how the home is broken up into four basic sections: the first floor of the main house, the second floor of the main house, the rear addition, and the "L" section. The important dimensions are also provided for each of the sections including length, width, and height. If the house is not divided, the calculations would be much more complicated.

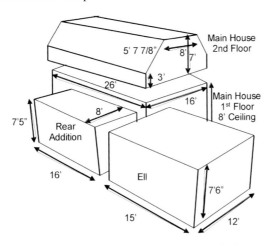

**Figure B-3. More Complicated Calculations**

Notice that the distraction of roofs has been removed since attic spaces are usually not conditioned areas. The sections of the home are easy to calculate for area because they are simple rectangles:

$$Ell = 15 \text{ ft} \times 12 \text{ ft} = 180 \text{ ft}^2$$
$$Addition = 16 \text{ ft} \times 8 \text{ ft} = 128 \text{ ft}^2$$
$$Main, \text{ 1st floor} = 26 \text{ ft} \times 16 \text{ ft} = 416 \text{ ft}^2$$
$$Main, \text{ 2nd floor} = 26 \text{ ft} \times 16 \text{ ft} = 416 \text{ ft}^2$$

If you need to know the total square footage, it would be: 180 ft² + 128 ft² + 416 ft² + 416 ft² = 1140 ft²

Calculating the volumes of these areas can be somewhat more complicated, especially for the second floor. However, the first floor section volumes are easy to calculate because they are simple rectangular boxes, Figure B-4:

$$\text{Ell} = 180 \text{ ft}^2 \times 7.5 \text{ ft} = 1350 \text{ ft}^3$$
$$\text{Addition} = 128 \text{ ft}^2 \times 7.42 \text{ ft} = 950 \text{ ft}^3$$
$$\text{Main, 1st floor} = 416 \text{ ft}^2 \times 8 \text{ ft} = 3328 \text{ ft}^3$$

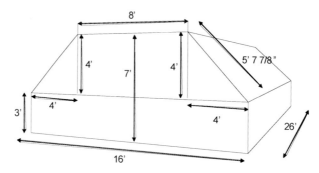

**Figure B-4. Simply the Calculations**

The volume of the second floor section of the main house is complicated because the knee wall attics and the sloping ceiling make it an odd shape. Breaking the shape down to even more pieces takes something complicated and makes it much easier. First, calculate the lower portion volume of the second floor. From the above calculations, the area of the second floor is 416 ft². It is known that the area's knee walls go up 3 feet from the floor. Now take out the rectangular box that is represented by the lower portion of the second floor, marked with an "A." This is a simple calculation:

$$(\text{floor area}) \times (\text{wall height}) = 416 \text{ ft}^2 \times 3 \text{ ft} = 1248 \text{ ft}^3$$

You will come back to this number later when you total all the different volumes for the second floor.

The next section to separate out is the area that angles from the upper attic flat straight down to 3 feet above the floor. This area is 8' x 26' x 4' because you have to subtract out the 3 foot high area from the 7 foot ceiling height which leaves 4 feet of height for this new box marked with a "B." Thus:

$$8' \times 26' \times 4' = 832 \text{ ft}^3$$

The last two sections to evaluate are the two triangular shapes next to each side of the volume just calculated. They are marked as a "C" on both sides of the "B" box. These are 26 feet long but have a triangular cross-section. As complicated as it may initially seem, it is actually easy to combine these two triangular shapes together to form a rectangular box. The two sides of the 90° or right angles on the inside of the triangle are both 4 feet long. You know this because the gable end of the second story is 16 feet long at the base. Since you have already taken out the middle 8-foot section and calculated its volume, you now have only a total of 8 feet on the 16-foot long gable end for one horizontal side of each of the two triangles. The vertical side next to this right angle is 4 feet long because it only extends between the knee wall height at 3 feet and the ceiling height at 7 feet: 7 ft – 3 ft = 4 ft.

Thus, if two triangular shaped sections of the second floor are combined, the rectangular box formed is:

$$4' \times 4' \times 26' = 416 \text{ ft}^3$$

So totaling all the pieces of the second floor:

$$1248 \text{ ft}^3 + 832 \text{ ft}^3 + 416 \text{ ft}^3 = 2496 \text{ ft}^3$$

The final step is to total the volumes of all four primary sections of the home to find its total volume:

| | |
|---|---|
| "Ell" : | 1350 ft³ |
| Addition: | 950 ft³ |
| Main, 1st floor : | 3328 ft³ |
| Main, 2nd floor: | 2496 ft² |
| Total Volume: | 8124 ft³ |

Therefore, the total volume in this home is 8124 ft³.

## Complicated Shapes

It may not always be easy to combine shapes to make rectangular boxes. It is important to know formulas for area and volume of some of the more complicated shapes.

*Triangles*

Triangle Area = ½ x Base x Height

Triangular Volume: ½ x Base x Height x Length

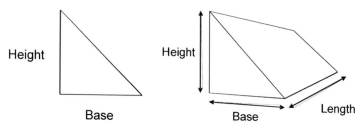

Height          Height          Length

Base                    Base

**Figure B-5. Triangle**          **Figure B-6. Triangularly Shaped**

*Circles*

Circle Area = π x r x r or πr2
π = 3.14 (this is always the same)
And r = radius

*Cylinders*

Cylinder = π x r x r x length or πr2L
where π = 3.14 (this is always the same)
r = radius and r x r = r2
and L = the length of the cylinder

In the previous example of calculating the area of both "C" sections, you could have calculated the area of the two triangular sections of the second floor by using the formula for triangularly shaped sections. In this case, calculate the volume of each triangle separately:

R = radius
of the circle

**Figure B-7. Circle**

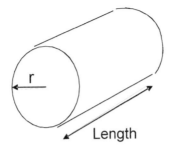

**Figure B-8. Cylinder Dimensions**

½ x Base x Height x Length =
½ x 4′ x 4′ x 26′ = 208

Then double that number to obtain the total for both triangles:

208 ft³ x 2 = 416 ft³

This is the same number calculated earlier when you combined the two triangles to form a rectangular shaped section before you did the calculation.

**Roof Slope**
There are some occasions where you may need to calculate roof slope. Roof slope is typically indicated by how many inches the roof rises vertically for every 12 inches (1 foot) of horizontal run. Thus, a typical 3 in 12 roof, rises 3 inches straight up for every 12 inches across on the level while the roof itself is an angle between the two:

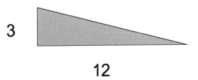

3

12

The pitch of a roof can be easily measured using a 12-inch level at a roof eave and measuring how many inches up the roof it travels as it runs 12 inches along the level. You can also measure the slope with a "slope gauge" (readily available from a hardware store) from the ground.

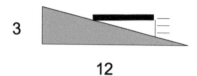

You may need to calculate the surface area of the roof since it may be part of the envelope of the home. In this case, use a table to find a number to multiply what the roof area would be if it were flat to find out what its true surface area is as a sloped roof. This multiplier is represented in Table B-1 depending upon the slope of the roof.

Using this multiplier, you simply multiply the square footage of the top floor below the roof by the multiplier to obtain the surface area of the roof on the slope. For instance, if the roof pitch were 3 in 12, also designated as "3:12," and the area of the top floor below the roof was 1000 ft²; check the table to find that the multiplier for finding area on a 3:12 roof is 1.031. Thus, 1.031 x 1000 ft² = 1,031 ft². This assumes that the slope on each side of the ridge was 3:12. Thus, the surface area for the entire roof on both sides of the ridge is 1,030 ft².

## Accuracy

Under some standards, such as BPI, the house only needs to be measured within 10% accuracy. This can take some of the meticulousness out of measuring. For the most part, you may only need to record the measurements to the nearest foot unless you are measuring a length less than

**Table B-1. Roof Surface Area Multiplier Table**

| Roof Pitch | Slope Multiplier (ft) |
|------------|-----------------------|
| 1 in 12    | 1.003                 |
| 2 in 12    | 1.014                 |
| 3 in 12    | 1.031                 |
| 4 in 12    | 1.054                 |
| 5 in 12    | 1.083                 |
| 6 in 12    | 1.118                 |
| 7 in 12    | 1.158                 |
| 8 in 12    | 1.202                 |
| 9 in 12    | 1.250                 |
| 10 in 12   | 1.302                 |
| 11 in 12   | 1.357                 |
| 12 in 12   | 1.414                 |

10 feet long. Even then, you may still be able to round off to the nearest foot. For instance, in the above example, one of the ceiling heights of 7'6" could have been rounded off to 8 feet. This will give you two out of four with 8 ft high ceilings and two other areas with a 7-foot high ceiling to create the following numbers:

$$[(180 \text{ ft}^2 + 416 \text{ ft}^2) \times 8] + [(128 \text{ ft}^2 + 416 \text{ ft}^2) \times 7] =$$
$$4768 \text{ ft}^3 + 380 \text{ ft}^3 = 8576 \text{ ft}^3.$$

Another ceiling height is 7'5" and can be rounded down to a 7-foot high ceiling. Thus, doing the calculation, assuming that the two ceiling heights are 8 ft and the third and fourth are 7 ft, the volume is 8576 ft³

Total area x ceiling height = Volume

This is close to the accurate measurement of 8124 ft³, and within the 10% accuracy standard, even without taking into account the triangular nature of parts of the upper floor ceiling.

8124 x 0.10 = 812
8124 – 812 = 731 ft³
8124 + 812 = 8936

Thus, the range of +/- 10% is from 7312 ft³ to 8936 ft³ and 8576 ft³ falls within this range.

# Appendix C

# Conversion Tables / Charts

Example: To convert Btu to Joules, go to the left hand column and find Btu, then move to the right to Joules to find the conversion factor. Therefore, to convert 3 Btus to Joules, it would be 3 x 1055 to obtain 3165 Joules.

## Energy

|  | British Thermal Unit (Btu) | Joule | Kilowatt Hour (kWh) | Natural Gas 100 Cubic Feet CCF | Therm | Cal | Ft-lb |
|---|---|---|---|---|---|---|---|
| British Thermal Unit | 1.0 | 1055 | 0.000293 | 0.0000098 | 0.00001 | 251.9 | 777.6 |
| Joule | 0.0009478 | 1.0 | 0.000000277 | 0.000000000292 | 0.00000000009478 | .239 | .739 |
| Kilowatt Hour | 3413 | 3,600,000 | 1.0 | 0.0334 | 0.03413 | 833,333 | 2631578 |
| Natural Gas, 100 Cubic Feet | 102,000 | 105,500,000 | 29.89449 | 1.0 | 1.02 |  |  |
| Therm | 100,000 | 105,506,000 | 29.30722 | 0.98 | 1.0 |  |  |
| Cal | .00397 | 4.184 | .0000012 |  |  | 1.0 | 3.085 |
| Ft-lb | .001286 | 1.353 | .00000038 |  |  | .3241 | 1.0 |

## Power

|  | Foot-Pounds per second | Horse-power | Calories per second | Kilowatts | Watts | Btu per hour |
|---|---|---|---|---|---|---|
| Foot-Pounds per second | 1 | 0001818 | 0.3238 | 0001356 | 1.35 | 4626 |
| Horsepower | 550 | 1 | 178.1 | 0.746 | 746 | 2544.4 |
| Calorie Per second | 3.088 | 0005615 | 1 | 0004187 | 4.187 | 14.289 |
| Kilowatt | 737.6 | 1.341 | 238.8 | 1 | 1000 | 3412.1 |
| Watt | 0.7376 | 0.001341 | 0.2388 | 0.001 | 1 | 3.412 |
| Btu per hour | 0.216158 | 0.000393 | 0.06998 | 0.00029307 | 0.29307 | 1 |

One ton of refrigeration is roughly equal to 12,000 Btuh

## Flow

|  | Cubic feet/minute | Gallons/minute | Cubic meter/second |
|---|---|---|---|
| Cubic feet/minute | 1 | 7482 | 0.0004719 |
| Gallons/minute | 0.1337 | 1 | 0.0000631 |
| Cubic meter/second | 2119 | 15,850 | 1 |

**Note**: FPM or feet per minute identifies the velocity or speed of airflow.

## Weight

|  | Pounds | Kilogram | Grains | Ounces |
|---|---|---|---|---|
| Pounds | 1 | 0.4536 | 7000 | 16 |
| Kilogram | 2.205 | 1 | 15,432 | 35.27 |
| Grains | 0.000143 | 0.0000648 | 1 | 0.00229 |
| Ounces | 0.0626 | 0.02835 | 437.5 | 1 |

## Pressure

|  | Pascals | Kilopascals | Pounds/square inch | Inches of water | Inches of mercury |
|---|---|---|---|---|---|
| Pascals | 1 | 0.001 | 0.000145 | 0.00401 | 0.0003 |
| Kilopascals | 1000 | 1 | 0.145 | 4.014 | 0.2952 |
| Pounds/square inch | 6895 | 6.895 | 1 | 27.68 | 2.036 |
| Inches of water | 249.1 | 0.2491 | 0.0361 | 1 | 0.0735 |
| Inches of mercury | 3388 | 3.388 | 0.4912 | 13.60 | 1 |

## Btu Content of Fuels

| Fuel | Btu |
|---|---|
| Natural Gas, cubic foot | 1030 |
| Propane, cubic foot | 2500 |
| Ethanol, gallon | 76,000 |
| Fuel oil, gallon | 135,000 (Kerosene) to 153,000 (#6) |
| Gasoline, gallon | 125,000 |
| Wood, cord | 10,000 (softwood) to 24,000 (hardwood) |
| Coal, lb | 10,000 to 15,000 |

# Appendix D
# Ventilation Calculations

## MVR STANDARDS AND CALCULATIONS

MVR is calculated using one of two standards: ASHRAE 62.1—1989 or ASHRAE 62.2—2007. The ASHRAE calculations tell the size of the fans needed to maintain MVR. In determining the amount of mechanical ventilation that must be added to the home, you can take into account already existing exhaust fans, etc. if you know their flow rates. While retrofitters typically do not need to know how to do these calculations, it is still good to know the reasoning.

ASHRAE 62.1—1989 was originally written for large commercial buildings that typically had forced ventilation from outside to inside. It was imposed on small residential buildings because it was the only ventilation standard available. Small buildings are defined as buildings having six or fewer units and are under three stories. Under this standard, indoor air quality pollutants are to be dealt with at the source of the pollutant. Fans or operable windows must provide the required ventilation. This standard is based on the home volume using calculations based on a profile of the vertical area above grade. To meet the ASHRAE 62.1-1989 requirement, choose the largest number among three calculations: the "people" calculation, the "bedroom" calculation, and the ACH calculation. The standard uses a target of 15 CFM per person with a maximum of 0.35 ACHnatural. In small buildings, the person count equals the greater of the number of bedrooms plus 1 or the actual occupancy number for the home. If any building has at least five occupants, the absolute minimum ventilation rate rises to 75 CFM since each person requires 15 CFM.

The ASHRAE 62.2—2007 standard applies to multifamily residential buildings and single-family buildings of three stories or fewer above grade. This includes manufactured and modular buildings as well. This standard is specifically designated as a performance standard, rather than just a prescriptive standard. This means that you have to actually test the fan delivering CFM for adequate flow rate before considering the work complete and not just assume that the fan is pumping the CFM that shows on the manufacturer's fan unit labels (see "Exhaust Fan Flow Meter" procedures). One of the important requirements of this standard is that every building must have a forced mechanical ventilation system to provide whole building ventilation. The standard also requires that local exhaust fans be installed in all kitchens and bathrooms. These bathroom fans must provide 20 CFM continuous or 50 CFM on-demand. The kitchen fans must provide five air changes per hour (ACH) continuous based on the kitchen volume, or they must provide 100 CFM on-demand. If the continuous fan provides less than 5 ACH, then a vented range hood is also required. This standard requires mechanical ventilation at 7.5 CFM per person plus 1 CFM for every 100 ft².

Here are the basic steps in calculating the Minimum Building Airflow Standard (BAS) using formulas:

See page 5 of BPI Building Analyst Standards to come up with the LBL 'N' Factors for your area.

Step 1: Calculate Ventilation Required for People
    Airflow (cfm) = 15 x occupants (occupants = bedrooms +1)
      = 15 x _____ = _____cfm

Step 2: Calculate Ventilation Required for Building
    Airflow (cfm) = 0.35 x Volume/60
      = 0.35 x _____/60 =_____cfm

Step 3: Using Higher Airflow Requirement, Convert to CFM50
    Minimum CFM50 = Airflow (cfm) x N
      = _____ x ____ = _____CFM50

Step 4: Multiply Step 3's BAS x 0.7 for Acceptable Range
    BAS x .7 = _____
    BAS Range: _____ to _____CFM50
           70%      100%

The following example shows how the standard can affect the outcome. Using ASHRAE 62.1-1989, assume a 1000 ft², three-bedroom house with 8 foot ceilings occupied by five people. Also, assume that the house is already retrofitted and the final blower door test reading is 800 CFM50. As for existing ventilation, there is only one kitchen exhaust fan, which is rated at 40 CFM, but when tested only operates at 20 CFM upon demand. Assume the house is a 1.5 story, well shielded home in Chicago, Illinois. From the climate zone map and LBNL chart shown earlier, this location falls in climate zone 2 with an N factor of 20. Divide the blower door reading by the N factor to get the CFM natural of the home:

$$800 \text{ CFM50}/20 = 40 \text{ CFM natural}$$

For the "people" calculation, multiply the number of people in the home by 15 CFM natural:

$$5 \text{ people} \times 15 \text{ CFM natural} = 75 \text{ CFM}$$

For the "bedroom" calculation, add one to the total number of bedrooms in the home and multiply that total by 15 CFM natural:

$$(3 \text{ bedrooms} +1) \times 15 \text{ CFM natural} = 60 \text{ CFM}$$

To complete the "volume" of ACH calculation, multiply the volume of the home by 0.35 ACH and then divide by 60. Since the volume is 1000 ft² x 8 ft = 8000 ft³, the volume-based calculation is:

$$(8000 \text{ ft}^3 \times 0.35 \text{ ACH})/60 = 47 \text{ CFM}$$

Adopt the largest of these three calculations, which turns out to be the people calculation at 75 CFM. From this 75 CFM deduct both the CFM natural calculation for the blower door results and the 20 CFM provided through the kitchen exhaust fan:

$$75 \text{ CFM} - 40 \text{ CFM} - 20 \text{ CFM} = 15 \text{ CFM}$$

15 CFM needs to be added to the house as mechanical ventilation to provide a minimum air exchange so the occupants do not experience negative health effects because of the tightness of the home. It should be noted that a 2004 Wisconsin study showed that existing exhaust fans

were actually operating at 50% or less of their rated capacity because of poorly sealed ductwork, improperly sized ducts, etc. Do not be surprised if the kitchen exhaust fan in the example that had a manufacturer's rating of 40 CFM was only operating at 20 CFM when tested using the exhaust fan flow meter procedure.

Now, make calculations for this same house using ASHRAE 62.2—2007. Under this standard, multiply the number of people (5) by 7.5 CFM natural and add 1 CFM for every 100 ft² in the home:

$$5 \text{ people} \times 7.5 \text{ CFM natural} = 37.5 \text{ CFM}$$

Since this a 1000-square-foot home, it has ten 100 ft² sections, the square foot equation looks like this:

$$1 \text{ CFM} \times 1000 \text{ ft}^2 / 100 \text{ ft}^2 = 1 \text{ CFM} \times 10 = 10 \text{ CFM}$$

Add the people calculation to the square footage calculation to get the MVR under this standard:

$$37.5 \text{ CFM} + 10.0 \text{ CFM} = 47.5 \text{ CFM}$$

Thus, the additional required mechanical ventilation for this home is the MVR of 47.5 CFM minus the 20 CFM actually provided by the existing kitchen fan:

$$47.5 \text{ CFM} - 20 \text{ CFM} = 27.5 \text{ CFM}$$

Under ASHRAE 62.1, 15 CFM must be added to this house while under ASHRAE 62.2, 27.5 CFM must be added.

## Building Tightness Limits

Building Tightness Limit (BTL) is minimum ventilation. It uses a number of factors to determine what the lowest blower door CFM50 can be without adding mechanical ventilation. Those factors include the degree of exposure to wind, the N-factor, the heated square footage of the home, the special climate zone, the height of the building and stories, and the volume of the house.

Building tightness limits are the higher of two calculations that are based on 1) the number of occupants, and 2) the number of bedrooms.

If you assume a minimum of 1200 CFM50 is required for a home, the occupant based calculation begins with a minimum of 1200 CFM50 + an additional 300 CFM50 for every occupant over four occupants. Thus, using the "occupant" based calculation for a home with six occupants shows the following CFM50 requirements:

1200 CFM50 + (300 CFM50 x occupants >4) =
1200 CFM50 + (300 CFM50 x (6-4)) = 1800 CFM 50

The "bedroom"-based calculation requires a starting point of 1200 CFM50 + an additional 300 CFM50 for every bedroom over three bedrooms. The bedroom-based calculation for a sample home with four bedrooms shows:

1200 CFM50 + (300 CFM50 x bedrooms >3) =
1200 CFM50 + (300 CFM50 x (4-3)) = 1500 CFM50

The building tightness limit for this home is the higher of the two, 1800 CFM50

BTLa is also a building tightness limit. It uses the same factors as the BTL but also considers the CFM 50, the blower door airflow exponent, the story height, a weather factor, the heated height to produce the threshold CFM50, an estimate of the natural ACH, and the necessary amount of mechanical ventilation.

BTLa is also called minimum ventilation guideline (MVG), also referred to as the MVR. The name change takes away the emphasis from a tightening threshold and towards how much ventilation needs to be added after the building is tightened to the maximum cost-effectively. Thus, the MVG is the required fan flow of air from the outside in CFM to make up for the tightness of the building. It only needs to be calculated if the building shows a CFM below the BTL.

**Attic Ventilation**

Calculating attic-venting requirements is important. Almost every ceiling below an attic is painted and therefore the attic space is treated as having a vapor barrier. Thus, the maximum ventilation required is 1 square foot of "net free area" (NFA) of the vent for every 300 ft² of attic space. However, if there is a combination of high vents and low vents in the attic, you are allowed to go down to the more liberal requirement of

1 square foot NFA for every 600 ft² of attic area. Here is an example assuming that the attic area has low venting at the soffits and high venting at the gable ends.

$$1000 \text{ ft}^2 / 600 \text{ ft}^2 = 1.67 \text{ ft}^2 \text{ of vent needed}$$

Converting to square inches would look like this:

$$1.67 \text{ ft}^2 \times 144 \text{ in}^2 / \text{ft}^2 = 240.5 \text{ in}^2$$

Thus, if the two vents (the lower soffit vent and the higher gable end vent) are the same size, each vent must be about 120 in² or the equivalent of a 12" x 10" opening. However, this does not mean you can put in two 12" x 10" vents. You must correct for the louvers in the grille, so this usually means the vent must be larger than originally calculated. If these are metal vents and no correction factor is available from the manufacturer, assume that you need to deduct 25% of the vent area to make up for the space the louvers take up in the vent, thereby reducing airflow through it. Thus, the 120 in² requirement per vent represents 75% of the actual vent size. Divide 120 in² by 75% or 0.75 to determine the real vent size needed for a metal vent covered by that metal grille:

$$120.25 \text{ in}^2 / 0.75 = 160.3 \text{ in}^2$$

Taking the square root of 160 inches, the minimum size of a square vent that could be used would be 12.7" x 12.7." Rounding up to the next size vent, you would probably have to get a 12" x 14" vent, if not a 14" x 14" vent, for each of the vents on this home. Since you are using the "high/low" provision (if you were not utilizing any lower soffit vents, but rather two gable vents), make sure that one of the gable vents is significantly higher than the other gable vent.

If these are wood vents and no correction factor is available from the manufacturer, assume that you need to deduct 75% of the vent area to make up for the space the louvers take up in the vent, thereby reducing airflow through it. Thus, the 120 in² requirement per vent represents 25% of the actual vent size we need to install. Divide 120 in² by 25% or 0.25 to determine the real vent size needed for a metal vent covered by that metal grille:

$$120.25 \text{ in}^2 / 0.25 = 481 \text{ in}^2$$

Taking the square root of 481 square inches, the minimum size of a square vent that could be used would be 21.9" x 21.9." Rounding up to the next size vent, you would probably have to get a 22" x 22" vent, if not a 24" x 24" vent. You might also use roughly 12" x 40" sized vent for each of the vents on this home. Since you are using the "high/low" provision (if you were not utilizing any lower soffit vents, but rather two gable vents), make sure that one of the gable vents is significantly higher than the other gable vent.

**Foundation Venting**

Many experts believe that ventilating crawlspaces is not only counterproductive from the energy savings standpoint, but also offers no significant benefit. However, if it is required by code, the ventilation requirement must be followed. Assume that the ground will be covered with a vapor retarder in the crawlspace. Then, you only need to provide one square foot ($ft^2$) of NFA vent for every 1500 $ft^2$ of crawlspace. If there is a 1000 $ft^2$ crawlspace area, the equation would look like this:

$$1000 \text{ ft}^2 / 1500 \text{ ft}^2 = 0.67 \text{ ft}^2 \text{ of vent needed}$$

Converting it to square inches gives—

$$0.67 \text{ ft}^2 \times 144 \text{ in}^2/\text{ft}^2 = 96.5 \text{ in}^2 \text{ NFA}$$

In addition, correcting for the screening or louvers in the grille and assuming it is a metal vent, the calculation will look like this:

$$96.5 \text{ in}^2 / 0.75 = 128.7 \text{ in}^2$$

Taking the square root of 128.7 $in^2$, the minimum size of a square vent that could be used would be 11.3" x 11.3." Rounding up to the next size vent, you would probably have to get a 12" x 12" vent for the crawlspace vent on this home.

If the municipality requires splitting the NFA requirement between two vents, each vent would have to provide a minimum of 64.3 $in^2$. Since you have already corrected for the louvers or screens in the grille or vent, do not make this correction again. The square root of 64.3 $in^2$ is roughly 8 $in^2$ which means that you need to provide two 8" x 8" vents.

## Furnace Makeup Air

If outside air is not used to provide oxygen for the furnace burners, the CAZ must communicate with enough volume in the home to provide oxygen (for example, by using door grilles). For example, if 50 cubic feet (ft³) are needed for every 1000 Btu, how many total cubic feet are needed to satisfy the makeup air requirement for a 120,000 Btu furnace? Multiply 50 ft³ x 120 kBtu to get 6100 ft³—remember the ratio is based on 1000 Btu (equal to 1 kBtu) not 1 Btu. If there is an 8-ft ceiling in the home, then 763 ft² of space is needed to communicate air to the appliance (6100 ft³/8 ft = 763 ft²). If the appliance is in the basement, bedroom volume cannot be used to satisfy this requirement—you cannot put grilles in bedroom doors to get additional volume. Often vents in the bottoms and tops of doors to the basement are needed so the basement communicates with the upstairs. By the way, include all the appliances in the CAZ, including water heaters, to obtain the total Btus in the calculations. Thus, if you had a 40,000 Btu water heater in the CAZ, you would need an additional 2,000 ft³ (50 ft³ x 40 kBtu) to satisfy this requirement. If you add this 2,000 ft³ to a family room (FR) in the basement with a 15' x 20' dimension, a living room (LR), dining room (DR), kitchen (K) and upstairs with a 15' x20', 12' x 12', and 12' x 12' dimensions with 8 foot high ceilings, respectively, you have: FR ft³ + LR ft³ + DR ft³ + K ft³ = (15 x 20 x 8) + 15 x 20 x 8) + (12 x 12 x 8) + (12 x 12 x 8) = 2400 + 2400 + 1152 + 1152 = 7,104 ft³. Thus, in this example you would not have enough cubic footage in the home to meet the make-up air requirement in the home and would have to provide make-up air from outside to the CAZ.

## ELA

Here is a sample calculation of the hole required to get 1 ACH in a sample home:

Assume 1320 ft² of envelope, 4800 ft³ of volume, and a change of 1 ACH between pre and post repair/retrofit.

$$CFMnatural = ACHnatural/60 \text{ minutes x Volume}$$
$$= 1/60 \times 4800 = 80 \text{ CFMnatural}$$

$$flow \text{ (CFM)} = \sqrt{\Delta P} \times ELA \text{ (in}^2) \times 1.0755$$

$$and \text{ } Flow/(\Delta P \times 1.0755) = ELA \text{ (in}^2)$$

If it is assumed that the pressure difference under natural conditions in a typical 2-story building is 4 Pascals

80 CFMnatural/(4 Pa x 1.0755) = 18.5 in² ELA

To convert to square feet:
     18.5 in²/144 in² per ft² = 0.129 ft²

To determine the ratio of the ELA to the total envelope area:
     Total envelope was 1320 ft² so that 0.129 ft²/1320 ft² = 9.78 x 10-5 which would be 9.78 x 10-3%.

## Equivalent Duct Length Fan Sizing Calculations

This example uses equivalent duct length (EDL) to determine the fan size needed for a particular application. Assume that calculations showed that you need a fan that can move 90 CFM. Also assume there is a 12 foot length of four-inch flexible aluminum ductwork that has one 90° elbow in it, as well as an end cap or grille where the duct opens to the outside air.

With four-inch flexible aluminum duct, the manufacturer's table (Table D-1) requires using a factor of 1.25 for resistance to flow. You must multiply the length by a correction factor of 1.25 to obtain the EDL.

**Table D-1. Manufacturer's Table for Equivalent Duct Length**

| Equivalent Duct Length | | | | | |
|---|---|---|---|---|---|
| | | Duct Diameter | | | |
| | | 3 " | 4" | 6" | 8" |
| Duct | Smooth Metal | Same as measured duct length | | | |
| Material | Flex Aluminum | 1.25x duct length | 1.25 x duct length | 1.5 x duct length | 1.5 x duct length |
| | Insulated Flex | 1.5 x duct length | 1.5 x duct length | 1.75 x duct length | 1.75 x duct length |
| Terminal | Wall Cap | 30 feet | 30 feet | 40 feet | 40 feet |
| Device | Roof Jack | 30 feet | 30 feet | 40 feet | 40 feet |
| Elbow | Adjustable | 15 feet | 15 feet | 20 feet | 20 feet |

The manufacturer's table also suggests that for every elbow in four-inch diameter duct you need to add an additional 15 feet to the EDL. (In other words, using a terminal end grille is like adding 15 feet of straight duct.)

Furthermore, the manufacturer's table requires adding an additional 30 feet because you are using an end cap on the duct where the air exits to the outside. (In other words, placing an end cap is like adding the same amount of resistance as 30 feet of duct.) Thus, the calculations for EDL would be as follows:

$$\text{material (flex duct) factor} = (12 \text{ ft} \times 1.25)$$
$$\text{elbow correction} = 15 \text{ ft}$$
$$\text{end cap correction} = 30 \text{ ft}$$
$$(12 \text{ ft} \times 1.25) + 15 \text{ ft} + 30 \text{ ft} = 60 \text{ ft EDL}$$

Now that you know the equivalent duct length is actually 60 feet rather than the 12 feet, go to Table D-2 (provided by the manufacturer) to find the proper model of that manufacturer's fan to use in the bathroom. The manufacturer will have a special table that assumes eight air changes per hour as recommended by the Home Ventilation Institute. Knowing that the bathroom is 90 ft$^2$, you can check the table for the number closest to 60 feet of EDL to find which model of that manufacturer's fans, at a minimum, you must use. Assuming you had a 12' x 7' (84 sq. ft) bathroom, you would at least need a 15 VQ3 model fan for this bathroom because you need to round up to 90 square feet.

## Table D-2.

BATHROOM FAN : NUMBER OF FANS - MODEL NUMBER (to achieve 8 ACH under the Home Ventilation Institute Standards)

CALCULATION OF EQUIVALENT DUCT LENGTH NEEDED

| Area – Room Sq.Ft. | 10 ft. | 20 ft. | 30 ft. | 40 ft. | 50 ft. | 60 ft. | 70 ft. | 80 ft. | 90 ft. | 100 ft. |
|---|---|---|---|---|---|---|---|---|---|---|
| 50 | 05VQ2 | 05VQ2 | 05VQ2 | 0SVQ2 | 05VQ2 | 05VQ2 | 05VQ2 | 05VQ2 | 05VQ2 | 05VQ2 |
| 60 | 05VQ2 | 05VQ2 | 05VQ2 | 0SVQ2 | 05VQ2 | 07VQ2 | 07VQ2 | 08VQ2 | 08VQ2 | 08VQ2 |
| 70 | 07VQ2 | 07VQ2 | 08VQ2 | 08VQ2 | 11VQ2 | 11VQ2 | 11VQ2 | 15VQ3 | 15VQ3 | 15VQ3 |
| 80 | 08VQ2 | 08VQ2 | 08VQ2 | 08VQ2 | 11VQ2 | 11VQ2 | 11VQ2 | 15VQ3 | 15VQ3 | 15VQ3 |
| 90 | 08VQ2 | 08VQ2 | 11VQ2 | 11VQ2 | 11VQ2 | **15VQ3** | 15VQ3 | 15VQ3 | 20VQ3 | 20VQ3 |
| 100 | 11VQ2 | 11VQ2 | 11VQ2 | 15VQ3 | 15VQ3 | 15VQ3 | 20VQ3 | 20VQ3 | 20VQ3 | 20VQ3 |
| 140 | 15VQ3 | 15VQ2 | 20VQ3 | 20VQ3 | 20VQ3 | 20VQ3 | 20VQ3 | 20VQ3 | 20VQ3 | 20VQ3 |
| 180 | 20VQ3 or 2-11VQ2 | 20VQ3 or 2-11VQ2 | 20VQ3 or 2-11VQ2 | 30VQ3 or 2-11VQ2 | 30VQ3 or 2-11VQ2 | 30VQ3 or 2-15VQ3 | 30VQ3 or 2-15VQ3 | 30VQ3 or 2-15VQ3 | 30VQ3 or 2-20VQ3 | 30VQ3 or 2-20VQ3 |
| 240 | 30VQ3 or 2-15VQ3 | 30VQ3 or 2-15VQ3 | 30VQ3 or 2-15VQ3 | 30VQ3 or 2-15VQ3 | 30VQ3 or 2-20VQ3 | 30VQ3 or 2-20VQ3 | 30VQ3 or 2-20VQ3 | 30VQ3 or 2-20VQ3 | 40VQ3 or 2-20VQ3 | 40VQ3 or 2-20VQ3 |
| 300 | 30VQ3 or 2-15VQ3 | 40VQ3 or 2-20VQ3 | 40VQ3 or 2-20VQ3 | 40VQ3 or 2-20VQ3 | 40VQ3 or 2-20VQ3 | 40VQ3 or 2-20VQ3 | 40VQ3 or 2-20VQ3 | 2-20VQ3 | 2-20VQ3 | 2-20VQ3 |

Notice that one model often covers of number of different Equivalent Duct Lengths until it no longer has adequate capacity.

# Appendix E

# Using a Psychrometric Chart

Many years ago, the air-conditioning and refrigeration industry developed the psychometric method or chart (Figure E-1) to evaluate relative humidity based on the results of the wet and dry bulb temperatures from a sling psychrometer. Notice the dry bulb or ordinary temperature is found on the lower horizontal axis in degrees Fahrenheit while the wet bulb temperature is shown along the curved line on the upper left limit of the chart in degrees Fahrenheit as well.

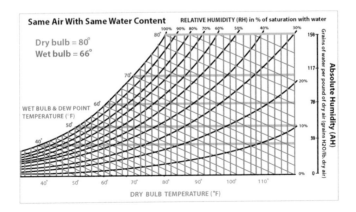

**Figure E-1. Psychrometric Chart**

In addition, note that percent relative humidity is shown across the top horizontal axis and that the absolute humidity is shown on the far right vertical axis in grains (a weight unit) of the water in the air.

To use a psychrometric chart, first go to the lower horizontal axis to locate the dry bulb temperature. The dry bulb temperature is the same temperature as measured with an ordinary thermometer. In this case, assume the dry bulb temperature is 80°F (Figure E-2).

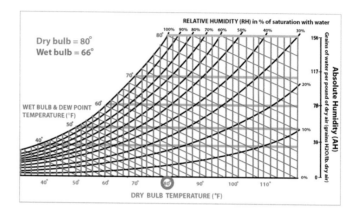

**Figure E-2. Psychrometric Chart 80°F**

Start the psychrometric analysis by drawing an imaginary line up vertically from the temperature (Figure E-3).

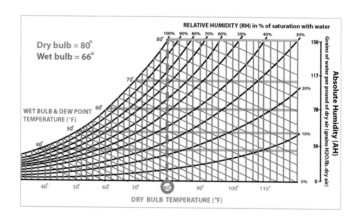

**Figure E-3. Psychrometric Chart with Imaginary Line**

Look at the curved line at the left and identify where to place the wet bulb temperature, in this case 66°F. Then, draw a vertical line from the dry bulb temperature on the horizontal axis and a slanting line at an angle from the wet bulb temperature point to find the point where they intersect. (Figure E-4)

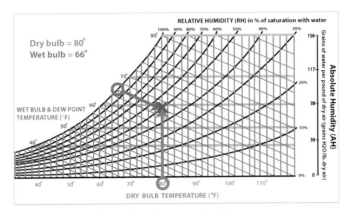

**Figure E-4. Intersecting Points**

Going from this intersection point horizontally to the left tells what the dew point is with this combination of dry and wet bulb temperatures. In this case, the dew point is 60°F, Figure E-5.

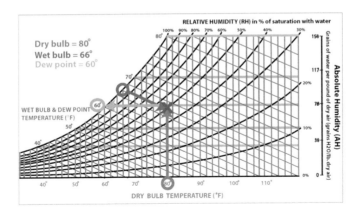

**Figure E-5. Dew Point Intersection**

If you start at the intersection point again but this time follow the curved line to the upper right up to the top of the graph representing the relative humidity, you can find the relative humidity, which in this case is 50%, Figure E-6.

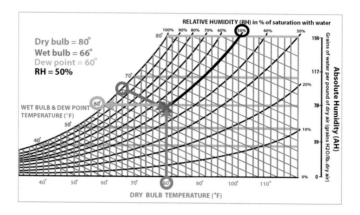

**Figure E-6. Relative Humidity Intersection**

Then, if you follow the dew point line to the right of the intersection, you can obtain the absolute humidity in grains of water per pound of dry air, Figure E-7.

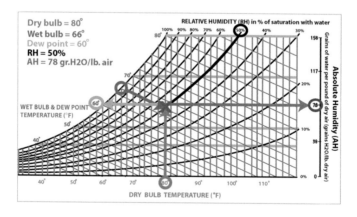

**Figure E-7. Absolute Humidity Intersection**

Absolute humidity is the weight or amount of water in a given volume or weight of air, and is not dependent on the temperature of the air. Thus, if you were able to remove all the water vapor from the air, the absolute humidity is what that removed water vapor would weigh in grains if it were condensed from a vapor into a liquid for a given volume or weight of air. Notice that the range of the absolute humidity starts at zero at the bottom and ends at 156 grains at the top, the most grains of water that can be contained in 80% relative humidity air. In this case, this sample of air has an absolute humidity of 78 grains of water per pound of air. Perhaps it is good to note that a grain of water is a very, very small weight of water such that:

$$1 \text{ grain} = 0.000143 \text{ lb or pounds} = 0.0648 \text{ g or grams.}$$

Warm air can hold more moisture than cold air. With a psychrometric chart, you can see the relationship between air temperature and relative humidity of the same sample of air. Thus, if the air temperature or dry bulb temperature of the same air rises from 80°F to 90°F, the relative humidity will go from 50% down to 38%. The following diagrams show this in stages similar to the stages in the first example:

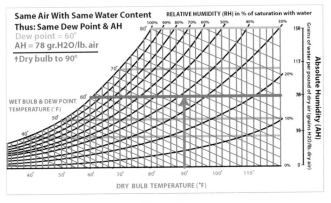

**Figure E-8**

Since this is the same air sample as in the previous example, find the intersection point of the temperature and the horizontal dew point line established for the air in the first example. The dew point stays the same in a given sample of air if the water content or absolute humidity of the air sample is not changed. By following the straight lines to the upper left from this intersection, find that the expected wet bulb temperature will be 70°F.

From this, follow the curved line towards the upper right of the chart to find the relative humidity at 38%.

On the other hand, if you cool the air from 80°F to 70°F, the relative humidity of the air rises to 70%. The following charts show the same steps of finding this as the previous examples (see the steps and numbers in the upper left corner of each graph box):

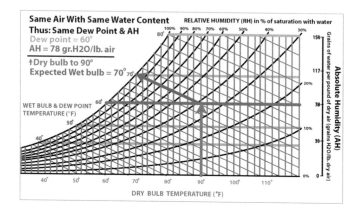

**Figure E-9**

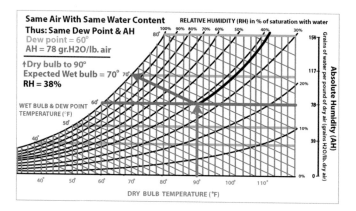

**Figure E-10**

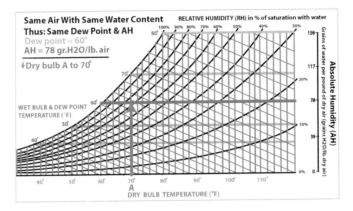

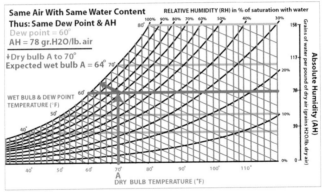

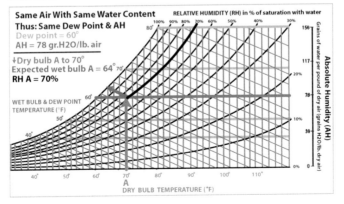

If it is cooled even further down to 60°F, the relative humidity increases to 100%. The following graphs show this in the same sequence as the previous examples (see the steps and numbers in the upper left corner of each graph box.):

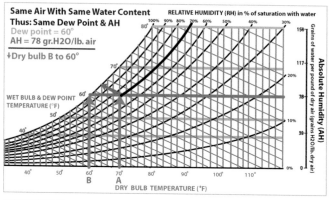

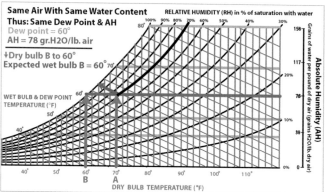

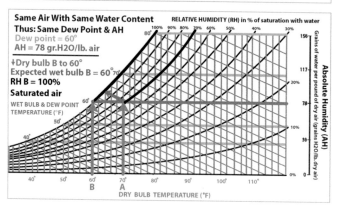

It is good to note that both the absolute humidity in grains of water per pound of dry air and the dew point temperature remained constant for the air even though each of the air temperature changes.

The next question might be what if the air is cooled even more and it goes down below the dew point temperature shown on the chart? The answer is that all the water cannot be held any longer in the air. Thus, if the air is cooled to 50°F, water will condense on surfaces, or rain in the case of a weather situation, until enough moisture is released from the air that the dew point moves down to match the air temperature. The following diagram shows this in one chart:

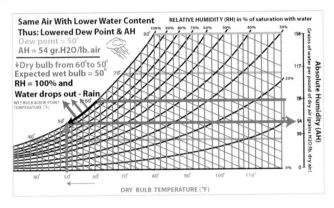

Notice that the sample of air has now changed. This is shown by shifting the dew point/absolute humidity line down on the chart. Notice also that the new dew point temperature is 50°F while the absolute humidity has dropped to 54 grains. The sample of air has now changed and the analysis of this new air sample must follow the new horizontal dew point/absolute humidity line. In addition, any further analysis of changes in temperature of this new air sample must be evaluated based on the new dew point/absolute humidity line, as has been done in the chart below using 80°F dry bulb temperature. Remember, anytime the temperature of an air sample drops below the dew point line, condensation of the water is the result. Thus, in the example below, when this new sample of air is at 80°F dry bulb (bottom horizontal axis), the wet bulb temperature (curved upper left axis) would be 62°F, while the relative humidity (upper horizontal axis) would be 37%. Interestingly, as a result, this new sample of air at 80°F has roughly the same relative humidity as the original sample of air at 90°F. This is a confirmation that this sample is a different sample, has a different amount of moisture in it, and has a lower absolute humidity than the first sample.

Where do you find a situation similar to this in the real world? One example occurs when heating a home in a cold weather area in the winter, and air leaks into the attic through bypasses in the upper floor ceiling. When the warm, moist air is released into the attic through the bypasses, it cools down, often well below its dew point temperature. As a result, condensate forms on surfaces inside the attic, on anything from the insulation and ceiling joists to the even colder bottom side of the roof deck. Remember that condensate must have a surface in order for it to form, such as wood. In the case of the weather, condensate begins to form on some of the microscopic particles in the air in order for water vapor to condense and rain to be created.

Another example also occurs in a cold weather area in winter when air leaks into an insulated wall cavity from the warmer inside and cools as it passes towards the colder materials closer to the outside. In this case, the water often condenses inside the insulation layer and reduces the effectiveness of the insulation.

In a warm weather area in the summer, the reverse of these two real life examples can occur, as warm air leaking in towards the colder air conditioned materials inside of the home often condenses in the insulation layer on its way in. Thus, even just a slight negative pressure in the home and cause warmer outside air to be drawn into the house through leaks in the wall and reduce the effectiveness of insulation.

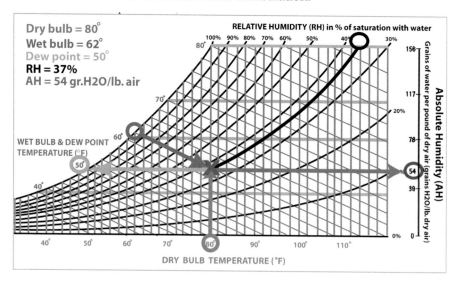

# Appendix F

# Special Weatherization Calculations

## VENTILATION CREDITS

When calculating the amount of ventilation needed for a home that is being weatherized, the weatherization program allows for less ventilation (referred to as a ventilation credit) if the following applies:

CFM natural (or the CFM50/N factor) > (2 x floor area/100)

If the ventilation credit can be applied, the CFM of the fan can be reduced by:

0.5 x ((CFM50/N factor) – (2 x area/100))

In this example, assume the following conditions: an 1150 square-foot, 1.5 story, three-bedroom home in zone 2 with a normal exposure (N factor = 16.7)`and a blower door reading of 1500 CFM50.

To determine if the ventilation credit can be applied, the question is "Is 1500 CFM50/16.7 > (2x1150/100)?" or by following through on the calculation "Is 1500/16.7 >2300/100?"

The answer is yes, 89.9 > 23. Thus, the ventilation credit can be taken in this case. Now that you know you can take the ventilation credit, you need to calculate the amount of the credit. To calculate the allowed credit, use

0.5 x ((CFM50/N factor) – (2 x area/100))

which in this case is represented by:

0.5 x((1500/16.7) – (2 x 1150/100)) = 0.5 (89.8 – 23) = 33.4 CFM Ventilation Credit

To calculate the amount of ventilation needed, the following equation is used:

Qfan = (.01 x floor area) + 7.5 (Number of bedrooms + 1)

Qfan = (.01 x 1150) + (7.5 x 4) = 11.5 + 30 = 41.5 CFM required

Then, to take this credit:
Required ventilation – ventilation credit = amount of ventilation actually needed
41.5 – 33.4 = 8.1 CFM actually needed

It should be noted that the BPI standards refer to the MVR as the Building Airflow Standard or BAS (often referred to as 62-89).

**Target Percent Infiltration Reduction**
Some weatherization programs use tables to guide air sealing that is based on a target percent infiltration reduction. A sample percent infiltration reduction goal is shown in Table F-1.

**Table F-1. Percent Reduction Goals**

| Home ACH 50 | % Reduction Goals |
|-------------|-------------------|
| 11-17       | 25%               |
| 18-22       | 35%               |
| >23         | 40%               |

For example, a home has been tested and found to have an ACH 50 of 14.77. The table above indicates that the goal is a 25% infiltration reduction. If the pre-retrofit blower door test indicated 2000 CFM50 leakage in the home, to reduce that infiltration leakage by 25%, you need to go down to at least 75% of the original 2000 CFM50 for the home. Thus, the equation would look something like this:

2000 CFM50 x 75% =

2000 CFM50 x 0.75 = 1500 CFM50

Under this weatherization program's target percent infiltration reduction standard, retrofitting the home ideally should bring the post-retrofit blower door reading down to at least 1500 CFM50, but not below the building tightness limit.

### "At Risk" Calculation for Indoor Air Quality

When doing a blower door test and tightening the house, make sure the house is not so tight that indoor air quality suffers. If you are not able to do a blower door test (vermiculite insulation was found in the attic space, etc.), some state weatherization programs require the weatherizers to determine if the home is "at risk" for indoor air quality and moisture problems in lieu of being able to test the home for tightness using a blower door. This involves a calculation. Start with a minimum standard requirement of 8400 ft$^3$ of actual heated volume in a home. Then add a factor of 2100 ft$^3$ for each occupant over 4 occupants. Using the same sample home as before and if the home has five occupants, the following equation would be used:

8400 ft$^3$ + (2100 ft$^3$ x # occupants > 4) =
8400 ft$^3$ + (2100 ft$^3$ x (5-4)) =
8400 ft$^3$ + (2100 ft$^3$ x 1) = 10,500 ft$^3$

The sample home only had a volume of 8124 ft$^3$, which is less than the required 10,500 ft$^3$. Therefore, the sample home is considered "at risk." In this case, most weatherization programs would not air seal this home because it falls outside of the standards for indoor air quality using the "at risk" calculation.

# Appendix G

# R-Values and Benefits in Reality

## INTRUSION AND WIND WASHING

There are some situations such as intrusion and wind washing where the effective R-value of insulation is significantly reduced. Intrusion occurs when air moves into or out of insulation without going through the wall or ceiling assembly. This can occur even when there is a greater air barrier present on one surface. The best home insulation has an air barrier on all six sides if the insulation is in a box shape. Wind washing occurs when air enters and leaves the attic through the attic vent openings, blowing through the flat fiberglass attic insulation and removing heat as it goes. Wind washing is peculiar to fiberglass insulation. Both wind washing and intrusion can reduce insulation's effective R-value. In fact, if wind washing and intrusion are combined, they can reduce the insulation's effectiveness up to 50%.

## THE EFFECT OF FRAMING ON R-VALUE

In reality, the expected R-value of a wall is not simply the clear wall R-Value—the R-value of the insulation inside the wall. There are wall studs, headers over openings, a concentration of studs at the corners, and bottom and top plates that fill the wall cavity. These wood-framing members are considered "thermal bridging." They allow the conduction of heat or cold through the wall faster than it passes through insulation. In addition, many walls have windows and doors. Thus, when calculating the true R-value of a wall, you must take into account the R-value of the wood framing, windows, and doors as well.

Here is an example to help with the concept. Figure G-1 is a cross-section of a typical outside wall in a home. To calculate heat flow or losses through the wall, use the R-value of the materials and the square footage of the wall. The table shows the different components in the wall and their R-values. Since framing cannot coexist in the wall in the same spot as insulation, the table separates the calculation for the framing R-value from the R-value of the rest of the wall. Calculating the R-values of just one part of a wall that is consistently the same, represents a calculation of a serial path for heat through a wall. For instance, the independent calculation of the R-value of the framing only looks at the R-value for heat transfer through the air film, drywall, framing, sheathing, etc. where there is framing in the wall. Likewise, if you only look at the components of the portion of the wall with insulation in it, you are looking at a serial path for heat conduction. However, when you combine relative strengths for each of these two portions of the wall to get the average U- or R-Value, you are evaluating the heat flow through two parallel paths to get the overall wall R-value. This value is known as Area Weighted R-Value, Whole Wall R-Value, or Weighted Average Wall R-Value.

Notice the air film R-value is higher when the air is assumed to be still (interior), versus when it is assumed to be exposed to a 15 mph wind

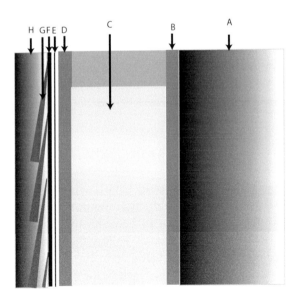

**Figure G-1. Components in Series of the Portion of a Typical Wall With Insulation in It**

(exterior). An air film that is being disturbed by wind allows more heat transfer than a film in still air. Thus, under an equation called Newton's Law of Cooling, the film coefficient or factor will be larger when the film is disturbed by wind than when not, much like the phenomenon of wind chill. At the same time, since R-value is lower, more heat transfer occurs. The R-value of the wind-exposed film outside will be lower than the inside (0.17 vs. 0.68)

Once the R-value is totaled (I in Table G-1), convert it to a U-value, which is 1/R (J) because calculating the average R-value of the wall (taking into account the different materials) can only be done using U-values. Multiply the U-value by the portion of the total area of the wall represented by framing. In standard 2x4 wall construction, this is 23% or .23. That means the rest of the wall represents 77% or .77 of the total wall area. Thus, in the case of wood framing, multiply the U-value for the wood framing by 0.23 and multiply the U-value for the rest of the wall by 0.77. Then add those two proportioned U-values together to come up with an overall U-value for the wall. Take this overall U-value and convert it back to an R-value to obtain an overall or average wall R-value. By adding in the effects of the studs' R-value, the calculated R-value for the wall drops from 14.6 to 11.6, a reduction of 3.

The calculated R-value of a wall is typically higher than the true R-value if it were to be tested in the laboratory. Therefore, discount the R-value by 9% or .09 if there is 2 x 4 framing in the wall, or by 14% or .14 if there is 2 x 6 framing. Once the R-value of the wall is discounted to fit reality, it is found that the true R-value or Corrected Wall R-Value, is actually only 10.6. This is a significantly lower R-value then first thought: 14.6.

## THE EFFECT OF WINDOWS AND DOORS

Almost every wall on a home could have a window or door. The Table G-2 takes into account not only the framing but also the windows and doors that typically exist in a home. Estimate an approximate R-value for windows and doors for a combined 3.3. Follow the same steps in calculating the true or corrected wall R-value as in the previous example. The only difference is that the wall area fraction for the framing and the rest of the wall is lowered to accommodate a roughly 10% or .10 wall area fraction for windows and doors. Thus, the other two wall area fractions for the framing and the rest of the wall are reduced by 10%.

Table G-1. Wall R Value When Take Framing Into Account

| Lettered/Wall Segment | Framing R | Other Wall R |
|---|---|---|
| A. Interior Air Film[1] | 0.68 | 0.68 |
| B. ½" Drywall[2] | 0.45 | 0.45 |
| C. 2x4 Wood Frame-3½"[3] | 4.38 | ---- |
| D. 3 ½" Fiberglass Batt Insulation[4] | ---- | 12.0 |
| E. 7/16" OSB Sheathing[5] | 0.51 | 0.51 |
| F. House Wrap Air Seal | 0.00 | 0.00 |
| G. Wood Lapped Siding, ½"x8", Beveled[6] | 0.81 | 0.81 |
| H. Exterior Air Film[7] | 0.17 | 0.17 |
| I. Total R (sum A - G) | 7.0 | 14.6 |
| J. U-Value (1/R) | 0.143 | 0.069 |
| K. Fraction of the Wall by Area (without windows)[8] | 0.23 | 0.77 |

**Calculate the Area Weighted Average U-Value** by multiplying each U-value (appropriate column in J) by its corresponding Fraction of the wall by area (K):

$0.143 \times 0.23 = 0.0329$; $0.069 \times .77 = 0.053$

and then add them together:

$0.0329 + 0.053 = 0.0859$ = Overall Wall U-value

**Calculate the Average Wall R value** from the Overall U Factor by dividing the U-value by 1:

$1/0.0859$ = R-value of 11.64 when the framing insulation values are also taken into account

| Reality Correction Factor[9] (9% or .09) | $0.09 \times 11.64 = 1.05$ |
|---|---|
| Actual Corrected Wall R-value | $11.64 - 1.05 = 10.6$ |

[1]Assumes still air; [2]ASHRAE Fundamentals; [3]R value of softwoods as used in framing is 1.25 per inch; [4]Assumes and R value of 3.5 per inch; [5]As per Structural Board Association; [6]Insulation Handbook, Bynum 2001; [7]Assumes a 15 mph wind; [8]Assumes Standard Framing – 2x studs, double top plate, single bottom plate, 3 stud corners, 2 stud openings, headers 2x or 4x with space between header and exterior sheathing, 2 stud interior/exterior partition intersections; [9]ORNL - Correction factor for 2x6 construction is 0.14 or 14%

Therefore, a wall that started out with an R-value of 14.6 when only the insulated portions were counted, has now dropped to a corrected R-value of 8.43 when the framing, windows, and doors in the home are taken into account. This is a drop of 42% from the original calculations, almost half of what was generally estimated to be the R-value in the wall. This is an even more dramatic reduction in R-value than when only the framing is taken into account.

If a similar evaluation were completed for an uninsulated wall assembly, it would have an R-value of R-3. An uninsulated upper ceiling with no attic floor would have an R-value of R-1. An uninsulated upper ceiling with an attic floor would have an R-value of R-2.

Another factor relating to R-values in reality is the effect of the radiant surface temperature on comfort levels. As mentioned before, the radiant surface temperature of the walls, windows and other surfaces in a home effect whether we feel warmer or colder. The better the insulation in the wall, the U factor of windows, etc. the closer these surface temperatures come to our own body temper-

**Table G-2.**
**R-Value When Corrected For Window Area**

| Lettered/Wall Segment | | | Windows and Doors R |
|---|---|---|---|
| I. Window R: 3 Pane with Low e | | | 3.3 |
| J. U-Value (Window) | | | 0.303 |
| K. Fraction of the Wall by Area (corrected for 10% windows added) | Frame[1] 0.21 | Wall[2] 0.69 | Window & Doors 0.10 |

**Calculate the Area Weighted Average U-Value** by multiplying each U-value by its corresponding Fraction of the wall by area (K):

.143x0.21=.03; .069x.69=.0476; .303x.1=.0303

and then add them together:

.03+.0476+.0303=.1079 = Overall Wall U-Value

**Calculate the Average Wall R value** from the Overall U-value by dividing the U-value by 1:

1/.1079 = R-value of 9.26 when both framing and window values are taken into account

| Reality Correction Factor[3] (9% or .09) | 0.09 x 9.26 = 0.834 |
|---|---|
| Actual Corrected Wall R value | 9.26 – 0.834 = 8.43 |

[1]Assumes Standard Framing – 2x studs, double top plate, single bottom plate, 3 stud corners, 2 stud openings, headers 2x or 4x with space between header and exterior sheathing, 2 stud interior/exterior partition intersections; [2]Wall fraction is everything else that is not framing or windows and therefore should have insulation in the walls; [3]ORNL - Correction factor for 2x6 construction is 0.14 or 14%

ature and the more comfortable we feel given a specific air temperature, humidity, etc. in the house. Thus, one of the reasons justifying high performance homes, the passive house concept, etc. is to not only save on utility bills, but also to help us feel more comfortable in the home.

# Appendix H

# Insulation Versus Air Sealing

## CALCULATING ENVELOPE ENERGY LOSS

Energy moves by conduction, convection, and radiation. Insulation handles conduction and radiation by trapping small air pockets within the insulation to slow conductive and radiant energy flow. However, insulation generally has a minimal effect on convective energy flow. Air sealing must be completed to control air transported energy loss. While many people say they can install insulation properly, few outside of the energy retrofit field see the value of air sealing and know how to do it. Heat can be lost through surface area (insulation problem) and through air infiltration (sealing problem). This appendix will demonstrate why both insulation and air sealing are important.

To make appropriate comparisons, the envelope losses on a home have to be quantified in terms of heating load (Btu/hr). Typically, solar and internal heat gains are not taken into account when calculating the heating load. Equations for surface heat loss and air transported heat loss are given below. Both use a concept called heating degree days (HDD), a term originally developed by the oil heating industry to help predict future fuel use. You can calculate a heating degree day by dividing the sum of the day's high and low temperatures by two and subtracting the result from 65°F (the heating balance point). 65°F is used as a standard because when the outside temperature is 65°F a house will stay at 70°F because of internal heat gains from occupants and mechanical systems, and most people feel comfortable at 70°F. The annual HDD can be found by calculating each heating degree-days in a year and adding them together.

For example, if a day has a high temperature of 75°F and a low of 45°F, add these numbers and divide by 2 to get 60°F. Subtract 60°F from 65°F to get 5°F HDD for that particular day of the year. As another example, look at a day with a high temperature of 25°F and a low temperature of -5°F. Add these numbers and divide by 2 to get 25°F + -5°F = 20°F/2 = 10°F. Subtract 10°F from 65°F and get 55°F for the HDD for that much colder day. A similar method is used for calculating cooling degree-days (CDD) using a balance point of 78°F. Because of the influence of conditions such as relative humidity and shade in determining the comfort level in a home, calculating the CDD is much less reliable for predicting cooling costs than HDD is for predicting heating costs.

**Surface Heat Loss Equation**

The surface heat loss equation in Btu/hr is given as $(A \times \Delta T \times t)/R$

Where:

    A=   Area in square feet
    $\Delta T$  =  Difference in temperature in °F
    t =   Time in hours
    R=   R-Value—the total resistance of assembly to heat flow

Remember that $R = 1/U$. R-values can be added, U values cannot.

To calculate the surface heat loss for an entire season, use the following equation:

$(A \times \#HDD \times 24 \text{ hours})/R$

Where:

    A=   Area in square feet
    #HDD  =  number of heating degree days in the season for the area
            the house is located
    R=   R-Value-the total resistance of assembly to heat flow

## Air Transported Heat Loss Equation

Air transported heat loss is always calculated in Btu per cubic foot of air per hour. The air transported heat loss equation per hour is

$$V \times ACH \times (0.0182 \text{ Btu/ft}^3,°F) \times \Delta T$$

Where:

     V    = Volume of the building
  ACH   = Air change per hour
  $0.0182 \text{ Btu/ft}^3,°F$ = Specific heat of air
    $\Delta T$ temperature difference between inside and outside

To calculate the air transported heat loss for an entire season, use the equation

$$V \times ACH \times 0.0182 \times \#HDD \times 24 \text{ hrs}$$

Where:

  V = Volume of the building
  ACH = Air change per hour
  $0.0182 \text{ Btu/ft}^3,°F$ = Specific heat of air
  #HDD = number of heating degree days in the season for the area the house is located

The specific heat of air constant is the number of Btus that are needed to raise one cubic foot of air 1°F. It is a "constant" because it never changes.

Comparison of Insulation (surface heat loss) versus Air Sealing (air transport loss)

Here are examples of different types of heat losses in a variety of homes under different conditions.

## Example 1. Surface Heat Loss of an Insulated Wall.

The first example is an 8' x 12' wall with no windows. The square footage of the wall is 96 ft². For this example, it is 70°F inside the house and 30°F outside. This equals a temperature differential of 40°F. For this example, the house is located where there are 7,200 heating degree-days or HDDs and the walls have an R-value of R-11. Remember that theoretical does not equal reality, so discount the R-11 by 9%. Therefore, treat the R-value in the wall as R-10 rather than R-11. What is the heat loss in Btu per hour and how many Btus are lost during an entire heating season?

Heat Loss Equation
    (A x ΔT x t)/R or (96 sq. ft x 40°F x 1 hr)/10 = 384 Btu/hour

Btu per heating season
    A x #HDD x 24 hours/R

    (96 ft² x 7,200 HDD x 24 hrs)/10 =

    1,658,880 Btu per heating season

## Example 2. Surface Heat Loss and Air Transported
## Heat Loss Uninsulated Home

A home with no insulation in the walls or ceiling has an R-value of 1 in the upper ceiling/attic and an R-value of 3 in the walls. This is a 20′ x 30′ home on a slab with 8-foot high walls, 10% of the wall area is windows and doors. The temperature inside is 70°F while the temperature outside is 30°F and it also has 7,200 heating degree-days. In this example, the home actually experiences 1.25 air changes per hour. With two sides of the house at 20 feet, the other two sides at 30 feet, the perimeter of the house is 100 feet around: (20x2) + (30x2) = 100. Multiple by 8 feet high to get 800 ft² of rough wall area. Windows and doors represent 10% of the wall area, so the windows and doors represent 800x0.1 or 80 ft². To calculate the actual wall area where there are no windows and doors, subtract the 80 ft² of windows from the 800 ft² of total wall area to get 720 ft² of actual wall area. This 20′ x 30′ home would have 600 ft² of ceiling area. Go a step further and calculate the volume of this home: 30′ x 20′ x 8′ = 4,800 ft³. How many Btus per hour of surface losses are there?

    (A x ΔT x t)/R

In the walls: (700 ft² x 40°F x 1 hr)/3 = 9600 Btu/hour
In the ceiling: (600 ft² x 40°F x 1 hr)/1 = 24,000 Btu/hr
    This gives a total of 33,600 Btu/hour

How many Btu of losses per heating season?

    A x #HDD x 24 hours/R

Walls = (720 ft² x 7,200 HDD x 24 hrs)/3 = 41,472,000 Btu/yr
Ceiling = (600 ft² x 7,200 HDD x 24 hrs)/1 = 103,680,000 Btu/yr
      For a total of 145,152,000 Btu/yr

To calculate the heat loss through air infiltration, use the following analysis.

Air infiltration per hour:
      V x ACH x (0.0182 Btu/ft³,°F) x $\Delta$T

      4,800 ft³ x 1.25 ACH x (0.0182 Btu/ft³,°F) x 40°F = 4,368 Btu/hr

Air infiltration loss for heating season:
      V x ACH x 0.0182 x #HDD x 24 hrs

      4,800 ft³ x 1.25 ACH x (0.0182 Btu/ft³,°F) x 7,200 HDD x 24 hrs = 18,869,760 Btu/heating season

*Analysis*
      Examples 1 and 2 show the major differences that insulation makes to a home. A home with a total area of walls and ceiling that was roughly 13 times the size of a wall (1300 ft² versus 96 ft²) but with roughly 1/5 the insulation (R-10 down to approximately and R-2) used about 87 times the energy (33,600 Btu per hour versus 384 Btu per hour). These are the surface losses due to no insulation. This means that the house used roughly seven times more energy per square foot because it had about 1/5 the R-value. At the same time, there are significant losses from air infiltration as well.

## Example 3. Surface Heat Loss and Air Transported—
## Heat Loss in a Minimally Insulated Home
      The next example is a minimally insulated home. Assume this house is similar to Example 2 with a 20' x 30' on slab home with 8 ft walls, 10% wall area as windows and doors, 7200 heating degree days, and inside temperature of 70°F and an outside temperature of 30°F, 3 1/2 inches of R-11 fiberglass in the walls along with 6 inches of R-19 fiberglass in the ceiling, and 1.25 air changes per hour.
      BPI standards require that voids in insulation must be taken into account using a BPI correction table. The "Effective R-Values for Batt In-

sulation" table in the building analyst standards of BPI shows effective R-values of 2.5 per inch if the gaps represent less than 2.5% of the area insulated ("Good" installation). An R-value of 1.8 per inch if the gaps represent more than 2.5% of the area insulated, ("Fair" installation). An R-value of 0.7 per inch when the gaps are over 5% of the area insulated, ("Poor" installation). It is worth noting that a 2.5% gap could be roughly equal to only a 3/8-inch space along the length a 14.5-inch wide batt.

Calculating the heat loss through surface area of the minimally insulated home gives:

Btu per hour of surface losses:
$(A \times \Delta T \times t)/R$

Walls = (720 ft² x 40°F) x 1 hour/10 = 2,880 Btu/hr
Ceiling = (600 ft² x 40°F) x 1 hours/17 = 1,412 Btu/hr
For a total of 4,292 Btu/hr

How many Btu of losses per heating season?

$A \times \#HDD \times 24 \text{ hours}/R$

Walls = (720 ft² x 7,200 HDD x 24 hrs)/10 = 12,441,600 Btu/yr
Ceiling = (600 ft² x 7,200 HDD x 24 hrs)/17 = 6,098,824 Btu/yr

This gives 18,540,424 Btu/yr, quite a bit lower than the 145,000,000 Btu/yr found in the same, but uninsulated home.

To calculate the heat loss through air infiltration, use the following analysis:

Air infiltration per hour:

$V \times ACH \times (0.0182 \text{ Btu}/ft^3, °F) \times \Delta T$

4,800 ft³ x 1.25 ACH x (0.0182 Btu ft³, °F) x 40°F = 4,368 Btu/hr

And heating season analysis of heat loss from air infiltration:

$V \times ACH \times 0.0182 \times \#HDD \times 24 \text{ hrs}$

4,800 ft³ x 1.25 ACH x (0.0182 Btu/ft³,°F) x 7,200 HDD x 24 hrs
= 18,869,760 Btu/heating season

Note that the losses due to air infiltration are the same as the uninsulated home since air infiltration has to do with leakage but not insulation and both these homes had an ACH of 1.25.

### Example 4. Surface Heat Loss and Air Transported Heat Loss in an Air-Sealed and Supplementally Insulated Home

In this final example, assume the same parameters as the minimally insulated home in the last example, but it has now been air sealed and insulated. So all the characteristics are the same as the minimally insulated home, except that the attic has been improved to a true R-38/40 and the ACH has been reduced from 1.25 down to 0.35. Calculating the heat loss through surface area of the minimally insulated home gives—

**Btu per hour of surface losses:**

$$(A \times \Delta T \times t)/R$$

Walls = (720 ft² x 40°F) x 1 hr/10 = 2,880 Btu/hr
Ceiling = (600 ft² x 40°F x 1 hr)/38 = 632 Btu/hr

This gives 3,512 Btu/hr, almost 20% lower than the previous minimally insulated home example.

How many Btu of surface losses per heating season?

$$A \times \#HDD \times 24 \text{ hours}/R$$

Walls = (720 ft² x 7,200 HDD x 24 hrs)/10 = 12,441,600 Btu/yr
Ceiling = (600 ft² x 7,200 HDD x 24 hrs)/38 = 2,728,421 Btu/yr

This gives 15,170,021 Btu/yr, almost 20% lower than the previous minimally insulated home example above.

Heat loss through air infiltration:

$$V \times ACH \times (0.0182 \text{ Btu}/\text{ft}^3,°F) \times \Delta T$$

4,800 ft³ x 0.35 ACH x (0.0182 Btu/ft³, °F) x 40°F = 1,223 Btu/hr

And heating season analysis of heat loss from air infiltration:

V x ACH x 0.0182 x #HDD x 24 hrs

4,800 ft³ x 0.35 ACH x (0.0182 Btu/ft³,°F) x 7,200 HDD x 24 hrs = 5,283,533 Btu/heating season

Note that the losses due to air infiltration have gone down significantly in proportion to the decrease in ACH. Thus, since an ACH of 0.35 is approximately 28% of an ACH of 1.25, the heat loss from air infiltration has been reduced to about 26% of that in the minimally insulated home example.


EFFECTS OF A 7% VOID IN INSULATION

**Example 1 where much insulation is expected (very cold weather area attic):**

> **1000 square foot attic.**
> 930 square feet at R-60
> 70 square feet void at R-2
>
> **Divide square feet of each area by R-Value**
> 930 square feet divided by 60 = 15.5
> 70 square feet divided by 2 = 35
>
> **Add the results together**
> 15.5+35 = 50.5

**Take total square feet of attic and divide by sum of previous calculation**
> 1000/50.5 = 19.8 Weighted R Value

This means that the R-value went from R-60 to R-19.8 due to the void = 19.8/60 = .33.

Note that reducing the R value to half its intended R value (R-30) occurs with a void of only 3.6% or 36 square feet in this example: 36 square feet divided by 2 = 17.8, 17.8 + 15.5 = 33.3 and 1000/33.3 = 30.0 Weighted R value

**Example 2 where far less insulation is expected (a wall in a warm weather area):**

> **1000 square foot wall.**
> 930 square feet at R-13
> 70 square feet void at R-2

> **Divide square feet of each area by R-Value**
> 930 square feet divided by 13 = 71.5
> 70 square feet divided by 2 = 35

> **Add the results together**
> 71.5+35 = 106.5

**Take total square feet of wall and divide by sum of previous calculation**
> 1000/106.5= 9.38 Weighted R Value

Which means that the R value went from R-13 to R-9.38 due to the void = 9.38/13 = 0.72 the R value: now only slightly more than 2/3 of the R value intended for the wall because of the 7% void in the insulation)

Note that reducing the R value to half its intended R value (R-6.5) occurs with a void of 16.72% or 165 square feet in this example: 164.72 square feet divided by 2 = 82.36, 82.36 + 71.5 = 153.9 and 1000/153.9 = 6.5 Weighted R value

**Example 3 in the middle range (warmer weather climate attic):**

> **1000 square foot attic.**
> 930 square feet at R-30
> 70 square feet void at R-2

> **Divide square feet of each area by R-Value**
> 930 square feet divided by 30 = 31
> 70 square feet divided by 2 = 35

**Add the results together**
31+35 = 66

**Take total square feet of Attic and divide by sum of previous calculation**
1000/66= 15.15 Weighted R Value

Which means that the R value went from R-30 to R-15.5 due to the void = 15.5/30 = 0.52 the R value: now only about 1/2 of the R value intended for the attic to have because there is a 7% void in the insulation)

In conclusion, the thicker you want to insulate an area with a given insulation, the smaller the void needed to bring the insulation value down to half of what it was intended it to be.

# Appendix 1

# Estimating Insulation Requirements

Here are examples of how to calculate how much insulation to put in a home that has no insulation.

## IN ATTICS

Assume that the attic should be brought up to an R-40 level. Do not include any knee wall insulation that could also be installed. Below is the area for each roof section on the home:

L x W = 15 ft x 12 ft = 180 ft²
Addition = 16 ft x 8 ft = 128 ft²
Main, 2nd floor = 26 ft x 8 ft = 208 ft²
Total Area = 516 ft²

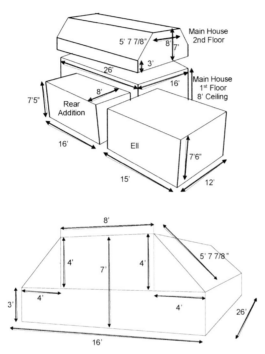

Use the coverage chart provided by the manufacturer for blown-in cellulose insulation. For this example, assume the chart shows that for an R-value of R-40, the minimum cellulose insulation thickness throughout the attic needs to be 12 inches. In addition, it also shows that if you correct for 2 x 6 framing that is 16 inches on center, each bag should be able to cover 18.1-ft² maximum. Many installers will also add 15% to this calculation to account for over packing and spillage.

Calculate how many bags are needed to adequately insulate the total attic area:

516 ft²/18.1 ft²/1 bag = 28.5 bags

To add 15% waste allowance to account for over packing and spillage, multiply the results by 1.15:

28.5 bags x 1.15 = 32.78 bags = 33 bags

Thus, the number of bags of cellulose insulation to bring the sample home from R-0 to R-40 in the attic is 33 bags.

It is very important to label all the numbers with units and keep track of those units. This includes making sure the units cancel out each other. For instance, if you find in the equation that there are cubic feet as a unit on the top part of the equation and cubic feet as a unit on the bottom part of the equation, you can cancel those out. Likewise, if you find that there are pounds as a unit on the top part of the equation and pounds as a unit on the bottom part of the equation, you can cancel those out as well, for example:

$$\frac{2500 \text{ lbs}}{50 \text{ lbs/bag}} = 50 \text{ bags}$$

IN WALLS

**With Manufacturers' Coverage Charts:** In this example, assume the manufacturers' coverage chart for installing wall insulation states that if there are 2 x 4 studs that are 16 inches on center you can cover 33.8 ft² of the wall per bag of R-13 insulation. If you are dense packing 1200 ft² of wall, use the following equation to calculate how many bags are needed:

Wall ft² ÷ ft²/Bag = Total Bags Needed
1200 ft² ÷ (33.8 ft²/Bag) = 35.5 = 36 bags

To correct for 15% waste allowance, multiply the result by 1.15:

35.5 bags x 1.15 = 41 bags

**Density**: Assume the chart shows that the weight per square foot of wall area is 0.758 lbs/ft². Divide that by the volume behind each square foot of wall space to calculate the density of the dense pack insulation. Since the room in the 2x4 wall space is about 3 1/2 inches (0.29 feet), you can calculate the volume behind each square foot area of wall space through the following equation:

1 ft x 1 ft x 0.29 ft = 0.29 ft³ in each square foot of wall or 0.29 ft³/ft²

Divide the 0.758 pounds in the wall per square foot by 0.29 ft³/ft² to get density in pounds per cubic foot:

$$\frac{0.758 \text{ lbs.}/\text{ft}^2}{0.29 \text{ ft}^3/\text{ft}^2} = 2.6 \text{ lbs.}/\text{ft}^3$$

While at first glance it appears that the density falls below the roughly 3.5 pounds per cubic foot guideline in determining whether dense pack has been packed densely enough, this is actually the number obtained accounting for the space taken up by framing. The framing actually reduces the volume being insulated per square foot of wall space, typically by about 23%.

**Without Manufacturers' Coverage Charts:** If there is no manufacturer coverage chart, calculations can still be made. The goal is to dense pack to 3.5 pounds per cubic foot and the bag states that each bag contains 36 pounds of insulation. First, calculate the volume in the 1200 ft² wall:

1200 ft² x 3.5 in/12 in per ft =
1200 ft² x 0.29 ft = 350 ft³

Multiply this volume in the wall by the dense pack standard to determine how many pounds should be installed in the wall for proper dense packing:

Volume x Density = Weight of Insulation
350 ft³ x 3.5 lbs/ft³ = 1225 lbs.

Divide this number by the weight per bag to find out how many bags are needed:
1225 pounds =34.03 bags = 35 bags
36 lbs/bag

Thus, even without having the manufacturers' coverage chart, you can calculate how many bags of insulation to put into a wall for it to be properly dense packed.

Taking a step back, calculate how much of a wall will actually need insulation after correcting for the windows and doors. First, calculate the total wall area without correcting for windows and doors:

First Flr = 2(16' x 8') + 2(26' x 8') = 672 ft2
2nd Flr = 2(3'x 26') + 2(6' x 26')* + 2(16' x 7') = 692 ft2
Addition = 2(8' x 7'5") + (16' x 7'5")** = 237 ft2
Ell = (12' x 7'6")***+ 2(15' x 7'6") = 315 ft2

* *Sloped ceiling, from walk-through notes*
** *Only count one long wall, other wall abuts the rest of home*
*** *Only count one short wall, other wall abuts the rest of home*

This gives 1916 ft² of wall area.

Note that this does not take into account the sloped cutouts that occur on the gable ends as a result of the sloped roof. To do so, subtract roughly two 4' x 4' triangles or 16 ft² of area from the second floor calculation. Since, 16 ft² does not even come close to taking it out of the 10% accuracy standard for BPI, it can be ignored for calculation purposes.

Now assume there are eight windows in this home that are 12.5 ft² each and two doors that are 20 ft² each. Totaling the square footage for these windows and doors:

8 x 12.5 ft² = 100 ft²
2 x 20 ft² = 40 ft²
Total ft² = 140 ft²

Subtract this total square footage for the windows and doors from the previously calculated total square footage for the walls:

1916 ft² – 140 ft² = 1776 ft²

The net wall area requiring insulation is 1776 ft² on the sample home. If you can cover 33.8 ft²/bag (with framing already taken into account) then:

$$\frac{1776 \text{ ft}^2}{33.8 \text{ ft}^2/\text{bag}} = 52.5 \text{ bags} = 53 \text{ bags}$$

# Appendix J

# DOE Major U.S. Climate Zones and Climate Data for Select Cities

In 2002, a DOE lab, Pacific Northwest National Laboratory, developed a climate classification that could be used in building codes. A county-by-county map was developed to identify the climate zone for each county in the United States. Eight zones were distinguished by the number of cooling degree-days or heating degree-days and humidity, among other factors. Table J-1 shows these now well-established eight zones with their Moist (A), Dry (B), and Marine (C) subcategories. Figure J-1 shows this information in a map format and Table J-2 shows data from various cities in the U.S.

These climate zones, the HDD, the CDD, and other factors are used in determining the heating and cooling requirements for a given home.

**Table J-1. Eight zones developed by the DOE**

| Zone Number | Thermal Criteria |
|---|---|
| 1 | More than 9000 CDD50°F |
| 2 | between 6300 and up to 9000 CDD50°F |
| 3A or B (Moist or Dry) | between 4500 and up to 6300 CDD50°F AND less than 5400 HDD65°F |
| 4A or B (Moist or Dry) | 4500 or less CDD50°F AND less than 5400 HDD65°F |
| 3C (Marine) | 3600 or less HDD65°F |
| 4C (Marine) | between 3600 and up to 5400 HDD65°F |
| 5 | between 5400 and up to 7200 HDD65°F |
| 6 | between 7200 and up to 9000 HDD65°F |
| 7 | between 9000 and up to 12600 HDD65°F |
| 8 | More than 12600 HDD 65°F |
| HDD65°F= heating degree days based on temperature below 65°F outside CDD50°F=cooling degrees day based on temperature above 50° F outside | |

These factors help in designing the correct capacity heating and cooling systems necessary for occupants to live comfortably while at the same time not paying for an oversized and less efficient system. Heating design and cooling design parameter subfactors also provide useful information. These include the following winter and summer design parameters (for more information, see the ASHRAE tables available through ASHRAE Standard 90.1 amongst other ASHRAE documents):

Winter Design Dry-Bulb (97.5%) in degrees Fahrenheit (°F): This temperature is the regular outdoor air temperature dividing line *below which* 97.5% of the temperatures taken during a typical winter season (hourly low temperatures considered) are found. In the United States, the winter season is defined as the months of December through February.

Summer Design Dry-Bulb and Mean Coincident Wet Bulb Temperature (2.5%) in degrees Fahrenheit (°F): This temperature is the regular outdoor air temperature dividing line *above which all but 2.5%* of the temperatures taken during a typical summer season (hourly high temperatures considered) are found. In the United States, the summer is defined as the months of June through September.

Cooling Degree Hours assuming a 74°F base (CDH74°F): This is similar to the Cooling Degree Days at 50°F definition above except that it is based on hours, not days, and it assumed outdoor temperature is 74°F rather than 50°F.

### Table J-2. Climate Data For Select Cities

(complete list and U.S. Map available on the above websites)

| City | Heating Degree Day | Summer Design Temp | Summer Mean Daily Range |
|---|---|---|---|
| Baltimore, MD | 4654 | 91/75 | 21 |
| Boston, MA | 5634 | 88/71 | 16 |
| Chicago, IL | 6155 | 89/74 | 20 |
| Dallas, TX | 2363 | 100/75 | 20 |
| Detroit, MI | 6232 | 88/72 | 20 |
| Houston, TX | 1396 | 95/77 | 18 |
| Los Angeles, CA | 2061 | 80/68 | 15 |
| Las Vegas, NV | 2709 | 106/65 | 30 |
| Miami, FL | 214 | 90/77 | 15 |
| New York City, NY | 4871 | 89/73 | 17 |
| Philadelphia, PA | 5144 | 90/74 | 21 |
| Pittsburgh, PA | 5987 | 86/71 | 22 |
| Salt Lake City, UT | 6052 | 95/62 | 32 |
| San Francisco, CA | 3001 | 77/63 | 20 |
| Washington, DC | 4224 | 91/74 | 18 |

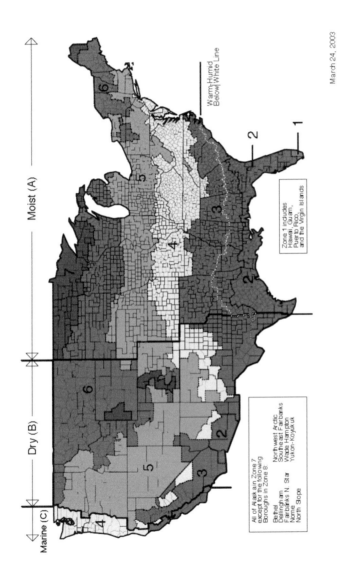

**Figure J-1. Map of the eight zones**

http://www1.eere.energy and http://www.energysavers.gov/tips/insulation

# Appendix K

# Meter Reading

### Real Btu/hour Appliance Calculation from Meter Dial Timing

| Seconds for one rotation of the dial | Size of the Meter Dial in Cubic Ft. | | |
|---|---|---|---|
| | ½ Cu. Ft. | 1 Cu. Ft. | 2 Cu. Ft. |
| | *Thousands of Btu/hr. @ 1,000 Btu/Cu. Ft.* | | |
| 12 | 150 | 300 | 600 |
| 13 | 138 | 277 | 553 |
| 14 | 129 | 257 | 514 |
| 15 | 120 | 240 | 480 |
| 16 | 113 | 225 | 450 |
| 17 | 106 | 212 | 424 |
| 18 | 100 | 200 | 400 |
| 19 | 95 | 190 | 379 |
| 20 | 90 | 180 | 360 |
| 21 | 86 | 171 | 343 |
| 22 | 82 | 164 | 327 |
| 23 | 78 | 157 | 313 |
| 24 | 75 | 150 | 300 |
| 25 | 72 | 144 | 288 |
| 26 | 69 | 139 | 277 |
| 27 | 67 | 133 | 267 |
| 28 | 64 | 129 | 257 |
| 29 | 62 | 124 | 248 |
| 30 | 60 | 120 | 240 |
| 31 | 58 | 116 | 232 |
| 32 | 56 | 113 | 225 |
| 33 | 55 | 109 | 218 |
| 34 | 53 | 106 | 212 |
| 35 | 51 | 103 | 206 |
| 36 | 50 | 100 | 200 |
| 37 | 49 | 97 | 195 |
| 38 | 47 | 95 | 190 |
| 39 | 46 | 92 | 185 |
| 40 | 45 | 90 | 180 |
| 41 | 44 | 88 | 176 |

| Seconds for one rotation of the dial | Size of the Meter Dial in Cubic Ft. | | |
|---|---|---|---|
| | ½ Cu. Ft. | 1 Cu. Ft. | 2 Cu. Ft. |
| | Thousands of Btu/hr. @ 1,000 Btu/Cu. Ft. | | |
| 42 | 43 | 86 | 171 |
| 43 | 42 | 84 | 167 |
| 44 | 41 | 82 | 164 |
| 45 | 40 | 80 | 160 |
| 46 | 39 | 78 | 157 |
| 47 | 38 | 77 | 153 |
| 48 | 38 | 75 | 150 |
| 49 | 37 | 73 | 147 |
| 50 | 36 | 72 | 144 |
| 51 | 35 | 71 | 141 |
| 52 | 35 | 69 | 139 |
| 53 | 34 | 68 | 136 |
| 54 | 33 | 67 | 133 |
| 56 | 32 | 64 | 129 |
| 57 | 32 | 63 | 126 |
| 58 | 31 | 62 | 124 |
| 59 | 30 | 61 | 122 |
| 60 | 30 | 60 | 120 |
| 61 | 30 | 59 | 118 |
| 62 | 29 | 58 | 116 |
| 63 | 29 | 57 | 114 |
| 64 | 29 | 56 | 112 |
| 65 | 28 | 55 | 111 |
| 66 | 28 | 54 | 109 |
| 67 | 27 | 54 | 107 |
| 68 | 27 | 53 | 106 |
| 69 | 26 | 52 | 104 |
| 70 | 26 | 51 | 103 |
| 71 | 25 | 50 | 101 |
| 72 | 25 | 50 | 100 |
| 73 | 25 | 49 | 99 |
| 74 | 24 | 49 | 97 |
| 75 | 24 | 48 | 96 |
| 76 | 24 | 47 | 95 |
| 77 | 23 | 47 | 94 |
| 78 | 23 | 46 | 92 |
| 79 | 23 | 46 | 91 |
| 80 | 23 | 45 | 90 |
| 81 | 22 | 44 | 89 |
| 82 | 22 | 44 | 89 |

| Seconds for one rotation of the dial | Size of the Meter Dial in Cubic Ft. | | |
|---|---|---|---|
| | ½ Cu. Ft. | 1 Cu. Ft. | 2 Cu. Ft. |
| | Thousands of Btu/hr. @ 1,000 Btu/Cu. Ft. | | |
| 83 | 22 | 43 | 87 |
| 84 | 21 | 43 | 86 |
| 85 | 21 | 42 | 85 |
| 86 | 21 | 42 | 84 |
| 87 | 21 | 41 | 83 |
| 88 | 21 | 41 | 82 |
| 89 | 20 | 40 | 81 |
| 90 | 20 | 40 | 80 |
| 92 | 20 | 39 | 78 |
| 94 | 19 | 38 | 76 |
| 96 | 19 | 38 | 75 |
| 98 | 18 | 37 | 74 |
| 100 | 18 | 36 | 72 |
| 102 | 18 | 35 | 71 |
| 104 | 17 | 35 | 69 |
| 106 | 17 | 35 | 68 |
| 108 | 17 | 35 | 67 |
| 110 | 16 | 33 | 66 |
| 112 | 16 | 32 | 64 |
| 114 | 16 | 32 | 63 |
| 116 | 16 | 31 | 62 |
| 118 | 15 | 31 | 61 |
| 120 | 15 | 30 | 60 |
| 122 | 15 | 30 | 59 |
| 124 | 15 | 29 | 58 |
| 126 | 14 | 29 | 57 |
| 128 | 14 | 28 | 56 |
| 130 | 13 | 28 | 55 |
| 135 | 13 | 27 | 53 |
| 140 | 13 | 26 | 51 |
| 145 | 12 | 25 | 50 |
| 150 | 12 | 24 | 48 |
| 155 | 12 | 23 | 46 |
| 160 | 11 | 22 | 45 |
| 165 | 11 | 22 | 44 |
| 170 | 11 | 21 | 42 |
| 175 | 10 | 21 | 41 |
| 180 | 10 | 20 | 40 |

# Appendix L

# Refrigerant Charge Testing

All refrigerant handling and testing requires that an individual be an EPA section 608 certified and qualified technician. WARNING: Do not do these tests if you are not certified. Before testing and adjusting the refrigerant charge on an air-conditioning unit, make sure the outdoor temperature is at least 60°F. Measure and then adjust the airflow in the air-conditioning unit to match the manufacturer's recommendations, which are typically around 400 CFM per ton ±20%. Test and seal the ductwork, clean the condenser coil and make sure it is completely dry. Make sure you have the required equipment for testing and adjusting. The equipment includes:

1)  a Department of Transportation (DOT) approved recovery cylinder for removing any excess refrigerant,
2)  a digital thermometer with a wet cloth or a sling psychrometer for determining a wet bulb reading, and
3)  a refrigeration gauge set.

Refrigerant technicians must use different tests depending upon which valve system is used on the air-conditioning unit. The temperature difference between the very middle of the condenser and the liquid service valve located outside is called the "subcooling." Use the subcooling test when there is a thermal expansion valve system. When conducting this test, make sure the outdoor temperature is at least 60°F and have the heat pump or air conditioner running in the cooling load a minimum of 10 minutes before starting the test.

The temperature difference between the evaporator and the compressor inlet is called the "superheat." Use the evaporator superheat test when evaluating a capillary tube or fixed orifice system. In this case, the airflow must be at least 400 CFM per ton. Once again, the outdoor temperature must be at least 60°F and the test should not be conducted when the outdoor temperatures are very high.

The steps to conduct the sub-cooling test include

1)   measuring and converting liquid pressure at the liquid service valve,
2)   measuring the temperature of the liquid refrigerant as it leaves the condenser,
3)   subtracting this liquid refrigerant temperature from the converted condensing temperature to obtain the measured sub cooling, and
4)   comparing this number with the correct sub cooling indicated on a permanent sticker in the condenser, the manufacturer's literature, or slide rule.

If there is a 3°F variation or more, adjust the levels of the refrigerant. After adjusting the levels, let the system run for 10 minutes and then repeat the test.

The steps to conduct the evaporator superheat test include

1)   verifying that there is adequate air flow,
2)   measuring the dry bulb temperature of air that enters the outdoor coil,
3)   measuring the wet bulb temperature of the return air at the air handler,
4)   using the superheat table to determine what the recommended superheat temperature should be, and
5)   measuring and evaluating the compressor-suction pressure at the suction service valve.

If the superheat temperature is low, remove the refrigerant, wait 10 minutes, and then run through the test again. If the superheat temperature is high, add refrigerant, wait 10 minutes and then run to the test again.

# Appendix M

# Certifications in the Home Energy Arena

## BPI, RESNET, AND DOE/NREL

BPI, RESNET, AEE and DOE/NREL. The Building Performance Institute (BPI) (www.bpi.org) Residential Energy Services Network (RESNET) (http://resnet.us) and the Association of Energy Engineers (AEE) are the leading organizations in providing certifications in the energy arena. Getting a certification from one certifier does not keep one from obtaining certifications from other providers. Regional preferences by the public may exist. Perhaps a significant difference between BPI and RESNET certifications is that RESNET requires the auditor, contractor, etc. to contract and affiliate with "Providers" – those that provide the ratings to the public. AEE allows you to work independently. In addition, more certifications are under development (See DOE/NREL Certifications below).

### New vs. Existing Buildings

It is important to differentiate new construction energy issues from existing home energy issues. Obviously, it is easier to make a home more energy efficient if the necessary components can be installed when the building is being built. RESNET was originally organized to cover new construction energy issues, while BPI historically concentrated on energy issues in existing homes. However, RESNET has recently established standards in the existing home arena and BPI standards can be used to evaluate new homes as well.

### Evaluation vs. Retrofit

In addition to the "new" and "existing" division, the energy industry also differentiates on whether you are evaluating/auditing the building for energy issues, or whether you will work as a contractor/retrofitter to make the home more energy efficient. BPI has designations for auditors ("Building Analysts" or BA) and retrofitters ("Envelope Professionals" or EP), while RESNET has historically concentrated on auditors ("Rater"). However, RESNET now has begun certifying contractors under their "Energy Smart" program.

### Inspector/Evaluator/Auditor/Analyst/Rater ("Inspector")

This group inspects or tests the house to determine the condition of the house from the energy standpoint. This person never does actual work on the home outside of testing and evaluation.

### Retrofitter/Contractor/Installer/Insulator/Air Sealer ("Retrofitter")

This group uses the information from the inspector to do the actual work on the building to make it more energy efficient. This work includes sealing leaks between the outside and inside of the home, adding attic, wall, and floor insulation, etc. Retrofitters do some testing, but typically only to see if their work is productive or to address possible safety issues. Retrofitters are required to "test out" at the end of each day and at the end of the job to ensure their work has not compromised occupant safety.

### Compliance to Standards/Quality Control

Both BPI and RESNET certifications involve some type of quality control over the evaluation and retrofitting process. One of the major differences between the two programs is the type of arrangement for providing the quality control. Under BPI, the quality of auditors and contractors are under the supervision of BPI. Under RESNET, both Raters and contractors must be evaluated for quality control by their "Providers" who also provide inspector services to the public.

### Difficulty to Achieve

Not only do the different certifications seem confusing, trying to determine the degree of difficulty of achieving each of the different levels of certification is daunting. In the below chart, an attempt has been made to list the different certifications in order of difficulty so that a more realistic evaluation can be made.

**Note:** Most states *do not require* licensing in order to offer energy auditing/inspection services to the public. However, many states may *require* licensing in order to do retrofit work on buildings.

## CERTIFICATIONS (ROUGHLY IN ORDER OF INCREASING DIFFICULTY OF ACHIEVING, DEPENDING ON THE CIRCUMSTANCES)

### Home Energy Survey Professional—HESP (RESNET) Type

Inspector—No specific training is required but you must pass an exam on specific topics relating to the Home Energy Survey. You must be affiliated with a RESNET provider to oversee quality control on your work (a contract with the provider is required). To find a HESP provider go to: http://www.resnet.us/programs/survey_providers

50 question open book written test

No Field Exam

### Residential Building Envelope Air Leakage Control Installer—ALCI—(BPI) Type: Retrofitter

This certification is a special testing circumstance—there is a very basic field exam with props that is directed solely at proving ones' skills at caulking, blocking bypasses, foaming, etc. Technically there is no "written exam" since questions are delivered verbally by the field examiner and the examinee answers them verbally. So, this certification involves both a question-based verbal exam as well as the field exam, but both are limited to the examinee proving knowledge and skills in the air sealing arena (no blower door or combustion appliance zone/closet (CAZ) testing, no auditor oriented inspection skills required, etc.)

### Diagnostic Home Energy Survey Professional—DHESP (RESNET) Type: Inspector

For this certification, you must first be certified as a Home Energy Survey Professional. In addition to having the knowledge base of the HESP, you must also prove an ability to conduct diagnostic testing on both the building envelope and ductwork. You must be affiliated with a RESNET provider to oversee quality control on your work (a contract with and fees to the provider are required). You are designated by your RESNET provider as a DHESP. In most cases, this probably means that your provider will require you to show him/her that you can properly conduct these diagnostic tests and show familiarity with any special concerns with your local climate.

### Building Analyst—BA (BPI) Type: Inspector

There is no specific training required for this certification, but some training is clearly still necessary, at least in the field to actually learn how to use a blower door, a combustion analyzer, etc. You pass a written exam as well as a field exam for this certification. This is the oldest and most common certification available for evaluating existing homes.

### Envelope Professional—EP (BPI) Type: Retrofitter

There is no specific training required but some training is clearly still necessary, at least in the field to actually learn how to use a blower door, a combustion analyzer, etc. There is a written exam as well as a field exam required for this certification. The written test and the field exam for this certification are almost the same as for the BPI Building Analyst. This is the oldest and most common certification available for retrofitting existing homes.

### Building Performance Auditor—BPA (RESNET) Type: Inspector

This certification is not yet currently available but will likely be similar to the BPI Building Analyst Certification.

### Home Energy Rater—HERs—Rater—RESNET Type: Inspector

It is recommended (but not required) to take a roughly 1-week course, pass the open book RESNET Energy Rater exam and, in the process, show proficiency:

— in the field in blower door testing
— in evaluating a home in accordance with the Mortgage Industry National Home Energy Rating System
— entering data collected on a home evaluation onto a computer program
— complete two energy ratings in the presence of the trainer as well 3 supplemental ratings (total of 5).

You must be affiliated with a RESNET provider to oversee quality control on your work (a contract with and fees to the provider are required). For a list of rater providers, see http://www.resnet.us/programs/search_directory

### Comprehensive Home Energy Rater—CHERS— (RESNET) Type: Inspector

For this certification, you must be a Rater and then show proficiency in:

— Providing a work scope according to BPI BA standards

      — Providing combustion testing according to BPI BA standards
      OR
Must be a BA and then also be able to accomplish:
      — Building simulation and performance analysis
      — Provide HERS ratings in accordance with the Mortgage Industry
        National Home Energy Rating System
      OR
      — Be certified by the Texas Home Energy Rating Organization
      — Must be affiliated with a RESNET provider to oversee quality
        control on your work (a contract with and fees to the provider
        are required).

### Other Specialized Certifications

Manufactured Housing Professional Type: Retrofitter
See BPI website

Heating Professional: Type Evaluator and Retrofitter
See BPI website

A/C or Heat Pump Professional: Type Evaluator and Retrofitter
See BPI website

Multifamily certifications
See BPI website

### Association of Energy Engineers Certification—
### Residential Energy Auditor—REA

The Residential Energy Auditor (REA) certification is available through the Association of Energy Engineers (AEE).

To qualify for candidacy for REA certification you must have:

**A two-year technical degree or higher** in engineering, building management, energy auditing, construction management, science, or a related specialty. No prior specific experience in residential energy auditing is required;

—OR—

**Currently working as a residential energy auditor** with at least three years of verifiable relevant experience.

—OR—

**The status of U.S. military veteran** with *at least three years of verifiable relevant technical experience*. No prior specific experience in residential energy auditing is required;

—OR—

**The current status of Certified Energy Manager (CEM) or Certified Energy Auditor (CEA)** in good standing.

After submitting an application for candidacy, to become certified you must attend an AEE preparatory REA training seminar, and complete and pass a four-hour written REA examination, proctored by an AEE-approved exam administrator. REA examination questions are drawn from concepts and experiences basic to residential energy auditing methodology, home energy use, residential energy systems, energy conservation measures, energy auditing methodology and calculations, and diagnostic and software tools. It is an open book exam, and the questions are a mixture of multiple choice and true/false. A passing score of at least 70% is required in conjunction with meeting all remaining eligibility requirements to become certified.

### DOE/NREL Certifications
Through BPI (pilot exams only are available June of 2012, certification available Fall 2012). In addition to the prerequisites shown below, all of these certifications require passing a written and field exam specifically designed for that certification, and having a GED or high school degree. "Years full-time" represents 1,000 hours of work, whether done full-time or part-time. "WAP" is the Weatherization Assistance Program. To be a "guinea pig" in the pilot/exam/certification development process go to: www.bpi.org/pilot

*Quality Control Inspector*
This certification is ideally suited for those who wish to evaluate the quality of new building or retrofit work of contractors in making homes more energy efficient and who have experience inspecting, including home inspections. The requirements are to obtain a minimum of 40 points from the following areas:

— Maximum of 20 points at the rate of 10 points per year of full time work as an inspector (including home inspections).

— Maximum of 10 points at the rate of 5 points per 1.5 years of full time work in the building arena, construction of any kind, repair and maintenance, etc.

— Maximum of 10 points at the rate of 5 points per 40 hours of training specific to the curriculum for Quality Control Inspectors.

— Maximum of 10 points from a minimum of 2 years experience energy auditing (10 points) or as a crew leader on a retrofit crew (5 points).

— Maximum of 10 points at the rate of 5 points per related certification—EPA HVAC, NATE HVAC, BPI, RESNET

As you can see, this certification seems to be the best fit for those with an inspection and construction background as they need only complete an 80-hour course to become a quality control inspector. The three other certifications, Energy Auditor, Retrofit Installer Technician, and Crew Leader, unlike the Quality Control Inspector, each require at least one year of experience in the energy arena.

*Energy Auditor*
This certification is for those who wish to evaluate existing buildings before they are retrofitted in making homes more energy efficient.

— Minimum of 1 year experience energy auditing or as a crew leader on a retrofit crew including proof of 15 home performance, RESNET, BPI, or WAP energy audits

AND

— Minimum of 20 points total from the following:

— Maximum of 10 points at the rate of 5 points per year of full time work in the building arena, construction of any kind, repair and maintenance, etc.

— Maximum of 10 points at the rate of 5 points per 40 hours of training specific to the curriculum for Energy Auditors.

— Maximum of 10 points at the rate of 5 points per related certification—EPA HVAC, NATE HVAC, BPI, RESNET

*Retrofit Installer Technician*

This certification is for those who wish to retrofit homes to make them more energy efficient. The requirements are to obtain a minimum of 35 points from the following areas:

— Maximum of 30 points at the rate of 10 points per each half year of full time work as an energy retrofitter.
— Maximum of 10 points at the rate of 5 points per year of full time work in the building arena, construction of any kind, repair and maintenance, etc.
— Maximum of 10 points at the rate of 5 points per 16 hours of training specific to the curriculum for Retrofit Installer Technician.
— Maximum of 10 points at the rate of 5 points per related certification—EPA HVAC, NATE HVAC, BPI, RESNET.

*Crew Leader*

This certification is for those who wish to be a crew leader of a retrofit crew in making homes more energy efficient. A crew leader must complete all of the following:

1.    OSHA 30 Certification
2.    Minimum of 4 years experience in any energy audit, retrofit, etc. field or 2 years experience if one year of that experience occurred after you were certified as a Retrofit Installer Technician.
3.    100 hours of training specific to the curriculum for both Retrofit Installer Technician and Crew Leader.
4.    Minimum of 15 points total from the following:
      — Maximum of 10 points at the rate of 5 points per year of full time work in the building arena, construction of any kind, repair and maintenance, etc.
      — Maximum of 10 points at the rate of 5 points per 50 hours of training in building science.
      — Maximum of 10 points at the rate of 5 points per related certification—EPA HVAC, NATE HVAC, BPI, RESNET.

If you would like more information on the requirements for these other three certifications visit the BPI website at:

http://www.bpi.org/news_expansion

| Air Conditioner(AC)/Heat Pump(HP) Efficiency Ratings and Moisture Removal Capacity | | | | | | |
|---|---|---|---|---|---|---|
| Rating Acronym | Name | Purpose | Application | Formula | Details | Example |
| SEER | Seasonal Energy Efficiency Ratio | Efficiency rating based on a season | Central AC's and AC sides of HP's | **Heat Removed Btu's / Watt-hours Used** | Heat removed by AC relative to electricity used | Operating a SEER of 7 will cost twice a 14 |
| EER | Energy Efficiency Ratio | Efficiency rating based on a season | Room/Unit AC's or AC side of HP's | **Heat Removed Btu's / Watt-hours Used** | Like SEER but is for Unit AC/HP's | Operating a EER of 7 will cost twice a 14 |
| HSPF | Heating Seasonal Performance Factor | Efficiency rating based on a season | Heat Pumps (Heat Side Only) | **Heat *Added* Btu's / Watt-hours Used** | Like SEER except is for heating | HSPF is about 10 or less |
| COP | Coefficient of Performance | Old Efficiency rating at 47° F | Heat Pumps (Heat Side Only) | **Heat *Added* in Btu's / Btu of electricity Used** | Use the Btu equivalent of kWHs | Not a season, 3 = 3x the heat as energy used |
| SHF | Sensible Heat Factor | Ability to remove moisture (water vapor) from air | Any AC or AC side of an HP | **Sensible Heat / Total Heat** **The units (for example Btu's) used should be the same on the top of the formula as on the bottom of the formula** | Ratio of heat in the air compared with total heat (Sensible + Latent Heat = Total Heat). Typically the SHF is between 0.5 and 1.0 | The part sensible heat increased to by removal of latent heat (water vapor). SHF of 1.0 = all energy is in the air and none in moisture (all moisture removed) |

# Appendix N

# Energy Efficiency in ACs and Heat Pumps

Conversion Factors for HVAC Equipment Efficiency Ratings
   HSPF = AFUE x 0.03413
   HSPF = COP x 3.413
   COP = AFUE / 100
   COP = HSPF / 0.03413
   AFUE = HSPF / 0.3413
   AFUE = COP x 100

A "rule of thumb" that some accept for getting an estimate of air condition sizing required for a given home is:
   — 1 ton per 400 square feet for older, less efficient homes
   — 1 ton per 700 to 1200 square feet in newer energy efficient homes
Cooling load is more difficult to calculate than heating load because of the variability of solar heat, internal gains, and air leakage. Thermal mass of the building has an effect on calculating both the cooling and heating load—it reduces daily temperature variation in the building and the energy demands. Air leakage has a double negative effect—it brings in warm air from outside but it also takes energy to remove the moisture in that air.

Note that Manual J is a more accurate method for determining air conditioner sizing that these estimates.

# Appendix O
# Multi-point Blower Door Pressure Testing

Blower door testing on homes typically involves testing the home at only one pressure setting, typically -50 Pascals. This is referred to as the "single point" test. However, blower door testing can also be conducted at more than one pressure setting. If so, this is referred to as the "multi-point" test. If you are willing to devote the additional time involved in testing a home at more than one single point pressure, the accuracy of the results will be increased. In fact, conducting a multi-point blower door test on a home helps determine statistically the accuracy level of the testing. More often than not, it is used for commercial or industrial building testing to obtain reliable results despite wind, stack and other factors. While setting up the home for the multipoint test varies little from the method of setting up the home for a single point test, many find the use of additional software useful in conducting multipoint tests. In fact, if you need to run two fans at once because you are evaluating a particularly leaky home, your results may be far more accurate using software that can even control multiple fans at the same time, such as the Retrotec FanTestic software and blower doors.

While some standards recommend conducting a multipoint test in 5 or 10 Pascal increments starting at about -25 Pascals and ending at about -75 Pascals, others suggest starting at -15 and proceeding through 8 to 10 pressure points until the house is finally tested at -60 Pascals. All seem to agree that testing beyond 100 Pascals could potentially damage air barrier systems, so this level should never be exceeded and many believe that testing should not exceed -75 Pascals. However, most agree that the greater the pressure differential during a test, the less the wind, etc. has an effect on the results and therefore the more accurate the results are expected to be. To add additional accuracy to the testing, some protocols require pressurizing the building (respect the cautions involved when pressurizing a building) at similar positive Pascal levels as conducted in the negative range. Under those circumstances, the building may be tested at 8 to 10 points in the negative Pascal range as well as 8 to 10 points in the positive Pascal range. Regardless of how many pieces of data are collected in evaluating the air leakage in the building, the data is then inputted into a software program that places the data points on a log-log graph showing airflow or leakage on the vertical y-axis and the interior/exterior pressure difference data points on the horizontal x-axis, as shown in Figure O-1.

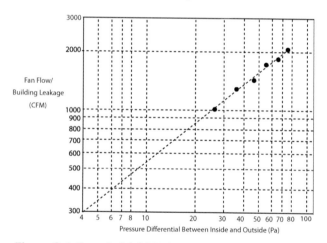

**Figure O-1. Sample Multi-Point Blower Door Test Graph**
Samples were taken between 25 and 75 Pascals at 10 Pascal increments. Notice that at 4 Pascals the line intersects the vertical y-axis at about 300 CFM, the natural air leakage value. From these data you can obtain a number that gives an indicator of the level of confidence that the data are good data.

Some software programs additionally allow automatically showing the statistical analysis for the data collected, including the correlation coefficient or "r2" value. Typically obtaining a correlation coefficient of 0.98 or better is considered a minimum under some testing standards. An example of the results from a software program that automatically controls the fan(s), automatically calculates the average flow rate at the different pressures, graphs these results, and conducts a statistical evaluation of the results is shown below (Retrotec).

The Power Law of Flow Through an Orifice is the equation that governs the flow of air through an orifice (bypasses in the case of a house) given a specific pressure differential between the inside and the outside of the home. This Power Law is expressed as follows:

$$Q = C\Delta P n$$

Where Q is the flow through the bypasses, C is the Air Leakage Coefficient and reflects how large the bypasses are, $\Delta P$ is the pressure differential between the inside and the outside of the house (for example, at -75 Pascals), and n is the pressure exponent or n exponent. Since the blower door has been tested and calibrated, you know its flow rate and its size in square inches and you can determine how much air is leaking through the bypasses in the home under different pressures. The pressure exponent tells whether the bypasses are shaped more like a round hole or like a long narrow crack. The closer the n exponent is to around 0.5, the closer the bypasses will be shaped like a round hole. The closer the exponent is to 1.0, the closer the bypasses will be shaped like a long narrow crack. From the practical standpoint however, the n exponent should be in the range of 0.45 to 0.8 or the results should be considered inaccurate.

Some protocols already exist (such as ASTM E779-03) and some are in the process of development for multipoint testing. Figure O-2 is a sample of the Retrotec Multi-Fan FanTestic Software showing only the Pressurization results:

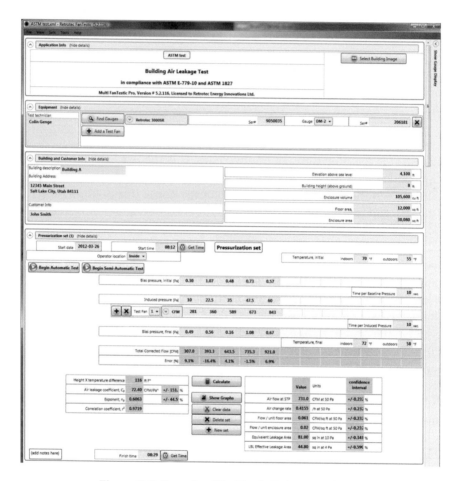

**Figure O-2. Sample of FanTastic Software Printout**

# Appendix P

# Utility Bill Interpretation

Typically any utility bill will have basic information on it such as name, address, account number, date of billing, dates billing covers/meter read, date payment due, amount currently due, etc. (see Figure P-1). This information is usually easy to find. However, understanding the rest of the bill is usually more challenging. These diagrams help explain the more challenging concepts.

A peak electrical demand charge is typically for using electricity at its highest usage rate. The highest electrical demands/peaks on a daily basis usually occur on weekdays in late afternoon and early evening, usually between 4 p.m. to 7 p.m. in the winter and noon to 8 p.m. in the summer. Annual peaks tend to occur on hot summer days due to the heavy use of electrical-powered air conditioning. This can also be shown as a "Peak Demand," "Demand Charge" or "Service Charge" on the bill. It is the electric load that corresponds to the highest level of electric usage by the utility customer in a specified time period (such as 15 minutes) during the month of billing. This charge or fee is usually billed in addition to the normal charges on the bill per kilowatt hour, etc. Some suggest that having a demand charge in the fee structure helps discourage electricity use when it is being used by a large number of other users at the same time. If the demand charge discourages electrical use during high demand times, it is thought that this can reduce the likelihood of "brown outs" where the supply of electricity cannot match the demand.

## Electrical Bill

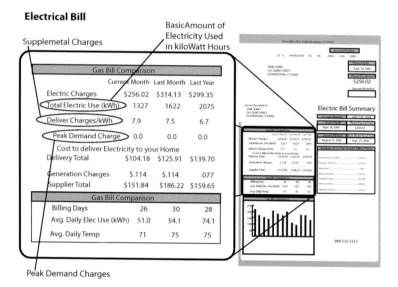

## Gas Bill

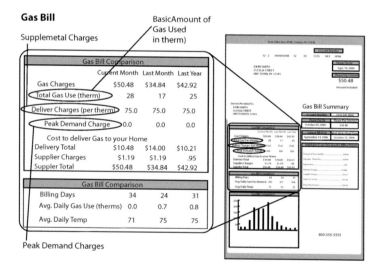

**Figure P-1. Utility Bill.**

# Appendix Q

# PHB Passive House

Europe seems to have been on the leading edge of energy conservation in buildings for many years. In 1988, Dr. Wolfgang Feist started the passivhaus or passive house concept that ultimately resulted in the completion of the first passive house in Darmstadt, Germany in 1991. These highly energy-conservative, high-performance buildings (HPBs) or passivhaus buildings (PHBs) use so little energy and are built in such a relatively cost-effective manner that they seem to have become the standard to which many buildings are compared. The program is similar to the RESNET program and uses software. The five most important aspects of PHBs are that they have:

1) very high R value thermal insulation (they must keep the energy use of the building below a maximum limit)

2) a highly airtight building envelope (≤ 0.6 air changes per hour at 50 Pascals)

3) thermal bridge-free construction (no thermal bridges in the building)

4) high R-value window frames with triple pane low E windows (solar heat gain coefficient or SHGC of 0.5 to 0.55)

5) highly efficient heat recovery ventilation (≥ 75% heat recovery efficiency with an electrical demand for the fans of 1.308 CFM/Watt or less)

Many of the PHBs involve the use of very efficient heating systems that often rely on geothermal-based heating and cooling systems and, in some cases, solar air as well (recognized around the world as one of the most cost-effective to build). A PHB cannot use more than 4.75 kBtu/ft²-yr of either heating energy or the same for cooling energy, nor more than 38.0 kBtu/ft²-yr total energy (including plug loads, etc.). The total energy limit is multiplied by an energy factor that allows for losses in generation and transmission. The building cannot have an indoor temperature above the maximum (77 °F temperature) more than 5% of the time. The concept of "operative" temperature is used as a better indicator of relative comfort by integrating the mean of the average surface temperatures of the floors, ceilings, and walls as well as the air temperature in the building. Thus, in traditional construction that is not thermal bridge free, you would feel colder than in a PHB because of the net radiation loss as a result of the colder walls in traditional construction. Because PHBs are thermal bridge free, the surface temperatures are not as significant in the evaluation of comfort. Surface temperatures and net radiation "coldness" to the occupants have much more of a nominal effect than in traditional construction.

PHBs typically can save up to 80% of space heating and cooling energy that would normally go into a building. Some of the big benefits for the occupants are that these buildings are much more comfortable and, because of extensive use of heat recovery ventilation that reduces even the CO2 levels, have much better indoor air quality than a typical building. These buildings tend to show less stratification in air temperature from floor to ceiling than less efficient buildings. This promotes greater comfort for the occupants. The comfort and indoor air quality issues are in addition to the savings in energy costs for the building over the lifetime of the building. The economic analysis for a PHB is based on the lifetime of the building and not on a much narrower time period, such as three to five years, as is common in the evaluation of some buildings in the United States. In many cases, the payback on PHBs has been as short as 8 years. In some cases, PHBs are costing only an additional 8% above the costs of normal construction, even in the United States.

The PHB principles can also be used to retrofit existing buildings, although the standards are more liberal in the case of a retrofit. Some new concepts have arisen out of some of the high-performance building research in the PHB program. For instance, because the purpose is to provide a thermal bridge-free construction, a psi value ($\psi$) was developed which helps confirm whether a building is thermal bridge-free. Using the operative temperature is also a concept not typically used in the United States.

For more information please visit the passipedia website at:
passipedia.passiv.de/passipedia_en/start
Or the Passiv Haus website at: www.passiv.de

# Appendix R

# Zone Pressure Diagnostic (ZPD) Calculations

This Appendix gives examples of ZPD calculations. See the ZPD section in the book on the procedures for these tests and how to use the appropriate charts to interpret the results. To utilize Excel files to complete the calculations (including calculating the likely uncertainty involved in the result), see the appropriate Excel files that may be available.

## OPEN-A-DOOR TESTS

Read Table R-1 instructions for Open-a-Door Tests before proceeding with these examples:

- *Opening made in H/Z barrier:* If you have opened a door from the house-to-zone (such as the house-to-garage door), use Column A of the Open-a-Door Chart below to find the house-to-zone pressure before opening the door (circle this pressure in Column A).
- *Opening made in Z/O barrier:* If you have opened a zone-to-outside door (such as the garage-to-outside door), use Column B of the Open-a-Door Chart below to find the house-to-zone pressure before opening the door (circle this pressure in Column B).
- *All Open-a-Door Tests:* Record the multipliers in the appropriate chart across to the right in the same row as the house-to-zone pressure you circled for the "Internal" (house-to-zone leakage), "External" (zone-to-outside leakage), and "Combined" (leakage when barriers combined) found:

Internal :_____ External: _____
Combined: _____

Now multiply each of the multipliers by the CFM difference recorded above to find out how much reduction, at a maximum, can be achieved by completely sealing both barriers (Combined):

Internal: \_\_\_\_\_ (multiplier) x _____ CFM Diff =
\_\_\_\_\_ CFM50 H/Z

External: \_\_\_\_\_ (multiplier) x _____ CFM Diff =
\_\_\_\_\_ CFM50 Z/O

Combined: \_\_\_\_ (multiplier) x \_\_\_\_ CFM Diff =
\_\_\_\_\_ CFM50 maximum reduction available if both barriers perfectly sealed

### Table R-1. Instructions for Open-a-Door Test

For <u>Door Opening in Barrier Between House to Zone</u>:
Use Column A to Match H/Z and Column B to Match Z/O
For <u>Door Opening in Barrier Between Zone to Outside</u>:
Use Column B to Match H/Z and Column A to Match Z/O

| Door closed pressure | | multiply CFM50 change by: | | |
|---|---|---|---|---|
| A | B | Internal | External | Combined |
| 48 | 2 | 0.14 | 1.14 | 0.14 |
| 47 | 3 | 0.20 | 1.19 | 0.19 |
| 46 | 4 | 0.25 | 1.24 | 0.24 |
| 45 | 5 | 0.31 | 1.29 | 0.29 |
| 44 | 6 | 0.37 | 1.34 | 0.34 |
| 43 | 7 | 0.43 | 1.39 | 0.39 |
| 42 | 8 | 0.49 | 1.44 | 0.44 |
| 41 | 9 | 0.56 | 1.49 | 0.49 |
| 40 | 10 | 0.63 | 1.54 | 0.54 |
| 39 | 11 | 0.70 | 1.60 | 0.60 |
| 38 | 12 | 0.78 | 1.65 | 0.65 |
| 37 | 13 | 0.87 | 1.71 | 0.71 |
| 36 | 14 | 0.96 | 1.78 | 0.78 |
| 35 | 15 | 1.06 | 1.84 | 0.84 |
| 34 | 16 | 1.17 | 1.91 | 0.91 |
| 33 | 17 | 1.29 | 1.98 | 0.98 |
| 32 | 18 | 1.42 | 2.06 | 1.06 |
| 31 | 19 | 1.56 | 2.14 | 1.14 |
| 30 | 20 | 1.71 | 2.23 | 1.23 |
| 29 | 21 | 1.88 | 2.32 | 1.32 |
| 28 | 22 | 2.07 | 2.42 | 1.42 |
| 27 | 23 | 2.27 | 2.52 | 1.52 |
| 26 | 24 | 2.50 | 2.64 | 1.64 |
| 25 | 25 | 2.76 | 2.76 | 1.76 |

Following Collin Olson and Anthony Cox, 2006, following Michael Blasnik
Internal = multiplier for zone to house leakage
External = multiplier for zone to outside leakage
Combined = multiplier for total/combined path leakage

### Open-a-Door Chart

OPEN-A-DOOR CHART

### Example 1. Open-a-Door Between House and Zone

This example shows how to determine the leakiness of the floor between the house and the basement. Set the blower door to run at -50 Pascals. Assume that you have determined the pressure differential across the floor to the basement to be 25 Pascals with the basement door closed and the CFM50 to be 1850. When the door to the basement is opened with the blower door still running to maintain -50 Pascals pressure, the blower door goes to 2200 CFM 50. As expected and required for the Open-a-Door test, you find the new pressure differential across the floor to the basement is 0 Pascals. This means, by opening the door, you have allowed free communication of air and pressure between the house and the basement. Using Table R-2 below, you can see the first pressure differential with the door closed—25 Pascals.

The zone-to-outside pressure differential should be 25 Pascals

#### Table R-2. ZPD Open-a-Door Calculations

| Calculation Preliminaries | | | |
|---|---|---|---|
| A<br>Closed Door Pressure: House WRT Zone | B<br>CFM50 Door Closed | C<br>CFM50 Door Open | D<br>CFM50 Difference (C-B) |
| 25 | 1850 | 2200 | 350 |
| Leakage from House-to-Zone | | | |
| (D) CFM50 Difference | Multiplier | CFM50 | Square Inches |
| 350 | 2.76 | 966 | 96.6 |
| Leakage from Zone-to-Outside | | | |
| (D) CFM50 Difference | Multiplier | CFM50 | Square Inches |
| 350 | 2.76 | 966 | 96.6 |
| Combined Path Leakage | | | |
| (D) CFM50 Difference | Multiplier | Maximum CFM50 Reduction Available | |
| 350 | 1.76 | 616 | |

with the door closed (the difference between 25 and 50). Record the 1850 CFM50 you saw before opening the door ("Door closed"), and the 2200 CFM50 you saw after opening the door ("Door open"). Calculate the difference between these two flows and put the number in the upper far right column of the table ("Difference"): 350 CFM50. In the next three rows of the table, record 350 CFM50 in the left column. Then, using the Open-a-Door chart above and following the instructions, find the house-to-zone pressure in Column A (since you opened a door in the house-to-zone barrier) and then go across to the right to see the multipliers. For an H/Z pressure differential of 25 Pascals, you can see that 2.76, 2.76, and 1.76 are the H/Z barrier ("Internal"), Z/O barrier ("External"), and Total for both H/Z and Z/O barriers combined ("Combined") multipliers respectively. Record these multipliers in the second column of the bottom three rows. (see above). Then calculate the CFM50 and the square inches (if applicable) for each leakage path:

The product of 350 x 2.76 is 966 CFM50. To get an idea of the effective leakage area of an air barrier, use an established constant of 0.10 square inches/CFM50. Multiply 966 CFM50 by 0.10 square inches/CFM50 to get about 97 in² (96.6 to be exact). Because the zone-to-outside pressure differential is also 25 Pascals, the exterior multiplier found in the table under the "External" column is also 2.76. The product of these two numbers is 966 CFM50. Once again, use the established constant of 0.10 and multiply the 966 CFM50 by 0.10 to obtain roughly 97 in² of effective leakage area from the zone to the outside of the house.

To calculate the total CFM50 due to both air barriers—house-to-zone and the zone-to-outside—use the multiplier from the "Combined" column (1.76) to give a total pathway leakage flow of 616 CFM50. Of the total 1850 CFM50 for the entire house, roughly 600 CFM50 is coming from outside through the basement and into the house. Thus, if you were to perfectly seal the basement walls or the house floor, you should reduce the CFM50 by about 600 CFM50.

### Example 2. Open-a-Door Between Zone and Outside

Use the same lookup chart as in Example 1 if you are opening a door between the zone and outside, such as a walk-out basement door to the outside or a garage vehicle door. Assume in this example that you have the same numbers as in the first Open-a-Door example except you are opening a door in the zone-to-outside barrier. In this case, as in all cases where a door between the zone and outside is opened, find the house-to-zone pressure in Column B on the table and then go across to the right to look up the multipliers. Doing this for an H/Z pressure differential of 25 Pascals, you see that 2.76, 2.76, and 1.76 are the H/Z barrier ("Internal"), Z/O barrier ("External"), and Total for both H/Z and Z/O barriers combined ("Combined") multipliers respectively. These are the same as in the earlier example because the H/Z and Z/O barriers are roughly equally leaky. Thus, the result is the same for this zone if you use the same pressures and flows as in the previous example because you are using the same numbers as in Example 1, but just applying them to a different barrier—zone-to-outside.

**Table R-3. ZPD Open-a-Door Calculations**

| Calculation Preliminaries | | | |
|---|---|---|---|
| A<br>Closed Door<br>Pressure:<br>House WRT<br>Zone | B<br>CFM50<br>Door<br>Closed | C<br>CFM50<br>Door<br>Open | D<br>CFM50<br>Difference<br>(C-B) |
| 25 | 1850 | 2200 | 350 |
| Leakage from House-to-Zone | | | |
| (D) CFM50<br>Difference | Multiplier | CFM50 | Square<br>Inches |
| 350 | 2.76 | 966 | 96.6 |
| Leakage from Zone-to-Outside | | | |
| (D) CFM50<br>Difference | Multiplier | CFM50 | Square<br>Inches |
| 350 | 2.76 | 966 | 96.6 |
| Combined Path Leakage | | | |
| (D) CFM50<br>Difference | Multiplier | Maximum CFM50<br>Reduction Available | |
| 350 | 1.76 | 616 | |

**Example 3. Open-a-Door Between House and Zone**

This example describes an Open-a-Door test on an attic with a full door access to a permanent stairway into the attic. The closed pressure differential in the attic with respect to the house was found to be 40 Pascals with a blower door flow of 2050 CFM50. When the door was fully opened as required, the open door pressure differential went down to 0 Pascals (as required for the Open-a-Door test), and the open door flow went up to 2550 CFM50. Using Table R-4, write in the first pressure differential with the door closed—40 Pascals.

As the lookup chart indicates, the attic zone-to-outside pressure differential should be 10 Pascals, (the difference between 40 and 50). Record (on the table) the 2050 CFM50 when the door was closed, and the 2550 CFM50 after opening the door. Calculate the difference between these two and put the number in the upper far right column: 500 CFM50. Then, using the Open-a-Door chart above and following the instructions, find the house-to-zone pressure in Column A (since you opened a door in the house-to-zone barrier) and then go across to the right to see the multipliers. When you do this for an H/Z pressure differential of 40 Pascals, you can see that 0.63, 1.54, and 0.54 are the H/Z barrier ("Internal"), Z/O barrier ("External"), and Total for both H/Z and Z/O barriers combined ("Combined") multipliers respectively. Record those multipliers in the second column of the bottom three rows.

### Table R-4. Calculation Preliminaries

| A Closed Door Pressure: House WRT Zone | B CFM50 Door Closed | C CFM50 Door Open | D CFM50 Difference (C-B) |
|---|---|---|---|
| 40 | 2050 | 2550 | 500 |
| Leakage from House-to-Zone | | | |
| (D) CFM50 Difference | Multiplier | CFM50 | Square Inches |
| 500 | 0.63 | 315 | 31.5 |
| Leakage from Zone-to-Outside | | | |
| (D) CFM50 Difference | Multiplier | CFM50 | Square Inches |
| 500 | 1.54 | 770 | 77.0 |
| Combined Path Leakage | | | |
| (D) CFM50 Difference | Multiplier | Maximum CFM50 Reduction Available | |
| 500 | 0.54 | 270 | |

Calculate the CFM50 and the square inches (if applicable) for each leakage path. In the left column on the next three rows, record 500 CFM50 and multiply it by the 0.63 interior multiplier found in the table under the "Internal" Column corresponding to 40 Pascals. Remember, when you open a door between the house-to-zone barrier, use Column A to find the pressure differential between house-to-zone. The product of 500 x 0.63 equals 315 CFM50. To get an idea of the effective leakage area of an air barrier, use the established constant of 0.10 square inches/CFM50 and multiply 315 CFM50 by 0.10 square inches / CFM50 to get 31.5 in². Because the zone-to-outside pressure differential is also 10 Pascals, the exterior multiplier found in the table under the "External" column is 1.54. The product of these two numbers (500 CFM50 x 1.54) is 770 CFM50. Using the established constant of 0.10, multiply the 770 CFM50 by 0.10 to get 77 in², which is the estimated leakage area from the zone to the outside of the house.

To calculate the total CFM50 due to both air barriers, the house-to-zone and the zone-to-outside air barriers combined, use the multiplier from the "Combined" column—0.54 to give a total pathway leakage flow of 270 CFM50. This means that of the total of 2050 CFM50 for the entire house, roughly 300 CFM50 is coming from outside through the basement and into the house. Thus, if you were to perfectly seal the basement walls or the house floor completely, you should reduce the CFM50 by about 300 CFM50.


OPEN-A-HATCH OR ADD-A-HOLE TEST

Read the following chart instructions on Open-a-Hatch Tests before proceeding with these examples:

Opening made in H/Z barrier: If you have opened a hatch or hole from the house-to-zone (such as the house-to-attic hatch, house-to-crawl-space hatch), use Column A of the Table R-5 Open-a-Hatch Chart to find the house-to-zone pressure before opening the hatch (circle this pressure in Column A). Then, use Row A to find the house-to-zone ending pressure after opening the hatch (circle this pressure in Row A). Find the multiplier at the intersection of these two circled items by going right across from the starting pressure and down from the ending pressure.

Opening made in Z/O barrier: If you have opened a hatch or hole from the zone-to-outside barrier (such as the roof to attic hatch, outside to crawlspace hatch), use Column B of the Open-a-Hatch Chart below to find the house-to-zone pressure before opening the hatch (circle this pressure in Column B). Then, use Row B to find the house-to-zone ending pressure after opening the hatch (circle this pressure in Row B). Find the intersection of these two circled items by going right across from the starting pressure and down from the ending pressure and record them in the equation below where the multiplier is expected to go.

All Open-a-Hatch Tests: Multiply the multiplier by the CFM difference you calculated above (the change in flow from before opening the hatch and after opening the hatch). The result gives the expected maximum reduction in leakage if you were to completely seal either the house-to-zone or the zone-to-outside barrier:

_____ (multiplier) x _____ CFM50 Diff =
_____ CFM max H/Z

## Example 1. Opening a Hatch or Adding a Hole in the House-to-Zone (H/Z) Barrier

Assume the zone to be the attic and assume the start house-to-zone (H/Z) differential pressure is 37 Pascals, then the zone-to-outside pressure differential is 13 Pascals. Also, assume that once the hatch is opened, the H/Z differential pressure goes down to 20. This represents a drop in

### Table R-5. Open a Hatch Chart

**Hatch/Hole Opened in House to Zone (H/Z) or Zone to Outside (Z/O) Barriers**

For Hatch/Hole Opening in House to Zone Barrier: Use A Column and Row to Match H/Z and Row B to Match Z/O
For Hatch/Hole Opening in Zone to Outside Barrier: Use B Column and Row to Match H/Z and Column and Row A to Match Z/O

Start Press — Ending Pressure After Opening Hatch or Adding Hole in Barrier

| A | B | 44/6 | 42/8 | 40/10 | 38/12 | 36/14 | 34/16 | 32/18 | 30/20 | 28/22 | 26/24 | 24/26 | 22/28 | 20/30 | 18/32 | 16/34 | 14/36 | 12/38 | 10/40 | 8/42 | 6/44 | 4/46 | 2/48 | 0/50 |
|---|---|---|---|---|---|---|---|---|---|---|---|---|---|---|---|---|---|---|---|---|---|---|---|---|
| 50 | 0 | 0.00 | 0.00 | 0.00 | 0.00 | 0.00 | 0.00 | 0.00 | 0.00 | 0.00 | 0.00 | 0.00 | 0.00 | 0.00 | 0.00 | 0.00 | 0.00 | 0.00 | 0.00 | 0.00 | 0.00 | 0.00 | 0.00 | 0.00 |
| 49 | 1 |  | 0.35 | 0.29 | 0.25 | 0.22 | 0.20 | 0.18 | 0.17 | 0.15 | 0.15 | 0.15 | 0.14 | 0.13 | 0.12 | 0.12 | 0.11 | 0.11 | 0.10 | 0.10 | 0.09 | 0.09 | 0.09 | 0.09 |
| 48 | 2 |  | 0.68 | 0.54 | 0.45 | 0.39 | 0.35 | 0.32 | 0.29 | 0.27 | 0.25 | 0.23 | 0.22 | 0.21 | 0.20 | 0.19 | 0.18 | 0.17 | 0.17 | 0.16 | 0.15 | 0.15 | 0.15 | 0.14 |
| 47 | 3 |  |  | 0.84 | 0.68 | 0.58 | 0.51 | 0.45 | 0.41 | 0.38 | 0.35 | 0.33 | 0.31 | 0.29 | 0.27 | 0.26 | 0.25 | 0.24 | 0.23 | 0.22 | 0.21 | 0.20 | 0.20 | 0.19 |
| 46 | 4 |  |  | 1.23 | 0.96 | 0.80 | 0.68 | 0.60 | 0.54 | 0.49 | 0.45 | 0.42 | 0.39 | 0.37 | 0.35 | 0.33 | 0.32 | 0.30 | 0.29 | 0.28 | 0.27 | 0.26 | 0.25 | 0.24 |
| 45 | 5 |  |  |  | 1.30 | 1.05 | 0.89 | 0.77 | 0.68 | 0.62 | 0.56 | 0.52 | 0.48 | 0.45 | 0.43 | 0.40 | 0.38 | 0.37 | 0.35 | 0.33 | 0.32 | 0.31 | 0.30 | 0.29 |
| 44 | 6 |  |  |  | 1.76 | 1.36 | 1.12 | 0.96 | 0.84 | 0.75 | 0.68 | 0.63 | 0.58 | 0.54 | 0.51 | 0.48 | 0.45 | 0.43 | 0.41 | 0.39 | 0.38 | 0.36 | 0.35 | 0.34 |
| 43 | 7 |  |  |  |  | 1.76 | 1.41 | 1.18 | 1.02 | 0.90 | 0.81 | 0.74 | 0.68 | 0.63 | 0.59 | 0.56 | 0.53 | 0.50 | 0.48 | 0.45 | 0.43 | 0.42 | 0.40 | 0.39 |
| 42 | 8 |  |  |  |  | 2.28 | 1.76 | 1.44 | 1.23 | 1.08 | 0.96 | 0.87 | 0.80 | 0.74 | 0.68 | 0.64 | 0.60 | 0.57 | 0.54 | 0.52 | 0.49 | 0.47 | 0.45 | 0.44 |
| 41 | 9 |  |  |  |  |  | 2.20 | 1.76 | 1.47 | 1.27 | 1.12 | 1.01 | 0.92 | 0.84 | 0.78 | 0.73 | 0.68 | 0.65 | 0.61 | 0.58 | 0.55 | 0.53 | 0.51 | 0.49 |
| 40 | 10 |  |  |  |  |  | 2.80 | 2.15 | 1.76 | 1.49 | 1.30 | 1.16 | 1.05 | 0.96 | 0.89 | 0.82 | 0.77 | 0.72 | 0.68 | 0.65 | 0.62 | 0.59 | 0.56 | 0.54 |
| 39 | 11 |  |  |  |  |  |  | 2.65 | 2.11 | 1.76 | 1.51 | 1.33 | 1.20 | 1.09 | 1.00 | 0.92 | 0.86 | 0.81 | 0.76 | 0.72 | 0.68 | 0.65 | 0.62 | 0.60 |
| 38 | 12 |  |  |  |  |  |  | 3.32 | 2.54 | 2.07 | 1.76 | 1.53 | 1.36 | 1.23 | 1.12 | 1.03 | 0.96 | 0.90 | 0.84 | 0.80 | 0.75 | 0.72 | 0.68 | 0.65 |
| 37 | 13 |  |  |  |  |  |  |  | 3.09 | 2.45 | 2.04 | 1.76 | 1.55 | 1.38 | 1.26 | 1.15 | 1.07 | 0.99 | 0.93 | 0.87 | 0.83 | 0.79 | 0.75 | 0.71 |
| 36 | 14 |  |  |  |  |  |  |  | 3.83 | 2.93 | 2.38 | 2.02 | 1.76 | 1.56 | 1.41 | 1.28 | 1.18 | 1.09 | 1.02 | 0.96 | 0.90 | 0.86 | 0.81 | 0.78 |
| 35 | 15 |  |  |  |  |  |  |  |  | 3.54 | 2.80 | 2.33 | 2.00 | 1.76 | 1.57 | 1.42 | 1.30 | 1.21 | 1.12 | 1.05 | 0.99 | 0.93 | 0.89 | 0.84 |
| 34 | 16 |  |  |  |  |  |  |  |  | 4.35 | 3.32 | 2.70 | 2.28 | 1.98 | 1.76 | 1.58 | 1.44 | 1.33 | 1.23 | 1.15 | 1.08 | 1.01 | 0.96 | 0.91 |
| 33 | 17 |  |  |  |  |  |  |  |  |  | 3.98 | 3.14 | 2.61 | 2.24 | 1.97 | 1.76 | 1.59 | 1.46 | 1.34 | 1.25 | 1.17 | 1.10 | 1.04 | 0.98 |
| 32 | 18 |  |  |  |  |  |  |  |  |  | 4.86 | 3.70 | 3.01 | 2.54 | 2.20 | 1.95 | 1.76 | 1.60 | 1.47 | 1.36 | 1.27 | 1.19 | 1.12 | 1.06 |
| 31 | 19 |  |  |  |  |  |  |  |  |  |  | 4.42 | 3.49 | 2.89 | 2.48 | 2.18 | 1.94 | 1.76 | 1.61 | 1.48 | 1.38 | 1.29 | 1.21 | 1.14 |
| 30 | 20 |  |  |  |  |  |  |  |  |  |  | 5.38 | 4.09 | 3.32 | 2.80 | 2.43 | 2.15 | 1.93 | 1.76 | 1.62 | 1.49 | 1.39 | 1.30 | 1.23 |
| 29 | 21 |  |  |  |  |  |  |  |  |  |  |  | 4.86 | 3.83 | 3.18 | 2.72 | 2.38 | 2.13 | 1.92 | 1.76 | 1.62 | 1.50 | 1.41 | 1.32 |
| 28 | 22 |  |  |  |  |  |  |  |  |  |  |  | 5.89 | 4.48 | 3.63 | 3.05 | 2.65 | 2.34 | 2.11 | 1.91 | 1.76 | 1.63 | 1.51 | 1.42 |
| 27 | 23 |  |  |  |  |  |  |  |  |  |  |  |  | 5.30 | 4.18 | 3.46 | 2.96 | 2.59 | 2.31 | 2.09 | 1.91 | 1.76 | 1.63 | 1.52 |
| 26 | 24 |  |  |  |  |  |  |  |  |  |  |  |  | 6.41 | 4.86 | 3.94 | 3.32 | 2.87 | 2.54 | 2.28 | 2.07 | 1.90 | 1.76 | 1.64 |
| 25 | 25 |  |  |  |  |  |  |  |  |  |  |  |  |  | 5.75 | 4.52 | 3.74 | 3.20 | 2.80 | 2.49 | 2.25 | 2.06 | 1.89 | 1.76 |
| 24 | 26 |  |  |  |  |  |  |  |  |  |  |  |  |  | 6.92 | 5.25 | 4.25 | 3.57 | 3.09 | 2.73 | 2.45 | 2.23 | 2.04 | 1.89 |
| 23 | 27 |  |  |  |  |  |  |  |  |  |  |  |  |  |  | 6.19 | 4.86 | 4.02 | 3.44 | 3.01 | 2.68 | 2.42 | 2.20 | 2.03 |
| 22 | 28 |  |  |  |  |  |  |  |  |  |  |  |  |  |  | 7.43 | 5.64 | 4.55 | 3.83 | 3.32 | 2.93 | 2.63 | 2.38 | 2.18 |
| 21 | 29 |  |  |  |  |  |  |  |  |  |  |  |  |  |  |  | 6.63 | 5.21 | 4.30 | 3.67 | 3.21 | 2.86 | 2.58 | 2.35 |
| 20 | 30 |  |  |  |  |  |  |  |  |  |  |  |  |  |  |  | 7.95 | 6.02 | 4.86 | 4.09 | 3.54 | 3.12 | 2.80 | 2.54 |
| 19 | 31 |  |  |  |  |  |  |  |  |  |  |  |  |  |  |  |  | 7.07 | 5.55 | 4.58 | 3.91 | 3.44 | 3.04 | 2.74 |
| 18 | 32 |  |  |  |  |  |  |  |  |  |  |  |  |  |  |  |  | 8.46 | 6.41 | 5.17 | 4.35 | 3.76 | 3.32 | 2.97 |
| 17 | 33 |  |  |  |  |  |  |  |  |  |  |  |  |  |  |  |  |  | 7.51 | 5.89 | 4.86 | 4.15 | 3.63 | 3.23 |
| 16 | 34 |  |  |  |  |  |  |  |  |  |  |  |  |  |  |  |  |  | 8.98 | 6.79 | 5.48 | 4.61 | 3.98 | 3.51 |
| 15 | 35 |  |  |  |  |  |  |  |  |  |  |  |  |  |  |  |  |  |  | 7.95 | 6.24 | 5.14 | 4.39 | 3.83 |
| 14 | 36 |  |  |  |  |  |  |  |  |  |  |  |  |  |  |  |  |  |  | 9.49 | 7.18 | 5.79 | 4.86 | 4.30 |
| 13 | 37 |  |  |  |  |  |  |  |  |  |  |  |  |  |  |  |  |  |  |  | 8.39 | 6.58 | 5.42 | 4.63 |
| 12 | 38 |  |  |  |  |  |  |  |  |  |  |  |  |  |  |  |  |  |  |  | 10.00 | 7.56 | 6.10 | 5.12 |
| 11 | 39 |  |  |  |  |  |  |  |  |  |  |  |  |  |  |  |  |  |  |  |  | 8.83 | 6.92 | 5.71 |
| 10 | 40 |  |  |  |  |  |  |  |  |  |  |  |  |  |  |  |  |  |  |  |  | 10.52 | 7.95 | 6.41 |
| 9 | 41 |  |  |  |  |  |  |  |  |  |  |  |  |  |  |  |  |  |  |  |  |  | 9.27 | 7.26 |
| 8 | 42 |  |  |  |  |  |  |  |  |  |  |  |  |  |  |  |  |  |  |  |  |  | 11.03 | 8.33 |
| 7 | 43 |  |  |  |  |  |  |  |  |  |  |  |  |  |  |  |  |  |  |  |  |  |  | 9.71 |
| 6 | 44 |  |  |  |  |  |  |  |  |  |  |  |  |  |  |  |  |  |  |  |  |  |  | 11.54 |

Following Anthony Cox and Collin Olson, 2006 following Michael Blasnik

pressure of 17 Pascals (37-20), far more than the minimum drop of 6 Pascals necessary to use this test.

Then, looking at the Open-a-Hatch chart above and following the instructions supplied, know that when you open a hatch in the house-to-zone (H/Z) barrier you find the starting house-to-zone pressure in Column A and the ending house-to-zone pressure in Row A. Thus, look for the intersection of the entry in Column A (37 Pascals) and the entry in Row A (20 Pascals) and find that the row with "37" in it under Column A intersects the column with "20" in it in Row A in the cell that reads a factor of "1.38" at that intersection.

Assume 1850 CFM50 before opening the hatch and 2500 CFM50 after opening the hatch, all while maintaining a 50 Pascal difference between the house and the outside (but letting the house-to-zone pressure differential change to its own new level). Record the CFM50 flow reading (1850 CFM50) and the H/Z pressure differential (37 Pascals) before the hatch was opened (see Table R-6).

**Table R-6. ZPD Open a Hatch Calculation Table**

| Before Hole | | Projected | |
|---|---|---|---|
| **CFM50** | | **Leakiness/ELA** | |
| 1850 CFM | | **CFM50 Diff** | |
| **H/Z Pressure** | | 650 CFM | |
| 37 Pascals | | **Multiplier** | |
| | | 1.38 | |
| **After Hole** | | **Max. Reduction** | |
| **CFM50** | | 897 CFM | |
| 2500 CFM | | **Square Inches** | |
| **H/Z Pressure** | | 89.7 | |
| 20 Pascals | | | |

Just below them, enter the ending CFM50 (2500 CFM50) and H/Z pressure differential (20 Pascal) in the appropriate boxes. Subtract the ending CFM50 from the starting CFM50 (2500 – 1850 = 650CFM50) and record that CFM50 flow reading in the box entitled "CFM50 Diff." Write the factor of 1.38 in the box entitled "Multiplier." Multiply the CFM50 Diff by the Multiplier (650 CFM50 x 1.38 = 897 CFM50) to get the approximate maximum air leakage due to this air barrier and place that number in the "Maximum Reduction" box. This maximum reduction is the most that the air barrier from the house to the zone (H/Z) behind the hatch would be tightened if you did a perfect job of air sealing either the H/Z or the Z/O barriers.

*Identifying Priorities in Air Sealing*

In your analysis, as an example, you could report the amount of air believed to be leaking out of the house as a whole (1850 CFM50) before opening the attic hatch. (This would be the usual blower door leakage number). You could report the numbers previously attributed to the floor between the house and the basement (see Open-a-Door Example 1) of 616 or approximately 600 CFM50, and to the ceiling between the house and the attic zone (this Open-a-Hatch Example 1) of 897 or approximately 900 CFM50. If you subtract the leakage numbers for the ceiling and floor from the total leakage (1850 – (600 + 900) = 350 CFM50), you get the leakage for the remainder of the house that is not part of the ceiling or floor, which is approximately 350 CFM50. This represents the leakage in the walls. As a result, you can suggest that the wall is not a problem while the ceiling to the attic represents the biggest priority. The basement is the middle priority in terms of leakiness. Thus, you are probably wise to concentrate the most cost-effective sealing work in the attic and basement.

## Example 2. Opening a Hatch or Adding a Hole in the Zone-to-Outside (Z/O) Barrier

This example of the Open-a-Hatch Test opens a hatch between the zone (in this case also an attic) and the outside. Assume that the start house-to-zone (H/Z) differential pressure is 45 Pascals, which leaves the zone-to-outside pressure differential at 5 Pascals. Assume that when opening the hatch, the H/Z differential pressure drops to 18. This represents a drop in pressure of 27 Pascals (45-18), far more than the minimum drop of 6 Pascals necessary to use this test and above the 15 to 25 Pascal drop required to feel comfortable relying on the result.

Then, looking at Figure R-6 Open-a-Hatch chart and following the instructions, you know that when opening a hatch in the zone-to-outside (Z/O) barrier you find the starting house-to-zone pressure in Column B and the ending house-to-zone pressure in Row B. Thus, look for the intersection of the entry in Column B (45 Pascals) and the entry in Row B (18 Pascals) and find that the row with "45" in it under Column B intersects the column with "18" in it in Row B in the cell that reads a factor of "0.43" at that intersection.

Assume 1200 CFM50 before opening the hatch and 1850 CFM50 after opening the hatch, all the while maintaining a 50 Pascal difference between the house and the outside (but letting the house-to-zone pressure differential change to its own new level). Record the CFM50 flow reading (1200 CFM50) and the H/Z pressure differential (45 Pascals) before the hatch was opened. Just below them, enter the ending CFM50 (1850 CFM50) and H/Z pressure differential (18 Pascal) in the appropriate box. Subtract the ending CFM50 from the starting CFM50 (1850 – 1200 = 650CFM50) and record that CFM50 flow reading in the box entitled "CFM50 Diff." Also, write the factor of 0.43 in the box entitled "Multiplier." Multiply the CFM50 Diff by the Multiplier (650 CFM50 x 0.43 = 280 CFM50) to get the approximate maximum air leakage and place that number in the "Maximum Reduction" box, see Table R-7.

**Table R-7. ZPD Open a Hatch Calculation Table**

| Before Hole | Projected |
|---|---|
| **CFM50** | **Leakiness/ELA** |
| 1200 CFM | **CFM50 Diff** |
| **H/Z Pressure** | 650 CFM |
| 45 Pascals | **Multiplier** |
|  | 0.43 |
| **After Hole** | **Max. Reduction** |
| **CFM50** | 280 CFM |
| 1850 CFM | **Square Inches** |
| **H/Z Pressure** | 28 |
| 18 Pascals |  |

This maximum reduction is the most that the air barrier from the house to the zone behind the hatch would be tightened if you did a perfect job of air sealing it or the zone –to-outside barrier.

*Identifying Priorities in Air Sealing*

In the final analysis, as an example, you can suggest that the number you believe is leaking out of the house as a whole is 1200 CFM50 before opening either the basement door or the attic hatch. If you had found the house to the basement leakage was 270 CFM (See Open-a-Door Example 3 above), you could report that the leakage from the house to the basement was about 270 or approximately 300 CFM50, and the leakage to the ceiling between the house and the attic zone (this Open-a-Hatch Example 2) was about 280 or approximately 300 CFM50. If you subtract the leakage numbers for the ceiling and floor from the total leakage (1200 – (300 + 300) = 600 CFM50), you get the leakage for the remainder of the house that is not part of the ceiling or floor of the envelope (the walls). That is, the leakage for the walls would be approximately 600 CFM50. As a result, you can now suggest that the wall represents the greatest worry while the ceiling to the attic and the basement represent the lowest priority. Thus, in this house, you are probably wise to concentrate the most cost-effective sealing work in the walls because the ceiling to the attic and the floor to the basement seem to be, relatively speaking, of lesser concern. You may find that dense packing the walls with cellulose may be the best approach to sealing and insulating these existing walls.

### Uncertainty Percentages

If you look at the small section of uncertainty percentages in the upper right of the hatch chart, you will notice that the uncertainty in the numbers calculated using this chart typically goes up because there are starting H/Z pressure differential numbers that are not roughly in the middle of 0 and 50 Pascals. As you get closer to the minimum closed/open drop of 6 Pascals, the higher the uncertainty, and the less you should rely on the calculated numbers in deciding where to concentrate the work.

# Appendix S

# Safety Items on Combustion Appliances

## Water Heaters (typical)

Temperature and pressure relief valve to keep the tank from exploding and from damaging home or things in the home if a discharge pipe to near the floor is provided; mixing hood, draft diverter, or bell housing to allow the combusted gases to cool down before exiting in the flue; thermocouple on the pilot so the burners will not be provided with gas unless the pilot is on; if of the enclosed variety ("unremovable" cover over burner compartment where it is normally open) this type will help reduce the risk of explosion from flammable gasses/vapors if they become present in the home (e.g. painting with a lead base paint, cleaning with a flammable liquid, etc.).

Direct vented water heaters (bring combustion air in directly from outside) also have the added safety feature of not changing the pressures in the house since all the air going out is replaced by air going in as a closed system much like high efficiency furnaces do.

## Furnaces

All efficiencies: thermocouple, low limit/high limit, light switch like switch on the side of the furnace for shutting off all electrical power to furnace/boiler, etc.

Standard/Low efficiency—cooling down the combusted air by mixing with air from the room at the draft hood or mixing box

Medium/high efficiency—fan to remove combusted gasses to outdoors, fan compartment safety switch so fan is shut off if cover removed, flame roll-out or backdrafting sensors near the burners to shut off in the event of roll-out, glow plug or spark generator in lieu of continuously burning pilot light, sensors to detect backdrafting down the flue

High efficiency—direct vent system so combustion air taken from outside and combusted air coming off the burners directed back out to outside with little if any risk of furnace effecting pressures in the home.

## Boilers—All Boilers

Electrical Shut-off Switch on wall/near door/near appliance, Temperature/Pressure relief valve, Pressure gauge, Pressure Feeder Valve, Regulator/Reducing Valve, Automatic Water Backflow Preventer on Water Feeder line,

## Hot Water

High Temperature Limit Switch, Water Temperature Gauge, Aquastat (a thermostat for water that shuts of and turns on the burners depending on the water temperature in the boiler), Expansion Tank, Air Scoop (Auto Air Bleeder). Optional Safety items: Low water cutoff

## Steam

Low water cutoff, Clear Glass Water Level Gauge, Hartford Loop, Equalizer Pipe (vertical pipe directly above Hartford Loop), Pressurtrol (High and low pressure cut-in and cut-off of firing of burners).

**Oil Fired Appliances**

Fire detector that shuts off fuel supply and is placed on the appliance, ceiling, or tank), Cadmium Cell Flame Detector (serves a similar purpose as the thermocouple does on gas pilot lights/burners to shut down fuel supply when fuel unignited), Oil Burner Combustion Control (flue mounted detector that shuts off oil supply when combustion not occurring), Vacuum actuated oil supply valve (shuts off the oil supply near the tank when the oil pump is no longer creating enough vacuum to keep it open, such as when the oil pump is off.

# Glossary

**- A -**

Abatement—A measure or set of measures designed to permanently eliminate a hazard (for example, lead based paint). Abatement strategies include removal of the hazardous materials, replacement of building components containing the hazardous material, enclosure or encapsulation. All of these strategies require proper preparation, cleanup, waste disposal post abatement clearance testing, and if applicable, record keeping and monitoring.

Absorption—Absorption is the process by which a substance can be readily taken into the body through the skin or membranes. The best defense is to have a protective barrier between the substance and the skin.

Air Changes per Hour at 50 Pascals (ACH50)—The number of times that the complete volume of a home is exchanged for outside air in one hour when a blower door depressurizes or pressurizes the home to 50 Pa.

Air Changes per Hour natural (ACHnat)—The number of times the indoor air is exchanged with the outdoor air in one hour under natural driving forces. It can be estimated with blower door use. Also designated as ACHn or ACHnatural.

Air exchange—The process where indoor air is replaced with the outdoor air through air leakage and ventilation. One CFM out equals one CFM in.

Air-Free Carbon Monoxide—A method used to be able to compare CO readings with varying amounts of dilution air (oxygen) mixed in. The air-free method adjusts air content (oxygen) to zero.

Air handler—A steel cabinet containing a blower with cooling and/or heating coils connected to ducts, which circulates indoor air across the exchangers and into the living space.

Air infiltration barrier—A spun polymer sheet (for example, house wrap) that stops almost all the air traveling through a building cavity, while allowing moisture to pass through it.

Altitude Adjustment—When a gas appliance is installed more than 2000 feet above sea level, its input rating must be reduced by approximately four percent per 1000 feet above sea level.

AFUE—Annual Fuel Utilization Efficiency—A laboratory derived efficiency for heating appliances, which accounts for chimney losses, jacket losses, and cycling losses, but not distribution losses or fan/pump energy.

AAMA—Architectural Aluminum Manufacturers' Association

Asbestos—A fibrous mineral with fireproof and insulation characteristics which may be shaped into a variety of building materials. Small, sharp asbestos fibers may cause damage to lungs if they are inhaled.

Ambient air—Air in the living space.

ANSI—American National Standards Institute, Inc.

Appliance Depressurization Limits—The ambient depressurization at which the venting of a combustion appliance is likely to become hazardous for those testing the appliance or for the usual occupants of the building.

ASHRAE—American Society of Heating, Refrigerating, and Air-Conditioning Engineers, Inc.

ASME—American Society of Mechanical Engineers

ASTM—American Society for Testing and Materials

Ampere—A unit of measurement that tells how much electricity flows through a conductor. It is like cubic feet per second to measure the flow of water. For example, a 1,200-watt, 120-volt hair dryer pulls 10 amperes of electric current (watts divided by volts).

Appliance Depressurization Limits—The ambient depressurization at which the venting of a combustion appliance is likely to become hazardous for those testing the appliance or for the usual occupants of the building.

Aquastat—A heating control that switches the burner or the circulator pump in a hydronic heating system.

Area Weighted R-Value (also Whole Wall R-Value or Weighted Average Wall R-Value)—a determination of the overall average R-Value of a wall taking into account the fact that the stud and other framing has a lower R-value than the insulated sections of the wall. By determining what fraction of the wall is insulated and the remaining fraction that is framing, you can calculate, first the overall weighted average conductivity of the wall (U value) and then, by taking the inverse of the overall U value, calculate the overall R-Value by the following equation:

Where:

    U is the area-weighted U factor

    A1 is the fraction of the wall represented by the insulated wall.

    U1 is the U factor of the insulated portion of the wall

    A2 is the fraction of the wall represented by framing

    U2 is the U factor of the framed portion

    $U = (A1 \times U1) + (A2 \times U2)$ And $R = 1/U$

As-Measured Carbon Monoxide—Measure of CO in air or flue gas that does not take into account excess air that dilutes the CO concentration.

Atmospheric appliances—A heating device that takes its combustion air from the surrounding room air. Also, know as open-combustion heater.

Atmospheric burner– A burner utilizing atmospheric combustion.

Atmospheric combustion—Combustion which takes place under atmospheric pressure at a given altitude.

Atmospheric pressure—The weight of air and its contained water vapor on the surface of the earth. At sea level this pressure is 14.7 pounds per square inch.

**- B -**

Backdrafting—Continuous spillage of combustion gases from a vented combustion appliance into the living space.

Backdraft damper—A damper, installed near a fan, that allows air to flow in only one direction and prevents reverse flow when the fan is off.

Backer rod—Polyethylene foam rope used as a backer for caulking.

Baffle – A plate or strip designed to retard or redirect the flow of flue gases.

Balance point—The outdoor temperature at which no heating is needed to maintain inside temperatures.

Balanced flue vent system—Term used for oil-fired systems to indicate a direct-vent appliance with positive pressure in the vent connector through which the gases of combustion pass.

Ballast—A coil of wire or electronic device that provides a high starting voltage for a lamp and limits the current from flowing through it.

Balloon framing—A method of construction in which the vertical framing members (studs) are continuous pieces running the entire height of the wall.

Barometric Damper—See Draft Regulator.

Barometric Draft Regulator—See Draft Regulator.

Band joist—See Rim joist.

Barometric vent damper—a device installed in the heating unit vent system to control draft. Usually used on oil-fueled units or gas units with power burners.

Batt—A blanket of preformed insulation, generally 14.5" or 22.5" wide and varying in thickness from 3.5" to 9."

Bay—the space between two studs in a wall (stud bay), between floor joists (floor joist bay) or other framing.

BDL—See Building Depressurization Limit.

Belly return—A configuration found in some mobile homes that uses the belly cavity as the return side of the distribution system.

Belt Tension (proper adjustment of)—Minimum of one-inch play per side. The belt should not slip on the pulleys.

Benefit-to-Cost Ratio (BCR)—See Savings-to-Investment Ratio (SIR).

Bimetal element—A metal spring, lever, or disc made of two dissimilar metals that expand and contract at different rates as the temperature around them changes. This movement operates a switch in the control circuit of a heating or cooling device.

Blocking—A building element or material used to prevent movement into or through building cavities.

Blow-down—Removing water from a boiler to remove sediment and suspended particulates.

Blower—The "squirrel-cage" fan in a furnace or air handler.

Blower door—A calibrated device to measure the air tightness of a building by pressurizing or depressurizing the building and measuring the flow through the fan.

Blown insulation—A loose-fill insulation that is blown into attics and building cavities using an insulation blowing machine.

Boiler—A space heating appliance that heats water with hot combustion gases.

Boot—A duct section that connects between a duct and a register, floor, or wall cavity or between round and square ducts.

Branch circuit—An electrical circuit used to power outlets and lights within a home.

Breeching or Breech—See Vent Connector.

Brightness—The luminous intensity of any surface in a given direction per unit of projected area of the surface as viewed in that direction.

British Thermal Unit (Btu)—The quantity of heat required at sea level to raise the temperature of one pound of water one degree Fahrenheit.

BTL—Building Tightness Limit calculation procedure, expressed in units of CFM50, based on the American Society of Heating, Refrigerating and Air-Conditioning Engineers Standard 62-1999, Ventilation for Acceptable Indoor Air Quality. This method was clearly explained in an article in Home Energy magazine (Tsongas 1993). The method closely follows the parameters set in ASHRAE 62-1999: For acceptable indoor air quality, 15 CFM per person (set minimum of five people) or 0.35 air changes per hour (ACH), whichever is greater, must be supplied by natural air leakage and/or continuously operating ventilation.

BTLa—Building Tightness Limit calculation procedure, expressed in units of CFM50, that is more complex than the BTL method and is based on ASHRAE Standard 62, Standard 119 (Air Leakage Performance for Detached Single-Family Residential Buildings), and Standard136 (A Method of Determining Air Change Rates in Detached Dwellings). This method closely follows the parameters set in ASHRAE 62-1999: For acceptable indoor air quality, 15 CFM per person or 0.35 air changes per hour (ACH), whichever is greater, must be supplied by natural air leakage and/or continuously operating ventilation. However, the BTLa method uses different calculation methods—based on ASHRAE 119 and 136—than the BTL method to arrive at the final tightness limits.

Btu—British thermal unit. Roughly the amount of energy expended when a kitchen match is completely burned

Btu content of fuels—the rough equivalence of energy in a given amount of a fuel in terms of the number of Btus, such as Btus per gallon of fuel oil, cord of wood, etc.

Btuh—British thermal units per hour.

Building cavities—The spaces inside walls, floors, and ceilings or between the interior and exterior sheeting.

Building Depressurization Limit (BDL)—BDL is a selected indoor negative pressure; expressed in Pascals, immediately around vented combustion appliances that use indoor air for combustion supply air. If a combustion appliance experiences a negative pressure of a greater magnitude than the BDL, it has the potential to backdraft, causing a hazardous condition for the occupants. The BDL for furnaces and boilers is often -5 Pascals and for stand-alone natural draft water heaters, -2 Pascals. Field studies have been done to determine the negative pressure at which these appliances will begin to backdraft.

Building science—An involved perspective on buildings, using contemporary technology to analyze and solve problems dealing with design, construction, maintenance, safety, and energy efficiency of the buildings.

Building Tightness Limit—A general term for a house-tightening limit, expressed in units of CFM50, used for ensuring adequate indoor air quality for the house occupants. Two building tightness limit procedures used in the North Dakota Weatherization Program are BTL and BTLa.

Burner—A device that facilitates the burning of a fossil fuel like gas or oil.

Bypass—An air leakage site that allows air to leak out of a building passing around the air barrier and insulation

## - C -

Carbon dioxide ($CO_2$)—A heavy, colorless, nonflammable gas formed by the oxidation of carbon, by combustion, and in respiration of plants and animals.

Carbon monoxide (CO)—An odorless, colorless, tasteless, and poisonous gas produced by incomplete combustion. To measure the CO emissions in flue gas, the CO sample must be taken before dilution air enters the vent system. CO emissions are measured in parts per million (ppm). See Air-Free Carbon Monoxide and As-Measured Carbon Monoxide.

Caulking—A mastic compound for filling joints and cracks.

Category I Gas appliance—An appliance that operates with negative static pressure in the vent and a temperature that is high enough to avoid condensation in vent.

Category I Fan-assisted gas appliance—An appliance that operates with negative static pressure in the vent, a temperature that is high enough to avoid condensation in vent, and an integral fan to draw a controlled amount of combustion supply air through the combustion chamber.

Category II gas appliance—An appliance that operates with negative static pressure in the vent and a temperature that is low enough to cause excessive condensation in the vent.

Category III Gas appliance—An appliance that operates with positive static pressure in the vent and a temperature that is high enough to avoid condensation in vent.

Category IV Gas appliance—An appliance that operates with positive static pressure in the vent and a temperature this is low enough to cause excessive condensation in the vent.

CAZ—See Combustion Appliance Zone.

Cellulose insulation—Insulation, packaged in bags for blowing, made from newspaper or wood waste and treated with a fire retardant.

Central Return—System of ducts or passages for distribution return air, which connect different areas of the house to a central location at the forced air furnace

Chimney—A building component designed for the sole purpose of assuring combustion by-products are exhausted to the exterior of the building.

Chimney flue—A passageway in a chimney for exhausting combustion gases to the outdoors.

Circuit breaker—A device that automatically disconnects an electrical circuit from electricity under a specified or abnormal condition of current flow.

Cleanout opening—An opening in a chimney (usually at its base) to allow inspection and the removal of ash or debris.

Coefficient of Performance (COP)—A heat pump or air conditioner's output in Watt-hours of heat removed divided by Watt-hours of electrical input.

Coil—A snakelike piece of copper tubing surrounded by rows of aluminum fins that clamp tightly to the tubing to aid in heat transfer.

Cold Air Return (Return side): Ductwork through which house air is drawn for reheating during furnace cycle.

Color rendering index (CRI)—A measurement of a light source's ability to render colors the same as sunlight. CRI has a scale of 0 to 100.

Color temperature—A measurement of the warmness or coolness of a light source in the Kelvin temperature scale.

Combustible –Susceptible to combustion, inflammable, any substance that will burn.

Combustible Gas Leak Detector—A device for determining the presence and general location of combustible gases in the air.

Combustion—The act or process of burning. Oxygen, fuel and a spark must be present for combustion to occur.

Combustion air—Air required to chemically combine with a fuel during combustion to produce heat and flue gases, mainly carbon dioxide and water vapor.

Combustion analyzer—A device used to measure steady-state efficiency of combustion heating units.

Combustion appliance—Any appliance in which combustion occurs.

Combustion Appliance Zone (CAZ)—The closed space or area, which holds one or more combustion appliances.

Combustion chamber—The area inside a heating unit where combustion takes place.

Common vent—The portion of the vent or chimney through which products of combustion from more than one appliance pass.

Compact fluorescent light (CFL)—A small fluorescent light engineered to fit conventional incandescent fixtures.

Compressor—A motorized pump that compresses the gaseous refrigerant and sends it to the condenser where heat is released.

Concentrically construction direct-vent—A direct-vent appliance that has an exhaust-gas vent and a combustion-supply air vent arranged in a concentric fashion, (one vent is inside the other with a space between the walls of each).

Condense—To change from a gaseous or vaporous state to a liquid or solid state by cooling or compression.

Condenser—The coil in an air conditioning system where the refrigerant condenses and releases heat, which is carried away by air moving across the coil.

Condensate—The liquid formed when a vapor is condensed.

Condensate receiver—A tank for catching returning condensate water from a steam heating system.

Conditioned Space—A heated or cooled area of a building. Conditioned space includes any area of a dwelling that is determined to be within the insulated envelope or shell.

Conductance—The quantity of heat, in Btu, that will flow through one square foot of material in one hour, when there is a one degree Fahrenheit temperature difference between both surfaces. Conductance values are given for a specific thickness of material, not per inch thickness.

Conduction—The transfer of heat energy through a material (solid, liquid or gas) by the motion of adjacent atoms and molecules without gross displacement of the particles.

Conductivity—The quantity of heat that will flow through one square foot of homogeneous material, one inch thick, in one hour, when there is a temperature difference of one degree Fahrenheit between its surfaces.

Confined space—A space with a volume of less than 50 cubic feet per 1,000 Btu per hour of the total input rating of all combustion appliances installed in that space.

Contractor—Any for-profit, not-for-profit, or government entity that provides services to the program under contract, not as a result of a grant of funds.

Control circuit—A circuit whose work is switching a power circuit or opening an automatic valve.

Convection—The transmission of heat by the actual movement of a fluid because of differences in temperature, density, etc.

Conventionally vented combustion appliance—Combustion appliances that are characterized by atmospheric burners or natural draft. Sealed or direct-vent appliances are not conventionally vented.

Cooling load—The maximum rate of heat removal required of an air conditioner when the outdoor temperature and humidity are at the highest expected level.

Cost-effective—Having an acceptable payback, return-on-investment, or savings-to-investment ratio.

Critical framing juncture—An intersection of framing members and envelope components that require special attention during prep and installation of insulation.

Cross section—A view of a building component drawn or imagined by cutting through the component.

CFM—Cubic Feet per Minute—A measurement of air movement in cubic feet past a certain point or through a certain structure per minute.

CFM50—The number of cubic feet per minute of air flowing through the fan housing of a blower door when the house pressure is 50 Pa (0.2 inches of water column). This figure is the most common and accurate way of comparing the tightness of buildings that are tested using a blower door.

CFMnat—The number of cubic feet of air flowing through a house from indoors to outdoors during typical, natural conditions. This figure can be roughly estimated using a blower door using the LBL (Lawrence Berkeley Labs) infiltration model

## - D -

Degree-days (DD)—A measure of the temperature element of climate produced by summing the temperature differences between the inside (65°F) and the daily average outside temperature for a one-year period.

Demand—The peak need for electrical energy. Some utilities levy a monthly charge for demand.

Density—The weight of a material divided by its volume, usually measured in pounds per cubic foot.

DOE—The United States Department of Energy.

Depressurize—To lower the pressure in an enclosed area with respect to a reference pressure.

Depressurization Tightness Limit (DTL)—A calculation procedure, expressed in units of CFM50, performed to estimate the building tightness level at which combustion appliances might backdraft when the house is under conditions of worst-case depressurization. A BDL must be selected for the calculation of the DTL. The DTL sets a low limit for air sealing that may or may not be lower than the building tightness limit for the same house. Refer to Appendix C for a flow chart of this procedure.

Design temperature—A high or low temperature used for designing heating and cooling systems when calculating the building load.

Dilution air—Air that enters through the dilution device-an opening where the chimney joins to an atmospheric-draft combustion appliance.

Dilution device—A draft diverter, draft hood, or barometric draft control on an atmospheric-draft combustion appliance.

Direct-vent appliance—Appliances that are constructed and installed so that all combustion air is taken directly from and the flue gases are vented directly to the outside.

Distribution system—A system of pipes or ducts

DHW—Domestic Hot Water

Dormer—A framed structure projecting above a sloping roof surface, and normally containing a vertical window.

Downdraft—Air flowing down a chimney or vent during the appliance off-cycle.

Draft—A pressure difference that causes combustion gases or air to move through a vent connector, flue, chimney, or combustion chamber. May be natural draft, induced draft, or forced draft. Draft is often measure with a draft gauge.

Draft diverter—A device built into an appliance or made a part of the vent connector for an appliance that is designed to: 1) provide for the ready escape of the flue gasses from the appliance in the event of no draft, backdraft, or stoppage beyond the draft hood, 2) prevent a backdraft from entering the appliance, and 3) neutralize the effect of stack action of the chimney or gas vent upon the operation of the appliance.

Draft fan—A mechanical fan used in a venting system to augment the natural draft in gas- and oil-fired appliances. These electrically operated, paddle-fan devices are installed in vent connectors.

Draft hood—A nonadjustable deice built into an appliance or a part of the vent connector this is intended to 1) provide for escape of flue gases if blockage or backdraft occurs, 2) prevent a downdraft of outdoor air from entering the appliance, 3) neutralize the effect of stack action of the chimney, and 4) lower the dew point temperature of the flue gas by the infusion of ambient room air.

Draft regulator—An adjustable and self-regulating damper attached to a chimney or vent connector for the purpose of controlling draft. A draft regulator can reduce draft; it cannot increase draft.

Drywall—Gypsum interior wallboard used to produce a smooth and level interior wall surface and to resist fire. Also called gypsum wallboard and sheetrock.

Dry bulb temperature—Normal ambient air temperature measured by a thermometer.

DTL—See Depressurization Tightness Limit.

Duct blower—A blower-door-like device used for testing duct leakiness and air flow.

Duct zone—A building space or cavity that contains heating or cooling ducts.

- E -

Eave—The part of a roof that projects beyond its supporting walls. See also soffit.

Effective Leakage Area (ELA)—ELA was developed by Lawrence Berkeley Laboratory (LBL) and is used in their infiltration model. The Effective Leakage Area is defined as the area of a special nozzle shaped hole (similar to the inlet of the Blower Door fan) that would leak the same amount of air as the building does at a pressure of 4 Pascals.

Efficiency—The ratio of output divided by input.

Efficacy—The number of lumens produced by a watt used for lighting a lamp. Used to describe lighting efficiency.

Electric service—The electric meter and main switch, usually located outside the building.

Emittance—The rate that a material emits radiant energy from its surface. Also called emissivity.

Encapsulation—Any covering or coating that acts as a barrier between the hazard (such as lead-based paint) and the environment, the durability of which relies on adhesion and the integrity of existing bonds between any existing layers (for example paint) and the substrate.

Enclosure—The use of rigid, durable construction materials that are mechanically fastened to the substrate to act as a barrier between the hazardous material (such as lead-based paint) and the environment.

Energy—A quantity of heat or work.

Energy audit—The process of identifying energy conservation opportunities in buildings.

Energy consumption—The conversion or transformation of potential energy into kinetic energy for heat, light, electricity, etc.

Energy efficiency—Term describing how efficiently a building component uses energy.

EEM—Energy efficiency measure.

Energy efficiency ratio (EER)—A measurement of energy efficiency for room air conditioners. The EER is computed by dividing cooling capacity, measured in British Thermal Units per hour (Btuh), by the watts of power. (See also Seasonal Energy Efficiency Rating—SEER)

Envelope—The building shell. The exterior walls, floor, and roof assembly of a building.

Environmentally sensitive—A person who is highly sensitive to pollutants, often because of overexposure, is said to be environmentally sensitive.

Equivalent Leakage Area (EqLA): EqLA is defined by Canadian researchers at the Canadian National Research Council as the area of a sharp edged orifice (a sharp round hole cut in a thin plate) that would leak the same amount of air as the building does at a pressure of 10 Pascals. The EqLA is used in the AIM infiltration model (which is used in the HOT2000 simulation program).

Evaporation—The process of being changed into a vapor or gas at a temperature usually below the boiling point. Evaporation is a cooling process.

Evaporative cooler—A device for cooling homes in dry climates that cools the incoming air through the evaporation of water.

Evaporator—The heat transfer coil of an air conditioner or heat pump that cools the surrounding air as the refrigerant inside the coil evaporates and absorbs heat.

Exacerbate—To aggravate or make worse.

Exfiltration—Air flowing out of a building from its conditioned space through the holes in the shell.

**- F -**

Fahrenheit—A temperature scale for which water boils at 212° and freezes at 32°.

Fan-assisted combustion—A combustion appliance with an integral fan to draw combustion supply air through the combustion chamber.

Fan control—A bimetal thermostat that turns the furnace blower on and off as it senses the presence of heat.

Fan-off temperature—In a furnace, the supply air temperature at which the fan control shuts down the distribution blower.

Fan-on temperature—In a furnace, the supply air temperature at which the fan control activates the distribution blower.

Feeder wires—The wires connecting the electric meter and main switch with the main panel box indoors.

Fenestration—Window and door openings in a building's wall.

Fiberglass—A fibrous material made by spinning molten glass.

Fill tube—A plastic or metal tube used for its stiffness to blow insulation inside a building cavity and allows the insulation to be delivered at the extreme end of the cavity.

Fire stop—Framing member, usually installed horizontally between studs, designed to stop the spread of fire within a wall cavity.

Furring—Thin wood strips fastened to a wall or ceiling surface as a nailing base for finish materials.

Flame safety control—A control for avoiding fuel delivery in the event of no ignition.

Flammable/Inflammable—Combustible; readily set on fire.

Flashing—Waterproof material used to prevent leakage at intersections between the roof surface at walls or penetrations.

Floor joists—The framing members that support the floor.

Floor joist bay—the space between floor joists (floor joist bay).

Flue—A vent for combustion gases.

Foam board—Plastic foam insulation manufactured most commonly in 4' x 8' sheets in thicknesses of ½" to 3."

Foot-candle—A measure of light striking a surface.

Footing—The part of a foundation system that transfers the weight of the building to the ground.

Forced draft—A vent system for which a fan installed at the combustion appliance moves combustion gases to the outdoors with positive static pressure in the vent pipe. Because of this positive pressure, the vent connector must be air-tight.

Frost line—The maximum depth of the soil where water will freeze during the coldest weather.

Furnace—A space heating appliance that heats air with hot combustion gases.

## - G -

Gable—The triangular section of an end wall formed by the pitch of the roof.

Gable roof—A roof shape that has a ridge at the center and slopes in two directions.

GAMA—Gas Appliance Manufacturers' Association

Gas oven bake burner—Oven burner used for baking located just below the oven compartment floor.

Gas oven broiler burner—Oven burner used for broiling located at the top of the oven compartment.

Gasket—Elastic strip that seals a joint between two materials.

General heat waste—Weatherization measures for which savings or savings-to-investment ratios (SIR) are difficult or impossible to calculate. Examples include all air sealing work, ductwork sealing and insulation, pipe insulation, and dryer vent kit installation. No SIR values are required for these measures.

Glazing—Glass installation. Pertaining to glass assemblies or windows.

Glazing compound—A flexible, putty-like material used to seal glass in its sash or frame.

Gypsum board—A common interior sheeting material for walls and ceilings made of gypsum rock powder packaged between two sheets of heavy building paper. Also called sheetrock, gyprock, or gypboard.

Ground Fault Circuit Interrupter (GFI or GFCI)—An electrical connection device that breaks a circuit if a short occurs. These are required for all exterior use of electrical equipment or when an electrical outlet is located near a water source.

## - H -

Hazardous condition—A situation that is causing a danger to the client/crew/contractor that exists before, is created by, or is exacerbated by, weatherization. For example, a dwelling could have a moisture problem that is allowing biological hazards (molds, viruses, bacteria, etc.) to flourish. Another example would be allowing fiberglass to enter the living space due to improperly fastened or sealed ductwork.

Hazardous material—A particular substance that is considered a danger to the client/crew/contractor.

HHS—United States Department of Health and Human Services

Heat anticipator—A very small electric heater in a thermostat that causes the thermostat to turn off before room temperature reaches the thermostat setting, so that the house does not overheat from heat remaining in the furnace and distribution system after the burner shuts off.

Heat capacity—The quantity of heat required to produce a unit of temperature change in a given object. Specific heat capacity of one gram or pound of a pure specific material. Common units are in joules/degree Centigrade or Btu/degree Fahrenheit.

Heat exchanger—The area in a heating unit that separates the combustion process from the distribution fluid, with the sole purpose of transferring heat from the combustion process to the distribution fluid.

Heat loss—The amount of heat escaping through the building shell during a specified period.

Heat pump—A type of heating/cooling unit, usually electric, that uses a refrigerant fluid to heat and cool a space.

Heat rise—In a furnace, the number of degrees of temperature increase that air is heated as it is blown over the heat exchanger. Heat rise equals air supply temperature minus air return temperature.

Heating degree day (HDD)—Each degree that the average daily temperature is below the base temperature (usually 65°F) constitutes one heating degree-day.

Heating load—The maximum amount of heat needed by a building during the very coldest weather to maintain the designed inside temperature.

Heating seasonal performance factor (HSPF)—Rating for heat pumps describing how many Btus they transfer per kilowatt-hour of electricity consumed.

HVAC—Heating, Ventilating, Air-Conditioning

High limit—A bimetal thermostat that turns the heating element of a furnace off if it senses a dangerously high temperature.

Hip Roof—A roof that slants in four directions from a central peak.

Home energy index—The number of Btus of energy used by a home divided by its area of conditioned square feet and by the number of heating degree days during one year.

HVI—Home Ventilating Institute

WAP—Home Weatherization Assistance Program

House pressure—The difference in pressure between the inside and outside of the house.

HUD—United States Department of Housing and Urban Development

Humidistat—An automatic control that switches a fan, humidifier, or dehumidifier on and off based on the relative humidity at the control.

Humidity ratio—The absolute amount of air's humidity measured in pounds of water vapor per pound of dry air.

Hydronic—A heating system using hot water or steam as the heat-transfer fluid. A hot-water heating system (common usage).

## - I -

Illumination—The light level measured on a horizontal plane in foot-candles.

Incandescent light—The common light bulb found in residential lamps and light fixtures and sold in stores everywhere that is known for its inefficiency.

IAQ—Indoor Air Quality

Inches of Water Column (IWC)—A non-metric unit of pressure difference. 0.004 IWC is equal to one Pascal. One IWC is equal to approximately 249 Pascals.

Induced combustion fan—A mechanical device intended to pull a consistent volume of combustion supply air through the combustion
areas on a Category I, fan-assisted, gas-fired appliance.

Induced draft—A vent system for which a fan-installed at or very near
the termination point of the vent pipe- moves the combustion gases to the outdoors with negative static pressure in the vent pipe.

Infiltration—The uncontrolled movement of non-conditioned air into a
conditioned air space.

Infrared—Pertaining to heat rays emitted by the sun or warm objects
on earth.

Ingestion—Ingestion is the process by which a substance enters the
body by swallowing through the mouth. The best defense is to
wash your hands before eating or putting your fingers in your
mouth, keeping hazardous materials out of reach from small children, and guarding against splashing of hazardous materials into
your mouth.

Inhalation—Inhalation is the process by which a substance is breathed
into the body in the form of a gas, vapor, fume, mist, or dust. The
best defense is to use a proper filter to remove these contaminants
before they enter the body or to not create dust if possible.

Input rating—The designed capacity of an appliance usually specified
in Btus or units of energy.

Insulating glass—Two or more glass panes spaced apart and sealed in a
factory giving a higher R-value.

Insulation—A material used to retard heat transfer.

Isolated outdoor air supply—Term used with oil-fired systems to indicate a vent pipe through which outdoor combustion supply is
ducted to the oil burner.

Intermittent ignition device (IID)—A device that lights the pilot light on
a gas appliance when the control system calls for heat, thus saving
the energy wasted by a standing pilot.

Internal gains—The heat generated by bathing, cooking, and operating
appliances that must be removed during the summer to promote
comfort or will reduce the heating demand in the winter.

Interstitial—Space between framing and other building components.
These spaces are also referred to as interstitial cavities.

**- J -**

Joist—A horizontal wood-framing member that supports a floor or ceiling.

Joule—A unit of energy. One thousand joules equals 1 Btu.

**- K -**

Kilowatt—One thousand watts. A unit of measurement of the amount of electricity needed to operate given equipment.

Kilowatt-hour—The most commonly used unit for measuring the amount of electricity consumed over time. It means one kilowatt of electricity supplied for one hour.

Kinetic energy—Consisting of or depending on motion; distinguished from potential energy.

**- L -**

Lamp—A light bulb.

Latent heat—The amount of heat energy required to change the state of a substance from a solid to a liquid or from a liquid to a gas without changing the temperature of the substance.

Lath—A thin strip of wood or base of metal or gypsum board serving as a support for plaster.

Light quality—Good light quality is characterized by absence of glare and low brightness contrast.

Living space—A space in a dwelling that is lived in or regularly occupied. This space may be conditioned or unconditioned.

Low-water cutoff—A float-operated control for turning the burner off if a steam boiler is low on water.

Lumen—A unit of light output from a lamp.

Low-E—Short for low emissivity, which refers to the characteristic of a metallic glass coating to resist the flow of radiant heat.

**- M -**

Main panel box—The service box containing a main switch, and the fuses or circuit breakers located inside the home.

Makeup air—Air supplied to a space to replace exhausted air.

Manifold—A tube with one inlet and multiple outlets or multiple inlets and one outlet.

Manometer—A pressure differential gauge used for measuring gas and air pressures.

MHEA—Manufactured Housing Energy Audit, developed by DOE for WAP. Used to audit mobile homes.

Masonry—Construction of stone, brick, or concrete block.

Mastic—A thick creamy substance used to seal seams and cracks in building materials and especially useful on ductwork.

MSDS—Materials Safety Data Sheet

Mechanical draft—A combustion appliance with induced draft of forced draft.

Metabolic process—Chemical and physiological activities in the human body.

Mitigate—To make less severe.

Mortar—A mixture of sand, water, and cement used to bond bricks, stones, or blocks together.

## - N -

NBS—The National Bureau of Standards, Department of Commerce renamed the National Institute of Standards and Technology (NIST).

NEMA—National Electrical Manufacturers' Association

NEAT—National Energy Audit, developed by DOE for WAP. Used to audit single-family and low-rise multi-family buildings.

NFPA—National Fire Protection Association.

NWMA—National Woodwork Manufacturers Association.

Natural draft—A vent system that relies on natural draft (hot, buoyant air) to move combustion gases to the outdoors.

Natural ventilation—Ventilation using only natural air movement, without fans or other mechanical devices.

Net Free Vent Area (NFVA)—The area of a vent after that area has been adjusted for the restrictions caused by insect screen, louvers, and weather coverings. The free area is always less than the actual area.

Non-conditioned space—An area within the building envelope that is not heated or cooled and tends to be the same temperature as outside.

Nozzle—An orifice designed to change a liquid like oil into a mist to improve the combustion process.

## - O -

$O_2$—Oxygen.

Ohm—A unit of measure of electrical resistance. One volt can produce a current of one ampere through a resistance on one ohm.

Open cavity wall or floor—a wall or floor without any rigid cover over both faces, such as drywall over the face of the framing.

Open-combustion appliance—A combustion appliance that takes its combustion air form the surrounding room. Contrast this with a direct-vent appliance.

Open floor—a floor without any rigid cover over both faces, such as drywall over the face of the framing.

Open wall—a wall without any rigid cover over both faces, such as dry-wall over the face of the framing.

Orifice—A hole in a gas pipe where gas exits the pipe to be mixed with air in a burner before combustion in a heating device. The size of the orifice will help determine the flow rate.

OTL—See Overall Tightness Limit.

Output capacity—The conversion rate of useful heat or work that a device produces after waste involved in the energy transfer is accounted for.

Overall Tightness Limit (OTL)—The OTL is expressed in units of CFM50. The OTL considers both the building tightness limit and the DTL. For example, if the building tightness limit is 1300 CFM50 and the DTL is 1400 CFM50, the OTL for the house is 1400 CFM50, satisfying both the building tightness limit and the DTL.

Oxygen Depletion Sensor (ODS)—A safety device for unvented (vent-free) combustion heaters that shuts off gas when oxygen is depleted.

### - P -

Parts per million (ppm)—The unit commonly used to represent the degree of pollutant concentration where the concentrations are small.

Pascal (Pa)—A metric unit of measurement of air pressure. 2.5Pa = 0.01 inches of water column.

Payback period—The number of years that an investment in energy conservation will take to repay its cost in energy savings.

Peak Electrical Demand Or Peak Electrical Load—The electric load that corresponds to a maximum level of electric demand in a specified time period. In other words, the largest amount of electricity used during a set period (15, 30, or 60 minutes, etc.) during the month of billing. Also you could say it is the highest electrical demand within a particular period of time. The highest electrical demands/peaks on a daily basis usually occurs on weekdays in late afternoon and early evening, usually between 4 to 7 p.m. in the winter and 12 to 8 p.m. in the summer. Annual peaks tend to occur on occur on hot summer days due to the heavy use of electric powered air conditioning. Sometimes a supplemental "demand charge" or service charge, based on the usage during this peak electrical demand period, is added to the normal bill for larger buildings, including multi-family apartments

Perimeter Pull—A technique used in attics previously insulated with batt insulation. The batts are cut back 2 feet from the eaves and the area is insulated with blown insulation to ensure coverage over the outer wall top plate and to prevent wind washing of the insulation under the existing batts.

Perlite—A heat-expanded mineral used for insulation.

Perm—A measurement of how much water vapor a material will let pass through it per unit of time under a specified pressure difference. More specifically, permeance is a unit of measurement for the ability of a material to retard the diffusion of water vapor at 73.4 °F (24 °C). A perm or perm rating, short for permeance, is the number of grains of water vapor that pass through a square foot of material per hour at a differential vapor pressure equal to one inch of mercury. The smaller the perm rating or permeance of a vapor barrier, the better the water vapor resistance. A good vapor barrier should have a perm rating no greater than 1.0.

Permeability: The ease with which water vapor flows through a porous medium or material

Pilot Tube—A device for measuring fluid velocity. An instrument placed in a moving fluid and used along with a manometer to measure fluid velocity.

Plaster—A plastic mixture of sand, lime, and Portland cement spread over wood or metal lathe to form the interior surfaces of walls and ceilings.

Plate—A piece framing member installed horizontally to which the vertical studs in a wall frame are attached.

Plenum—The section of ductwork that connects the air handler to the main supply duct.

Plywood—Laminated wood sheeting with layers cross grained to each other.

Polyethylene—A plastic made by the polymerization of ethylene, used in making translucent, lightweight, and tough plastics, films, insulations, vapor retarders, air barriers, etc.

Polyisocyanurate—Plastic foam insulation sold in sheets, similar in composition to polyurethane.

Polystyrene insulation—rigid plastic foam insulation, usually white, blue, pink, or green in color.

Polyurethane—versatile plastic foam insulation, usually yellow in color.

Power burner—A burner for which air is supplied at a pressure greater than atmospheric pressure. Most oil-fired burners are power burners. Gas burners used to replace oil burners are usually power burners.

Power Draft—See Mechanical Draft.

Potential energy—Energy in a stored or packaged form.

Pressure—A force that encourages movement by virtue of a difference in some condition between two areas. High pressure moves to low pressure.

Pressure diagnostics—The practice of measuring pressures and flows in buildings to control air leakage, and to ensure adequate heating and cooling airflows and ventilation.

Pressure pan—A device used to block a duct register, while measuring the pressure behind it.

Pressure Relief Valve—A safety component required on a boiler and water heater, designed to relieve excess pressure buildup in the tank.

Pressuretrol—A control that turns a steam boiler's burner on and off as steam pressure changes.

Primary window—The main window installed on the outside wall. Not to be confused with a storm window.

Provider—Either a grantee or contractor.


**- R -**

R-value—A measurement of thermal resistance.

Radiant barrier—A foil sheet or coating designed to reflect radiant heat flow. Radiant barriers are not mass insulating materials.

Radiant temperature—The average temperature of objects in a home, including walls, ceiling, floor, furniture, and other objects.

Radiation—Heat energy that is transferred by electromagnetic or infrared light from one object to another. Radiant heat flow can travel through a vacuum and other transparent materials.

Radon—A radioactive gas that decomposes into radioactive particles.

Rafter—A beam that gives form and support to a roof.

Rated ventilation—A ventilation system that has been designed and installed under the guidelines established by the American Society of Heating, Refrigeration, and Air Conditioning Engineers (ASHRAE) Standard for Acceptable Indoor Air Quality (Standard 62).

Reflectance—The ratio of lamination or radiant heat reflected from a given surface to the total light falling on it. Also called reflectivity.

Refrigerant—Any of various liquids that vaporize at a low temperature, used in mechanical refrigeration.

Register—A grille covering a duct supply outlet used to diffuse the airflow and sometimes control the flow.

Relative humidity—The percent of moisture present in the air compared to the maximum amount possible at that given temperature. Air that is saturated has 100% relative humidity.

Relay—An automatic, electrically operated switch.

Reset controller—Adjusts fluid temperature or pressure in a central heating system according to outdoor air temperature.

RCS—Residential Conservation Service Program

Resistance—The property of a material resisting the flow of electrical energy or heat energy.

Retrofit—An energy conservation measure that is applied to an existing building. Also means the action of improving the thermal performance or maintenance of a building.

Return air—Air circulating back to the furnace or central air conditioning unit from the house, to be heated or cooled and supplied back to the living area.

Rim joist—The outermost joist around the perimeter of the floor framing.

Rocking on the High Limit– Refers to the gas burner being shut down by the high limit switch on a furnace, instead of being properly activated by the fan-on/fan-off control.

Room air conditioner—A unitary air conditioner installed through a wall or window, which cools the room by removing heat from the room and releasing it outdoors.

**- S -**

Sash—A movable or stationary part of a window that frames a piece of glass.

Savings-to-Investment Ratio (SIR)—For an energy saving measure, the ratio of the savings over the investment (cost), including the discounting the investment value and escalation of fuel costs.

Sealed-combustion appliance—An appliance that draws combustion air from outdoors and has a sealed exhaust system. Also called a direct-vent appliance.

Seasonal energy efficiency ratio (SEER)—A measurement of energy efficiency for central air conditioners. The SEER is computed by dividing cooling capacity, measured in Btuh, by the Watts. (See also Energy Efficiency Rating.)

Sensible heat—The heat required to change the temperature of a material without changing its form.

Sequencer—A bimetal switch that turns on the elements of an electric furnace in sequence.

Service wires—The wires coming from the utility transformer to the service equipment of the building.

Sheathing—structural sheeting, attached on top of the framing, underneath siding and roofing of a building. Any building material used for covering a building surface.

Sheetrock—See drywall.

Shell—The building's exterior envelope—walls, floor, and roof of a building.

Shingle—A modular roofing component installed in overlapping rows.

Short circuit—A dangerous malfunction in an electrical circuit, where electricity is flowing through conductors and into the ground without going through an electric load, such as a light or motor.

Sill—The bottom of a window or doorframe.

Sill box—The area bounded by the rim joist, floor joists, sill plate, and floor.

Site-built home—Includes a house built on the site from building supplies or manufactured homes assembled on the site from pieces shipped to the site on flatbed trucks. Does not include mobile homes and doublewides.

Sling psychrometer—A device holding two thermometers, one wet bulb and one dry bulb, which is slung through the air to determine relative humidity.

Slope—The roof section of a knee wall attic with the roof and ceiling surfaces attached to the rafters.

Smoke—referring to a number of different methods of generating a true or artificial smoky vapor that allows for checking for leaks in air barriers, backflow on combustion appliances, and other situations.

Some examples of smoke include using a "Wizard Stick" or comparable artificial smoke generator, incense that has been lit, and a blown out match, amongst others. Some artificial smoke generators can damage electronic equipment. The Wizard Stick and similar type technology does not fall in that category.

Soffit—The underside of a roof overhang or a small lowered ceiling, as above cabinets or a bathtub.

Solar gain—Heat from the sun that is absorbed by a building.

Solenoid—A magnetic device that moves a switch or valve stem.

Space heating—Heating the living spaces of the home with a room heater or central heating system.

Specific Heat Capacity—Amount of heat required to raise the temperature of one pound of a substance by one degree Fahrenheit.

Spillage—Temporary flow of combustion gases from a dilution device.

Stack effect—The tendency for warm buoyant air to rise and leak out of the top of the house and be replaced by colder outside air entering from the bottom of the house.

Stand-by Losses, Standby Heat Losses, or Storage Heat Losses—A term used to describe heat energy lost from a water heater tank. Thus, the stand-by losses are reduced, if not virtually eliminated, by using an instantaneous or tankless hot water heater because you do not have to burn fuel to keep idle water in a tank hot in the event it may be used.

Steady-state efficiency (SSE)—The efficiency of a heating appliance, after an initial start-up period and while the burner is operating, that measures how much heat crosses the heat exchanger. A combustion analyzer measures the steady-state efficiency.

Steam trap—An automatic valve that closes to trap steam in a radiator until it condenses.

Steam vent—A bimetal-operated air vent that allows air to leave steam piping and radiators, but closes when exposed to steam.

Storage Heat Losses—see Stand-by Losses

Stud—A vertical framing member used to build a wall.

Stud bay—the space between two studs in a wall.

Sub floor—The sheathing over the floor joists and under the flooring.

Supply air—Air that has been heated or cooled and is then moved through the ducts and out the supply registers of a home.

Suspended ceiling—Modular-ceiling panels supported by a hanging frame.

## - T -

Technical Waiver—a waiver of a requirement under the Weatherization program standards due to technical reasons.

Therm—A unit of energy equivalent to 100,000 Btus or 29.3 kilowatt-hours.

Thermal break—A piece of relatively low conducting material between two high conducting materials installed to reduce heat flow through the assembly.

Thermal bridging—Rapid heat conduction resulting from direct contact between thermally conductive materials like metal and glass.

Thermal bypass—An indirect penetration that tends to reduce the effectiveness of insulation by allowing conditioned air to move out of a structure, or allowing unconditioned air to move in, depending on the exerted pressures.

Thermal conductance—A material's ability to transmit heat; the inverse of the R-value.

Thermal resistance—Same as R-value, expressing ability to retard heat flow.

Thermocouple—A bimetal-junction electric generator used to control the safety valve of an automatic gas valve.

Thermostat—A device used to control a heating or cooling system to maintain a set temperature.

Through-the-wall vented—Combustion appliances that are vented through a wall rather than into a vertical-rise chimney or vent. Such appliances are usually Category III or IV, but might also be Category I.

Transformer—A double coil of wire that reduces or increases voltage from a primary circuit to a secondary circuit.

Truss—A braced framework usually in the shape of a triangle to form and support a roof.

## - U -

U-value—The total heat transmission in Btus per square feet per hour with a 1°F temperature difference between the inside and the outside; the thermal conductance of a material.

Ultraviolet radiation—Light radiation having wavelengths beyond the violet end of the visible spectrum; high frequency light waves.

Underlayment—Sheeting installed to provide a smooth, sound base for a finish material.

UL—Underwriter's Laboratory

**- V -**

Vapor barrier—A material that retards the passage of water vapor.

Vapor diffusion—The flow of water vapor through a solid material.

Vapor retarder—A vapor barrier.

Vaporize—Change from a liquid to a gas.

Vent connector—A pipe that connects the combustion appliance to a vent, chimney, or runs directly to the outdoors.

Vent damper—An automatic damper powered by heat or electricity that closes the chimney while a heating device is off.

Ventilation—The movement of air through an area for removing moisture, air pollution, or unwanted heat.

Venting—The removal of combustion gases by a chimney.

Venting system—A continuous passageway from a combustion appliance to the outdoors through which combustion gases can safely pass.

Vermiculite—A heat-expanded mineral used for insulation.

Volt—A unit of electromotive force. It is the amount of force required to drive a steady current of one ampere through a resistance of one ohm. Electrical systems of most homes in the United States have 120-volt systems.

**- W -**

Watt (W)—A unit measure of electric power at a point in time, as capacity or demand. One Watt of power maintained over time is equal to one joule per second.

Watt-hour—One Watt of power extended for one hour. One thousandth of a kilowatt-hour

Weatherization—The process of reducing energy consumption and increasing comfort in buildings by improving energy efficiency of the building and maintaining health and safety.

Weatherstripping—Flexible gaskets, often mounted in rigid metal strips, for limiting air leakage.

Weep holes—Holes drilled for allowing water to drain out of an area in a building component where it may accumulate.

Weighted Average Wall R-Value—See Area Weighted R-Value

Wet bulb temperature—The temperature of a dampened thermometer of a sling psychrometer used to determine relative humidity.

Whole Wall R-Value—See Area Weighted R-Value

Window films—Plastic films, coated with a metalized reflective surface, that adhere to window glass to reflect infrared rays from the sun.

Window frame—The sides, top, and sill of the window, which form a box around window sashes and other components.

Worst-case Depressurization—A condition created when 1) all exhaust appliances (bathroom exhaust, kitchen exhaust, vented dryers, etc.) are operating, 2) the interior doors of a house are in a position that causes the greatest negative pressure in the CAZ, and 3) the furnace air handler is operating if such operation causes increased negative pressure in the CAZ.

# Index